MATHEMATICAL TOPICS IN TELECOMMUNICATIONS VOLUME 2

Contributing Authors

R. M. Brooks (Chapter 16)
British Telecom Research Laboratories
Martlesham

Professor K. W. Cattermole (Chapters 1, 5, 9 and 14)
Department of Electrical Engineering Science
University of Essex

F. M. Clayton (Chapter 12)
GEC Hirst Research Centre
Wembley, Middlesex

Dr. P. Cochrane (Chapter 17)
British Telecom Research Laboratories
Martlesham

Professor B. G. Evans (Chapter 13)
Department of Electrical Engineering
University of Surrey

K. V. Lever (Chapter 11)
School of Mathematics and Physics
University of East Anglia

Dr. J. J. O'Reilly (Chapters 2, 3, 4, 7 and 10)
Department of Electrical Engineering Science
University of Essex

Dr. G. S. Poo (Chapter 15)
Computer Science Department
National University of Singapore

Dr. G. G. Pullum (Chapter 8)
Standard Telecommunications Laboratories
Harlow, Essex

D. J. Songhurst (Chapter 6)
Teletraffic Division
British Telecom Research Laboratories
Martlesham

Mathematical Topics in Telecommunications

Volume 2
PROBLEMS OF RANDOMNESS IN COMMUNICATION ENGINEERING

Edited by

Kenneth W. Cattermole
John J. O'Reilly
Department of Electrical Engineering Science
University of Essex

PENTECH PRESS
London : Plymouth

First published 1984 by
Pentech Press Limited
Estover Road, Plymouth
Devon PL6 7PZ

ISBN 0–7273–1308–8

British Library Cataloguing in Publication Data

Mathematical topics in telecommunications
Vol 2. Problems of randomness in communication engineering
I. Cattermole, K. W. II. O'Reilly, J. J.
621.38′01′51 TK5102.5

ISBN 0-7273-1308-8

Filmset by Mid-County Press, London SW15
Printed and bound in Great Britain by Billing & Sons Limited, Worcester

Preface

This is the second in a series of volumes on *Mathematical Topics in Telecommunications.* Its contents derive mainly from contributions to a series of colloquia under the same general title, organised by the editors and held at the University of Essex at various times since November 1980. The purpose of the series, like that of the colloquia, is to extend the mathematical resources of practising engineers, graduate students and researchers in the telecommunications and allied fields. Our aim has been to select those topics which, while not uncommon in specialist research literature, are less familiar to the communication engineering fraternity in general and to present these in such a way as to expose both the underlying principles and practical aspects of application. To this end we include, with about equal emphasis: (1) tutorial expositions of the basic mathematics; (2) several examples of applications to engineering problems. Each volume in the series is based on two or three of the colloquia with cognate themes. There have been some editorial changes, and some new material has been added to render each volume self-contained and coherent.

Almost all aspects of communication theory are fundamentally probabilistic; and much practical system design and dimensioning is based on statistical properties of signals, or traffic, or both. Most graduate engineers and advanced students will be acquainted with the more usual probability distributions such as the Gaussian and binomial; and will have some idea of random process theory as exemplified by the usual treatment of electronic noise, random digit streams, and the like. But many practical problems can hardly be tackled without going well beyond this repertoire of simple tools and concepts.

An all-pervading complication, in our experience, is that of compound randomness: a probability distribution or stochastic process has a parameter which itself is liable to random fluctuation. Consider for example the composite signal in a multiplex or multiple-access system: it is the sum of a random number of contributions each modelled by a random process. Similar problems arise in the analysis of such diverse matters as optical signal detection, burst errors in digital systems, or overflow traffic in telephone networks. Another frequent complication is non-linearity: consider the composite signal passing through a travelling-wave amplifier in a satellite transponder, or the recovery of timing in a digital repeater by application of a random signal to a square-law device. As this last example suggests, random processes in communications are not always stationary but may be cyclo-stationary. Many random variables of practical interest are the sum of several random contributions; this may be easy to treat if the contributions are independent, but is much more difficult if there is a complex set of correlations, as for example in the performance estimation of disparity-reducing line codes. All these topics and more have their place in this book.

Much of the current research literature uses mathematical methods which occur rather late, if at all, in a typical introductory course in probability theory; such as characteristic-function and other transform methods, and bounding or approximation techniques allied to them. In our teaching, we introduce these methods early on. Our belief is that, firstly, they become familiar through extensive application and so should be introduced in time to be used: secondly, they are in fact quite accessible to communication engineers, who, in general, are thoroughly familiar with transform methods in signal theory and can readily learn to use similar concepts in a different context. This approach also has its place in our book.

We believe that this volume has more coherence and unity of theme than is usually found in collective works and not only because the editors are themselves major contributors. There is a genuine intellectual unity to this subject, and the work of the eight other contributors, from six organisations, falls into place beside our own.

Each of the two editors would acknowledge a good deal of stimulus and influence from the other: this is a genuine joint work. Both of us are deeply indebted to our fellow contributors, alike, for their efforts in preparing and presenting colloquium material, for their willingness to allow it to be published in this form, and for the general advice and encouragement which many of them have given us.

Kenneth W. Cattermole
John J. O'Reilly

Contents

PART 1 THEORETICAL TOPICS AND METHODS

PART 2 COMPOUND RANDOMNESS

PART 3 NON-LINEAR OPERATIONS ON STOCHASTIC PROCESSES

PART 4 DIGITAL LINE TRANSMISSION

1

Generating-Function Methods in Probability Theory

K. W. Cattermole

1.1 INTRODUCTION

In modern probability theory, the concept of the generating function has the same central place as the concept of the linear transform in signal theory. The mathematical operations are very similar in the two fields, and so the generating function approach should be accessible to communication engineers who will already be familiar with the following concepts:

(1) Fourier and Laplace transforms.
(2) The z-transform.
(3) Discrete and continuous convolution.
(4) Relationship between operations in two domains, e.g. multiplication transforms into convolution.
(5) Relationship between discrete and continuous cases (discrete being a special case of continuous if one admits impulses and other generalised functions, and continuous being a limiting case of discrete).

The advantages of these approaches are much the same in the two fields, namely

(1) Coherent framework which subsumes many specific facts and relationships.
(2) Simple and compact derivations of many key theoretical results.
(3) Operations difficult in one domain may be simple in the other.
(4) Familiarity with the theory leads to intuitive grasp of qualitative properties (moreover, intuition becomes easier if one is familiar with both fields of application).

There are three further advantages in the field of probability, namely

(5) A wide range of practical problems concern the sum of many independent random variables, and this is particularly easy to handle by means of generating functions.
(6) In the most complicated probability problems, it is usual to calculate upper bounds if the exact analysis is intractable. One of the best known bounds (Chernoff) is closely related to the moment-generating function.
(7) Because of the integrability of probability distributions, the convergence problems common in other areas of transform theory do not arise.

Given the power of this approach, and its accessibility to communication engineers, it is a good basis for our probability course.

In this chapter we shall treat as closely-related concepts (1) characteristic functions, (2) Fourier transforms, (3) Laplace transforms, (4) probability-generating functions, (5) moment-generating functions. It will turn out that many common distributions and properties are most easily derived by means of this approach.

1.2 NOTATION AND DEFINITIONS

We denote a random variable by a capital letter, and a specific value in its sample space by a corresponding lower-case letter, for example X, x. The sample space for a *continuous variable* is the real line $-\infty < x < +\infty$ or a subset thereof: some variables (such as time durations) are inherently non-negative, in which case $0 \leqslant x < \infty$. We denote probabilities by $P(\ )$ with a suitable description of the event as argument. A continuous variable has a probability density $p(x)$ such that

$$P(x \leqslant X < x + dx) = p(x)\,dx \tag{1.1}$$

and a *cumulative distribution*

$$\mathscr{P}(x) = P(X \leqslant x) = \int_{-\infty}^{x} p(\xi)\,d\xi \tag{1.2}$$

The *expectation* is

$$E(X) = \int xp(x)\,dx \tag{1.3}$$

which can be thought of as a long-term average in most practical cases. A *statistic* of the random variable X is the expectation of some function of X.

$$E[f(X)] = \int f(x)p(x)\,dx \tag{1.4}$$

for example the *moments*

$$m_r \equiv E(X^r) = \int x^r p(x)\,dx \tag{1.5}$$

which include the mean $m \equiv m_1$. The *variance* is the mean squared deviation from the mean

$$\sigma^2 \equiv V(X) = E[(X - m_1)^2] = m_2 - m_1^2 \tag{1.6}$$

The *probability* of an event A is definable if A is a suitable subset of the sample space, for example an interval $a_1 < x < a_2$ or the union of a countable number of intervals. The *indicator function* $I_A(x)$ takes the values

$$\begin{aligned} I_A(x) &= 1, \qquad x \in A \\ &= 0, \qquad \text{otherwise} \end{aligned} \tag{1.7}$$

The probability can be expressed as the expectation of the indicator function

$$p_A \equiv P(X \in A) = E[I_A(x)] \tag{1.8}$$

If A is a simple interval, this can be written as

$$p_A = \int_{a_1}^{a_2} p(x)\,dx \tag{1.9}$$

A *discrete variable* has a sample space comprising a finite or countably infinite set of distinct points x_i each with an associated probability p_i. It can be considered as a density function containing impulses,

$$p(x) = \sum_i p_i \delta(x - x_i) \tag{1.10}$$

and with this interpretation all the operations defined in terms of continuous variables become quite general and applicable also to discrete or mixed distributions. A discrete sample space of very common occurrence is the set of non-negative integers $\{0, 1, 2 \ldots\}$, which will be assumed in connection with any random variable of the nature of a count. In this case a statistic takes the form

$$E[f(X)] = \sum_{r=0}^{\infty} p_r f(r) \tag{1.11}$$

and according to the problem the series may be infinite or terminating.

Two or more random variables have a joint distribution in a sample space which is the Cartesian product of their individual sample spaces (e.g. a plane, if the sample spaces are real lines). Random variables X, Y are *statistically independent* if assignment of a value to one has no influence on any statistic of the other. It follows that for independent variables

$$E[f(X)g(Y)] = E[f(X)]E[g(Y)] \tag{1.12}$$

The statistics which may be multiplied in this way include, most obviously, probabilities and mean values: we shall, however, use eqn. (1.12) more widely. If two variables are not independent, a specification of their relationship requires *conditional statistics.* Well-known statistics are *conditional probabilities* written in the form $P(X \in A \mid Y = y)$ where the event in question is $X \in A$ and the probability is evaluated under the condition that $Y = y$. We shall also use *conditional expectations* of the form $E[f(X) \mid Y = y]$, meaning that the expectation $E[f(X)]$ is evaluated only over the subset of joint sample space for which $Y = y$. General allusions to conditional distributions and statistics are sometimes written as $P(X \mid Y), E(X \mid Y)$ and similarly.

Conditional statistics are often a useful intermediary in the calculation of unconditional statistics. Let $\{y_1, y_2 \ldots y_n\}$ be an exhaustive set of mutually exclusive values of Y. Then

$$E[f(X)] = \sum_{i=1}^{n} E[f(X) \mid Y = y_i]p(Y = y_i) \tag{1.13}$$

It often happens that the conditional statistics are simple to derive: we shall then use eqn. (1.13) as a means of calculating unconditional probabilities, expectations and other statistics.

1.3 PROBABILITY GENERATING FUNCTIONS

Let a random variable X have a sample space consisting of non-negative integers, and a probability distribution which we shall symbolise as $\{g\}$, meaning that $P(X = r) = g_r$. Consider the statistic

$$E(z^X) = \sum_{r=0}^{\infty} g_r z^r \equiv G(z) \tag{1.14}$$

We can consider this in three ways. Firstly, it is the expectation of a function of X, and consequently obeys any general rules for statistics formed in this way. Secondly, it is a power series whose coefficients are probabilities: hence the name 'probability generating function'.

Thirdly, it is a function of an algebraic quantity z. Some texts on probability or combinatorics suggest that we can treat z as a dummy variable of no especial significance; its powers are merely a vehicle for carrying probability coefficients. It is more in line with the spirit of the present treatment, however, to consider $G(z)$ as a linear transformation of the distribution $\{g\}$, and z as a variable in a transform domain.

Since $\{g\}$ is a probability distribution, we have

$$g_r \geqslant 0 \quad \text{and} \quad \sum_{r=0}^{\infty} g_r = 1, \quad \text{hence} \quad \sum_{r=0}^{\infty} |g_r| = 1 \qquad (1.15)$$

Consequently the series (1.14) converges, and $G(z)$ exists, at least for $|z| \leqslant 1$ and possibly elsewhere in the z-plane. The questions of convergence which often arise in general transform theory are not usually a problem in the context of probability.

We shall use p.g.f.'s of the form (1.14) in four ways:

(1) to derive probability distributions meeting specified physical requirements
(2) to study the sums of independent random variables
(3) to study certain problems in compound randomness (such as the sum of a random number of random variables)
(4) to derive moments of probability distributions.

1.3.1 Distribution of sums

Let two independent random variables X, Y have distributions $\{f\}$, $\{g\}$. What is the distribution $\{h\}$ of the sum $X + Y$?

The first method uses conditional probabilities. The sum has a value j if $X = i$ and $Y = j - i$, for $0 \leqslant i \leqslant j$. The probability of this event if $f_i g_{j-i}$. Consequently, we sum over all relevant events:

$$h_j = \sum_{i=0}^{j} f_i g_{j-i} \qquad (1.16a)$$

This composition of $\{f\}$ and $\{g\}$ is known as *convolution*, and is written briefly as

$$\{h\} = \{f\} * \{g\} \qquad (1.16b)$$

The second method uses the expectation definition of the p.g.f.'s, together with the rule (1.12). Then

$$E(z^{X+Y}) = E(z^X z^Y) = E(z^X)E(z^Y) \qquad (1.17a)$$

which says that in the transform domain

$$H(z) = F(z)G(z) \tag{1.17b}$$

Expanding the p.g.f.'s and identifying coefficients of z^j gives (1.16a), so the two methods agree. Thus we have the three-cornered relationship

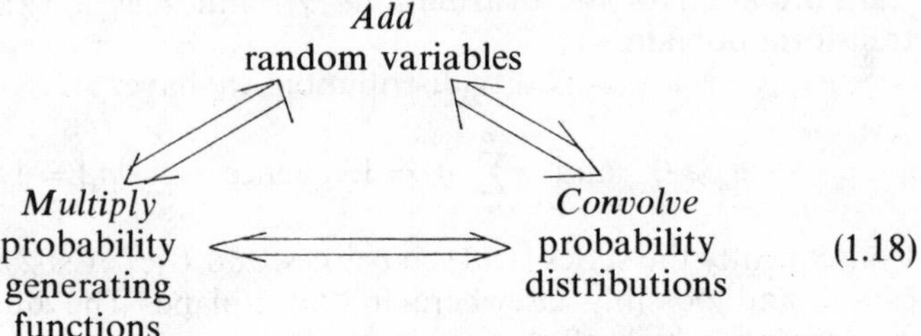

(1.18)

The multiplication/convolution relationship is reminiscent of the signal-theoretic transforms; it is not just coincidental but arises from a fundamental equivalence.

Example: The binomial distribution

A block of n binary digits arises from a random source: each digit is independently a 1 with probability p and a 0 with probability $q = 1 - p$. What is the distribution of the number of 1's?

For any one digit, the number of 1's is either 1 or 0: so the p.g.f. is

$$F_1(z) = q + pz \tag{1.19a}$$

Adding the contributions from n independent random digits with identical distributions, we multiply the p.g.f.'s to obtain

$$F_n(z) = [F_1(z)]^n = (q + pz)^n \tag{1.19b}$$

Expanding this in series gives the probability of r 1's as

$$p_r = \binom{n}{r} p^r q^{n-r} \tag{1.19c}$$

which is the binomial distribution. It is a useful exercise to confirm this without using p.g.f.'s.

1.3.2 A binary random process

The binary digits in the last example can be considered as a sequence, rather than a block: we can then ask such questions as 'how many 0's occur before the first 1?'. This waiting time is a random variable, which we denote by N_1. The event $N_1 = r$ implies r 0's followed by a 1, so its probability is $p_r = pq^r$: the probabilities being a geometric series. This is known as the *geometric distribution.* Its generating function is

$$E(z^{N_1}) = \sum_{r=0}^{\infty} pq^r z^r = \frac{p}{1 - qz} \tag{1.20}$$

What is the waiting time until the nth 1? This is preceded by $(n - 1)$ 1's together with a random number N_n of 0's. These 0's occur in n bursts, each with distribution similar to the first: so N_n is the sum of n random variables like N_1, and its p.g.f. is

$$E(z^{N_n}) = [E(z^{N_1})]^n = \left(\frac{p}{1 - qz}\right)^n \tag{1.21}$$

Expanding eqn. (1.21) in series shows that

$$p_r = \binom{-n}{r} p^n(-q)^r = \binom{n + r - 1}{r} p^n q^r \tag{1.22}$$

This is the *negative-binomial distribution.*

1.3.3 Moments

Differentiation of eqn. (1.14) gives

$$G'(z) = \sum_{r=1}^{\infty} g_r r z^{r-1} \tag{1.23}$$

Substitution of $z = 1$ gives

$$G'(1) = \sum_{r=1}^{\infty} g_r r = E(X) \tag{1.24}$$

In fact, there is no need to differentiate term by term: we could write

$$\frac{d}{dz} E(z^X) = E\left(\frac{d}{dz} z^X\right) = E(Xz^{X-1}) \tag{1.25}$$

which on substitution of $z = 1$ gives the same result. A second differentiation will show that

$$G''(1) = E(X^2 - X) \tag{1.26a}$$

whence the second moment is

$$E(X^2) = G''(1) + G'(1) \tag{1.26b}$$

and higher-order moments can be found by further differentiation. From eqns. (1.6), (1.24) and (1.26) the variance is

$$V(X) = G''(1) + G'(1) - [G'(1)]^2 \tag{1.27}$$

The statistics of our exemplary distributions are, by this method:

$$\text{Binomial} \begin{cases} \text{mean} = np & (1.28a) \\ \text{variance} = npq & (1.28b) \end{cases}$$

$$\begin{matrix}\text{Negative} \\ \text{binomial}\end{matrix} \begin{cases} \text{mean} = nq/p & (1.29a) \\ \text{variance} = nq/p^2 & (1.29b) \end{cases}$$

Note that means and variances increase in proportion to n, the number of random variables summed. This is in conformity with a more general rule that, when independent random variables are added, the mean of the sum is the sum of the means, and the variance of the sum is the sum of the variances. This rule can be proved either by multiplication of generating functions or by direct methods.

1.3.4 The Poisson distribution

Consider a binomial model applied to rare events, such as errors in a stream of binary digits in a rather good channel. A moderate number of errors will be observed only in a very long sequence, so n is large and p very small. Take the product $np = \lambda$ as a parameter, and consider the limiting behaviour as n increases. The p.g.f., eqn. (1.19b), becomes

$$F(z) = \operatorname*{Lim}_{n \to \infty} \left(1 + \frac{\lambda(z-1)}{n}\right)^n = e^{\lambda(z-1)} \tag{1.30a}$$

and the probability of r events is

$$p_r = \frac{\lambda^r}{r!} e^{-\lambda} \tag{1.30b}$$

This is the Poisson distribution, which is of very common occurrence. It can be shown from eqns. (1.24) and (1.27) that the mean and variance are both equal to λ.

1.3.5 The Poisson process

This is a point process on a continuous time scale: events may occur at any time equiprobably, in the sense that

$$P[\text{event in } (t, t + dt)] = \rho\, dt \tag{1.31}$$

for an infinitesimal interval dt: the parameter ρ is the *density* of the process. The probability of multiple events in dt is $O[(dt)^2]$ which vanishes in the limit.

What is the probability distribution for the number of events in an interval of length t? Let the probability of r events be $p_r(t)$: we define a generating function

$$F(z, t) = \sum_{r=0}^{\infty} p_r(t) \cdot z^r \tag{1.32}$$

which is similar in form to eqn. (1.14) except that the coefficients are functions of time. From the definition (1.31) we see that

$$F(z, dt) = 1 - \rho\, dt + \rho z\, dt \tag{1.33}$$

ignoring higher orders of infinitesimal. Now events in two non-overlapping time intervals are independent random variables: the p.g.f. of their sum is the product of the individual p.g.f.'s. So

$$F(z, t + dt) = F(z, t)\{1 - \rho\, dt + \rho z\, dt\} \tag{1.34a}$$

whence

$$\frac{F(z, t + dt) - F(z, t)}{dt} = (\rho z - \rho)F(z, t) \tag{1.34b}$$

and in the limit

$$\frac{d}{dt} F(z, t) = (\rho z - \rho)F(z, t) \tag{1.34c}$$

This differential equation defines $F(z, t)$. Taken with the obvious initial condition $F(z, 0) = 1$ it gives

$$F(z, t) = e^{\rho t(z-1)} \tag{1.35a}$$

whence the probabilities are

$$p_r(t) = \frac{(\rho t)^r}{r!} e^{-\rho t} \tag{1.35b}$$

This is Poisson distribution with mean ρt.

1.3.6 Compound randomness

Consider a digital transmission path subject to bursts of errors, due perhaps to fading or to impulsive interference. The number of bursts in a given period is a random variable with distribution $\{f\}$: the number of errors in any one burst is a random variable X with a distribution $\{g\}$. What is the distribution of the total number of errors Y?

This is readily derived using conditional statistics. Let the number of bursts be r: then the conditional generating function for Y is

$$E(z^Y \mid r) = [G(z)]^r \tag{1.36a}$$

since it is the sum of r variables each with p.g.f. $G(z)$. Summing over all values of r with the appropriate weight, as in eqn. (1.13), gives the unconditional p.g.f.

$$E(z^Y) = \sum_r f_r[G(z)]^r = F[G(z)] \tag{1.36b}$$

An important special case is the *compound Poisson process* in which each Poisson event has a random outcome (e.g. in an optical detector, each photon stimulates a burst of electrons by means of avalanche multiplication). In this case $F(z)$ is given by eqn. (1.30a) and

$$E(z^Y) = e^{\lambda[G(z)-1]} \tag{1.37}$$

The mean and variance can be found from eqns. (1.24) and (1.27): they are

$$E(Y) = \lambda E(X) \tag{1.38a}$$

$$V(Y) = \lambda E(X^2) \tag{1.38b}$$

We shall see later that similar expressions apply if X has a continuous sample space. An extension of this approach then provides a very complete theory of electronic noise.

1.4 LAPLACE TRANSFORM

The probability generating function (1.14) is essentially a discrete transform: its counterpart for continuous distributions must be a continuous linear transform. We consider first the non-negative random variable whose sample space is $0 \leqslant x < \infty$: practical examples are time durations or delays, as treated in queueing theory. The distribution is described by a probability density, say $g(x)$: the statistic commonly used is the Laplace transform

$$E(e^{-sX}) = \int_0^\infty g(x)e^{-sx}\,dx \equiv G(s) \tag{1.39}$$

Since

$$g(x) \geqslant 0 \quad \text{and} \quad \int_0^\infty g(x)\,dx = 1$$

hence

$$\int_0^\infty |g(x)|\,dx = 1 \tag{1.40}$$

the integral (1.39) converges and $G(s)$ exists, at least for $|e^{-s}| \leqslant 1$ and possibly elsewhere in the s-plane.

1.4.1 Waiting-time distributions

Consider the Poisson point-process defined in Section 1.3.5. The waiting-time from one event to the next is a random variable X_1. More generally, the waiting time to the rth event is a random variable X_r ($r = 1, 2, 3 \ldots$). What are the distributions of these variables?

A simple direct argument gives the densities $f^r(t)$ in terms of the Poisson probabilities (1.35b). The probability that the rth event falls in the interval $(t, d + dt)$ is the joint probability that $r - 1$ events occur in $(0, t)$ and one event in $(t, t + dt)$. By multiplication of independent probabilities,

$$f_r(t)\,dt = p_{r-1}(t)\rho\,dt \tag{1.41a}$$

which on substitution from eqn. (1.35b) gives

$$f_r(t) = \frac{\rho^r t^{r-1} e^{-\rho t}}{(r-1)!} \tag{1.41b}$$

This is the *gamma distribution.* The special case $r = 1$ gives the *negative exponential distribution*

$$f_1(t) = \rho e^{-\rho t} \tag{1.42}$$

which is the classical waiting-time to an event of uniform probability density.

1.4.2 Sums of continuous random variables

The discrete convolution (1.16) has its continuous counterpart

$$h(x) = \int f(\xi) g(x - \xi)\,d\xi \tag{1.43}$$

and by reasoning similar to the discrete case, we expect the probability density of the sum of two independent random variables to be the convolution of the two densities. In terms of the Laplace transform, however, it is clear that

$$E(e^{-s(X+Y)}) = E(e^{-sX}e^{-sY}) = E(e^{-sX})E(e^{-sY}) \qquad (1.44a)$$

so that in the transform domain

$$H(s) = F(s)G(s) \qquad (1.44b)$$

So again we have a three-cornered relationship

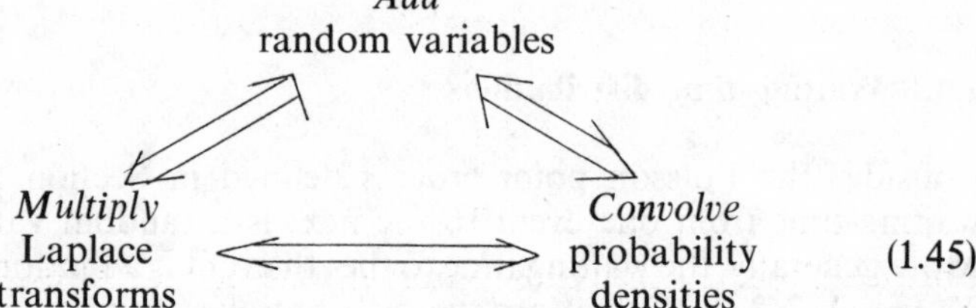

analogous to eqn. (1.18). The multiplication/convolution relationship is well-known in the general theory of Laplace transforms and, for example, in its application to electrical networks.

We can illustrate this result by reference to waiting-time distributions. Since X_r is the sum of r independent random variables each distributed as X_1, it follows that

$$f_r(t) = \{f_1(t)\}^{*r} \qquad (1.46)$$

where the notation indicates an r-fold convolution. It is easily shown by direct integration that

$$F_1(s) = \frac{\rho}{s+\rho} \qquad (1.47a)$$

and (perhaps a bit less easily) that

$$F_r(s) = \left(\frac{\rho}{s+\rho}\right)^r \qquad (1.47b)$$

which is consistent with the general relationship (1.45).

1.4.3 Poisson arrivals in a random time interval

Traffic problems often introduce a random variable which is the number of Poisson events in a time interval of random duration. For example, in a queueing system the queue comprises demands which have arrived during the holding times of calls in service: in an alternate-routing system, overflow traffic comprises calls which have arrived while the first-choice route is busy.

Let the time interval be X, with probability density $h(x)$ and Laplace transform $H(s)$. Let the arrivals be Poisson, with rate ρ: and denote the number of arrivals in the interval by Y. Again we invoke conditional statistics. For a given time interval x, the conditional generating function for Y is

$$E(z^Y \mid x) = e^{\rho x(z-1)} \tag{1.48a}$$

Averaging over the distribution of X according to eqn. (1.4) gives

$$E(z^Y) = \int_0^\infty e^{\rho x(z-1)} h(x)\, dx \tag{1.48b}$$

$$= H(\rho - \rho z) \tag{1.48c}$$

For example, if the time interval has a negative-exponential distribution with mean $1/\beta$, then

$$h(x) = \beta e^{-\beta x}, \quad H(s) = \frac{\beta}{s+\beta} \tag{1.49}$$

whence the p.g.f. is

$$E(z^Y) = \frac{\beta}{\beta + \rho - \rho z} \tag{1.50}$$

This is a geometric distribution as eqn. (1.20), with $p = \beta/(\beta + \rho)$.

Again, if the time-interval has a gamma distribution, the number of arrivals has a negative-binomial distribution. (We can now demonstrate this in at least three ways: by direct use of (1.48c), by addition of negative-exponential time intervals or by addition of geometric arrivals!) It is well known that overflow traffic approximates a negative-binomial distribution.

1.4.4 The departure process from a single-server queue

Finally, we sketch briefly a problem which is almost trivially simple using the Laplace transform, but much less obvious without it. A single-server queue has Poisson arrivals with rate ρ, and a negative-exponential holding time of mean $1/\beta$. What is the distribution of the intervals between departures? When there is a queue, the inter-departure time is just a holding time: when there is not, we have to wait for the next arrival to be served, so that the interdeparture time is the sum of the waiting time to arrival, and a holding time. How can be combine these distributions?

The answer is to take Laplace transforms conditional on the presence or absence of a queue (symbolised here by 1 and 0, respectively). A queue exists if a demand arrives while the server is occupied, which is for a fraction ρ/β of the time. Consequently

$$E(e^{-sX}) = p_1 E(e^{-sX} \mid 1) + p_0 E(e^{-sX} \mid 0) \tag{1.51a}$$

$$= \frac{\rho}{\beta}\left(\frac{\beta}{\beta + s}\right) + \left(1 - \frac{\rho}{\beta}\right)\left(\frac{\rho}{\rho + s}\right)\left(\frac{\beta}{\beta + s}\right) \tag{1.51b}$$

$$= \frac{\rho}{\rho + s} \tag{1.51c}$$

This is identical with the inter-arrival distribution. In fact, since the successive intervals are independent, departures are also a Poisson process: this is a fundamental factor in developing a theory of queueing networks.

1.5 MOMENT-GENERATING FUNCTIONS

There are several near-equivalent methods of calculating the moments of a distribution by means of linear transforms. We introduce one of them, after which the others will be obvious.

Consider the statistic

$$E(e^{sX}) \equiv F(s)$$

X is a random variable whose sample space is the real line or a subset thereof. By expanding the argument,

$$E(e^{sX}) = E\left\{\sum_{r=0}^{\infty} \frac{(sX)^r}{r!}\right\} = \sum_{r=0}^{\infty} \frac{s^r}{r!} E(X^r) \tag{1.52a}$$

we see that it is a power series whose coefficients include the moments m_r of eqn. (1.5). Comparison with the Taylor series expansion of $F(s)$

$$F(s) = \sum_{r=0}^{\infty} \frac{F^{(r)}(0)}{r!} s^r \tag{1.52b}$$

shows that

$$m_r \equiv E(X^r) = F^{(r)}(0) \tag{1.53}$$

The expression (1.52) is known as the *moment-generating function.* From the properties of a probability distribution, $F(s)$ must converge for a region in the s-plane whichincludes zero (compare (1.15) and

1.40)). However, it does not necessarily have derivatives of all orders, and certain distributions which vanish very slowly as x increases do not have finite moments. Most distributions of practical interest do have finite moments, so that we can write

$$F(s) = 1 + m_1 s + \tfrac{1}{2} m_2 s^2 + 0(s^3) \tag{1.54}$$

1.5.1 Sums of random variables

For independent random variables X and Y,

$$E(e^{s(X+Y)}) = E(e^{sX} e^{sY}) = E(e^{sX})E(e^{sY}) \tag{1.55}$$

so that the m.g.f.'s multiply (as for p.g.f.'s and Laplace transforms).

It is sometimes useful to define a function $\psi(s)$ such that

$$E(e^{sX}) = \exp\{\psi(s)\} \tag{1.56}$$

This is known as the log *m.g.f.* (for obvious reasons) or as the *cumulant generating function.* Taking the properties of the logarithm along with those of m.g.f.'s we have yet another three-corner relationship

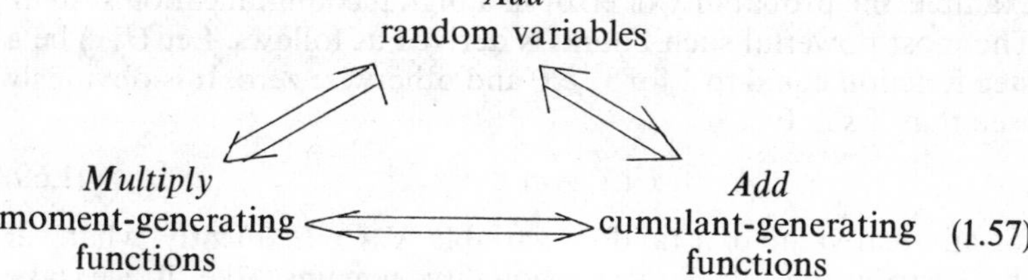

(1.57)

If the c.g.f. be expanded in power series

$$\psi(s) = \sum_{r=1}^{\infty} \frac{k_r}{r!} s^r \tag{1.58a}$$

then addition also corresponds to addition of the coefficients k_r which are known as *cumulants.* In general, k_r is a function of moments of all orders $\leqslant r$. Series expansion of the logarithm of expression (1.54) shows that

$$\psi(s) = m_1 s + \tfrac{1}{2}(m_2 - m_1^2)s^2 + 0(s^3) \tag{1.58b}$$

It follows that the first two cumulants are the mean and the variance, as we would expect from other evidence. Cumulants of higher order are sometimes useful when matching sum distributions to an analytical approximation.

If we know the Laplace transform of a distribution, the m.g.f. and c.g.f. follow readily. Foe example, the gamma distribution, eqn. (1.47b), gives

$$\psi(s) = -r \log(1 - s/\rho)$$
$$= r\left(\frac{s}{\rho} + \frac{1}{2}\frac{s^2}{\rho^2}\right) + 0(s^3) \tag{1.59a}$$

whence

$$\text{Gamma distribution}\begin{cases}\text{mean} = r/\rho & (1.59\text{b})\\ \text{variance} = r/\rho^2 & (1.59\text{c})\end{cases}$$

This is easily verified by calculating mean and variance directly for the negative-exponential ($r = 1$) and adding r similar contributions.

1.5.2 The Chernoff bound

When exact probabilities are difficult to calculate, it is often useful to calculate the upper bound to a tail probability representing, for example, the probability of error in a digital communication system. The most powerful such bound is derived as follows. Let $U(x)$ be a step function equal to 1 for $x \geqslant 0$ and otherwise zero. It is obviously true that if $s \geqslant 0$

$$U(X - c) \leqslant e^{s(X-c)} \tag{1.60}$$

for any real value of a random variable X. Consequently, whatever the distribution of X, the inequality remains true if we take expectations on each side. The step function is the indicator function for a tail probability, so

$$P(X \geqslant c) \leqslant E[e^{s(X-c)}] \tag{1.61a}$$
$$= e^{\psi(s) - sc} \tag{1.61b}$$

The bound is strongest if we minimise the exponent in (1.61b): there is obviously a turning point (which is in practice a minimum) when

$$\psi'(s) = c \tag{1.62}$$

For example, we take the average of n binary variables: what is a bound to the probability that this exceeds a value $d(>p)$? Consider a binomial distribution (eqn. (1.19)) with $c = nd$. The c.g.f. is

$$\psi(s) = n \log(1 - p + pe^s) \tag{1.63}$$

Optimising according to eqn. (1.62) and substituting in (1.61b) gives the bound as

$$P(X \geqslant d) \leqslant \left\{\left(\frac{1-p}{1-d}\right)^{1-d}\left(\frac{p}{d}\right)^{d}\right\}^{n} \tag{1.64}$$

which diminishes exponentially with n.

1.6 CHARACTERISTIC FUNCTIONS

Consider a random variable X whose sample space is the real line or any subset thereof; and let its probability density be $g(x)$. The *characteristic function* is the statistic

$$E(e^{j\theta X}) = \int_{-\infty}^{\infty} g(x)e^{j\theta x}\,dx \equiv G(\theta) \tag{1.65}$$

Since

$$g(x) \geqslant 0 \quad \text{and} \quad \int_{-\infty}^{\infty} g(x)\,dx = 1$$

hence

$$\int_{-\infty}^{\infty} |g(x)|\,dx = 1 \tag{1.66}$$

the integral (1.65) converges and $G(\theta)$ exists, at least for $|e^{j\theta}| \leqslant 1$. In particular, the characteristic function is defined for all real θ, whether the distribution is continuous or discrete (the density in the latter case being interpreted according to eqn. (1.10)). The integral (1.65) is recognisable as the Fourier transform (save for a scaling factor of 2π which is conventionally inserted in signal theory and various other applications, but omitted in probability theory).

1.6.1 Sums of random variables

For independent random variables X and Y

$$E(e^{j\theta(X+Y)}) = E(e^{j\theta X}e^{j\theta Y}) = E(e^{j\theta X})E(e^{j\theta Y}) \tag{1.67}$$

so that c.f.'s multiply (as for p.g.f.'s, m.g.f.'s and Laplace transforms). So again we have the three-cornered relationship

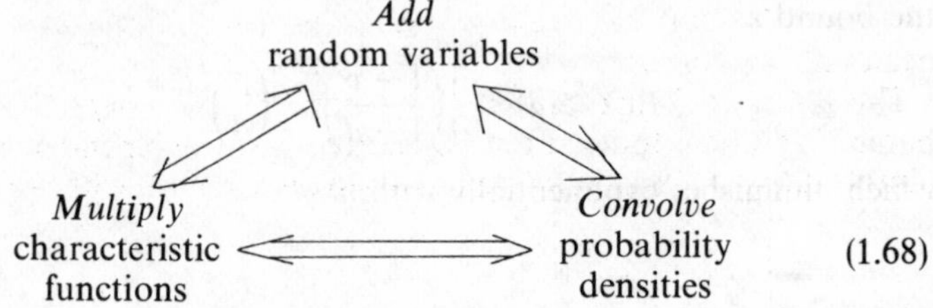

(1.68)

analogous to eqns. (1.18) and (1.45). The multiplication/convolution relationship is well known in the general theory of the Fourier transform.

The relationship between the c.f. and the m.g.f. means that moments and cumulants can also be derived from the c.f.; for example,

$$m_r = j^{-r}G^{(r)}(0) \tag{1.69}$$

(compare eqn. (1.53)).

1.6.2 The Gaussian distribution

The normalised form with zero mean and unit variance has the density and c.f.

$$g(x) = \frac{1}{(2\pi)^{1/2}} e^{-(1/2)x^2} \tag{1.70a}$$

$$G(\theta) = e^{-(1/2)\theta^2} \tag{1.70b}$$

The Gaussian (or normal) distribution is usually encountered as a two-parameter family, scaled and shifted as follows:

$$g(x) = \frac{1}{(2\pi)^{1/2}\sigma} \exp\left\{\frac{-(x-m)^2}{2\sigma^2}\right\} \tag{1.71a}$$

$$G(\theta) = \exp\{j\theta m - \tfrac{1}{2}\theta^2\sigma^2\} \tag{1.71b}$$

the relationship between these two functions conforming to the usual Fourier scaling and shifting rules. It is clear from (1.71b) that the cumulant generating function (1.56) is

$$\psi(s) = ms + \tfrac{1}{2}\sigma^2 s^2 \tag{1.71c}$$

so that the mean and variance are in fact m and σ^2, as the notation implies. It also appears that all higher-order cumulants (not, of course, the moments) are zero. Then the sum of any number of Gaussian random variables is itself Gaussian: for we can add terms of the form (71c), and still preserve the same form with the appropriate mean and variance.

The Gaussian distribution can be generalised to many dimensions: that is, several random variables (not necessarily independent) can be jointly Gaussian. The bivariate Gaussian distribution with zero mean is

$$g(x_1, x_2) = \frac{1}{2\pi\sigma_1\sigma_2(1-\rho^2)^{1/2}} \exp\left\{\frac{-1}{2(1-\rho^2)}\left(\frac{x_1^2}{\sigma_1^2} - \frac{2\rho x_1 x_2}{\sigma_1\sigma_2} + \frac{x_2^2}{\sigma_2^2}\right)\right\} \tag{1.72a}$$

$$G(\theta_1, \theta_2) = \exp\{-\tfrac{1}{2}(\sigma_1^2\theta_1^2 + 2\rho\sigma_1\sigma_2\theta_1\theta_2 + \sigma_2^2\theta_2^2)\} \tag{1.72b}$$

where σ_1, σ_2 are the marginal standard deviations and ρ the correlation coefficient. In any number of dimensions, all marginal and conditional distributions are Gaussian; the joint distribution is adequately specified by second-order statistics; and the sum of Gaussian variables is Gaussian.

1.6.3 Distributions related to the Gaussian

Consider two random variables X, Y related by the equation $Y = e^X$. If X is Gaussian (or normal), then the distribution of Y is *log normal.* Transmission over paths of random loss often yields lognormal distribution of amplitude or power (corresponding to normal distribution of level in decibels).

The lognormal distribution is usually defined by the parameters m, σ of the exponent. The median of Y is e^m: the mean follows readily from the c.f. (1.71b) or the c.g.f. (1.71c), and is

$$E(Y) = E(e^X) = e^{m+(1/2)\sigma^2} \tag{1.73a}$$

if natural logs are used. In decibel notation,

$$\text{Mean level in dB} = m + 0.115\sigma^2 \tag{1.73b}$$

where m and σ are also expressed in dB.

Consider a *random phasor* whose two quadrature components are each independently Gaussian, with mean 0 and variance $\frac{1}{2}$. The magnitude X of the phasor has the *Rayleigh distribution* with density

$$f(x) = 2xe^{-x^2} \tag{1.74}$$

which is well known in the theory of multi-path radio propagation. The squared magnitude $Y = X^2$ is proportional to power, if the phasor represents a physical sine-wave: putting $y = x^2$, $dy = 2x\,dx$

shows that

$$e^{-x^2}2x\,dx = e^{-y}\,dy \tag{1.75}$$

so the power has a negative-exponential density.

Distinguishing the form from the physical context, we could say that the variable here is the sum of two squared Gaussian random variables (the two quadrature-component amplitudes). It follows from eqn. (1.46) that the sum of $2r$ squared Gaussian random variables has a gamma distribution. Now the density (1.41b) can be written in a generalised form

$$f_\nu(y) = \frac{\rho^\nu y^{\nu-1} e^{-\rho y}}{\Gamma(\nu)} \tag{1.76}$$

where ν can have any real value for which the gamma function $\Gamma(\nu)$ is defined. Distributions with integer ν (as above) or with half-integer ν are of most common occurrence. The sum of n squared Gaussian variables has a gamma distribution of order $\nu = \frac{1}{2}n$, which is known in statistical theory as the *chi-squared distribution.*

1.6.4 The central limit theorem

If a large number n of independent random variables be added or averaged, in such a way that the total variance is finite and non-zero, with no one contribution dominating, then as n increases the sum distribution tends towards Gaussian form. We will demonstrate this for the special case of n identically distributed random variables with zero mean. Let each have variance $1/n$ (so that the sum has unit variance). Then from eqn. (1.54) the c.f. of each variable is

$$F_1(\theta) = 1 - \frac{\theta^2}{2n} + O(n^{-3/2}) \tag{1.77}$$

The c.f. of the sum is

$$F_n(\theta) = \left\{1 - \frac{\theta^2}{2n} + O(n^{-3/2})\right\}^n \tag{1.78a}$$

and as n grows

$$\operatorname*{Limit}_{n\to\infty} F_n(\theta) = e^{-(1/2)\theta^2} \tag{1.78b}$$

which defines a Gaussian distribution (1.70).

This explains why so many practical random variables are Gaussian: it is to be expected if the variable is the sum of many small contributions. Electronic noise is, of course, normally the sum of the effects of thousands (or more) of randomly moving charge carriers. Similarly, the product of many independent multiplicative effects will tend to have a lognormal distribution; this presumably applies to the transmission path attenuations, speech levels, etc., cited above.

1.6.5 Stable distributions

A distribution is *stable* if the sum of several variables, each of the given distribution, has another distribution of the same form (with, of course, different parameters). Clearly the Gaussian is a stable distribution. However, it is not the only one, as we can easily see by expressing the property in terms of characteristic functions.

The product of stable c.f.'s must give another c.f. of the same form: and this is obviously true for c.f.'s of the type

$$E(e^{j\theta X}) = e^{-c|\theta|^{\gamma}} \tag{1.79}$$

for some fixed index γ and any real positive parameter c.

It can be shown that eqn. (1.79) can be a characteristic function only if $\gamma \leqslant 2$: a higher value gives a sharper cut-off in the θ domain, which corresponds to overshoots in the x domain, whereas a probability density is non-negative. Several such distributions occur in various branches of probability theory: for instance, with γ values of $\frac{1}{2}$, 1 (Cauchy), $1\frac{1}{2}$ (Holtsmark) and 2 (Gaussian). Now the sum of random variables with a stable non-Gaussian distribution will not converge to a Gaussian limit; which implies some reservation to our proof of eqn. (1.78).

The clue is given by the moment theorem, eqn. (1.69). The second moment is finite only if the c.f. is twice differentiable at the origin: which is not true of (1.79) if $\gamma < 2$. The proof (1.78) assumes that variance is finite, and that higher-order moments become insignificant when scaled by powers of n. Thus the Gaussian is a limit distribution for sums of random variables with finite variance. For electrical or acoustic fluctuations, where variance is proportional to power, the other stable distributions are physically impossible, though they can arise in different contexts such as random position in space or time.

1.6.6 Stable properties of the Poisson process

The union of several Poisson point-processes is itself a Poisson point-process, with rate $\rho = \sum \rho_i$ (the latter being the individual rates). This is fairly clear from the basic derivation as a process of uniform probability density. Another approach, consistent with our treatment of stable distributions, is to multiply p.g.f.'s of the form (1.35a).

The Poisson process is also a *limiting point-process*, in the following sense. Consider the union of n independent point-processes, each with a stationary distribution of waiting times, with a finite total mean rate and no one dominant contribution. Then as n increases, the union tends towards a Poisson process. We illustrate this by superimposing processes each of which is individually as far from random as can be imagined. Let each contributory process have uniform intervals τ/n, so that the union of n similar processes has mean rate $1/\tau$. From an arbitrary time origin, the waiting time to the next event in one process of random phase has a uniform density $1/n\tau$ for $0 < t < n\tau$, so the tail probability of a wait exceeding t is $1 - t/n\tau$. On taking the union of n processes, the tail probability is

$$\left(1 - \frac{t}{n\tau}\right)^n \rightarrow e^{-t/\tau} \tag{1.80}$$

which is a negative-exponential distribution of mean τ: also as n increases the successive intervals are independent over an increasing range. It is not difficult to prove a similar result for a fairly general interval distribution: the brackets in eqn. (1.80) then include a term $0(n^{-2})$ whose effect vanishes in the limit. Calculation of a few finite examples shows that the interval distribution approaches a truncated negative-exponential fairly rapidly. Thus the union of many independent sources of point-events (very common in practical problems such as traffic flow, optical emission, electrical interference, etc.) is suitably modelled by a Poisson process.

1.6.7 The Gaussian random process

A random process whose state space and index set are both continuous is called a *Gaussian process* if any set of samples from it has a joint probability distribution of Gaussian form. Specifically, two samples from a stationary zero-mean process will have the joint distribution (1.72) with $\sigma_1 = \sigma_2$. The correlation coefficient ρ will be a function of the time separation τ between the samples. It is usual to describe a Gaussian process in terms of its autocorrelation function $\rho(\tau)$, and its power spectrum: these are related by the Fourier transform. This topic is covered thoroughly by textbooks of signal theory, and we shall not develop it here.

If a Gaussian process be applied as input to a linear filter, the output is also Gaussian. For a linear filter generates its output by convolution, i.e. by weighted summation of input values from different time epochs, all input samples are jointly Gaussian, the Gaussian distribution is stable, so the result follows.

1.6.8 The filtered Poisson process

Random photons incident on a photodetector have been cited as an example of a Poisson process. Suppose that each photon causes one charge carrier to flow in the detector circuit.* What are the statistics of the output current?

The near-instantaneous passage of one electronic charge at time t_i implies an impulsive current $e\delta(t - t_i)$. Let the impulse response of the circuit be $h(t)$ and the corresponding frequency response $H(f)$. Then the output at time t is a random variable

$$X(t) = e \sum_i h(t - t_i) \tag{1.81}$$

whose characteristic function we proceed to calculate. Consider the electrons flowing in a time interval $\Delta\tau$ centred on epoch τ. The number of electrons has a Poisson distribution with mean $\rho\,\Delta\tau$. Each electron contributes a term $\exp\{j\theta eh(t - \tau)\}$ to the c.f. Hence the c.f. in respect of the interval $\Delta\tau$ is

$$\sum_{r=0}^{\infty} e^{-\rho\Delta\tau} \frac{(\rho\Delta\tau)^r}{r!} e^{j\theta reh(t-\tau)} = \exp\left[\rho\Delta\tau\{e^{j\theta eh(t-\tau)} - 1\}\right] \tag{1.82a}$$

Now the events in different intervals $\Delta\tau$ are independent, so the total output is the sum of independent random variables. By eqn. (1.68), we must multiply the relevant characteristic functions. These are of exponential form (1.82a), so we add the exponents. Proceeding to the limit, this gives

$$\exp\left[\rho \int \{e^{j\theta eh(t-\tau)} - 1\}\, d\tau\right] \tag{1.82b}$$

Finally, if X is stationary then the time origin is immaterial and we can write

$$E(e^{j\theta X}) = \exp\left[\rho \int \{e^{j\theta eh(\tau)} - 1\}\, d\tau\right] \tag{1.82c}$$

which is the desired characteristic function.

† The optical physicist may enquire, what about quantum efficiency <1? The probability theorist can reply, that independent random deletions from a Poisson process leave a residual Poisson process of lower rate. Hint: use eqn. (1.37) to prove this.

Calculation of moments by means of eqn. (1.69) gives

$$E(X) = \rho e \int h(\tau)\, d\tau = \rho e H(0) \tag{1.83a}$$

$$V(X) = \rho e^2 \int |h(\tau)|^2\, d\tau = \rho e^2 \int_{-\infty}^{\infty} |H(f)|^2\, df \tag{1.83b}$$

The samples of a filtered Poisson process are analogous to a compound Poisson process, and the statistics (1.83) resemble (1.38) in an obvious way.

If we normalise to $H(0) = 1$ and define the mean current and the noise bandwidth as

$$I = \rho e \tag{1.84}$$

$$B = \int_0^{\infty} |H(f)|^2\, df \tag{1.85}$$

then the output statistics become

$$E(X) = I \tag{1.86a}$$

$$V(X) = 2IeB \tag{1.86b}$$

the latter being the classical expression for shot or quantum noise.

It 8s also possible to show (though we shall not prove it here) that

(1) The covariance of the filtered Poisson process is

$$\text{Cov}\,[X(t), X(t+\tau)] = \rho e^2 \int h(t)h(t+\tau)\, dt \tag{1.87}$$

the integral being the Fourier transform of the power spectrum of the filter. That is, the power spectrum of the filtered noise is purely that of the filter: the noise source can be considered as 'white'.

(2) The characteristic function (1.82c) tends toward that of a Gaussian distribution when the number of overlapping filtered events is large.

The theory we have presented, together with these extra items, constitutes a substantial link between point processes and continuous random processes.

1.7 PHILOSOPHICAL REFLECTIONS ON THIS APPROACH

A remarkable feature of the present theory is its coherence. The several transforms and generating functions are all minor variants of each other. The several distributions and processes presented (which include the most important of those commonly occurring in practide) turn out to be related by change of variable, change of parameter, a randomisation procedure, or a limiting process of some kind. The properties of the distributions emerge very naturally when we use the transforms.† Why is this?

In the author's opinion, the key is the threefold relationship which recurs in the context of each transform—see eqns. (1.18), (1.45), (1.57), (1.68) above. The generating function/linear transform technique is ideally suitable for manipulating the statistics of random variables which arise as the sum of many small contributions. But almost every random variable of interest to communication engineers is of this nature. Signals, and noise, are the summed effect of many electrons or photons. Traffic is the sum of demands from many sources. Transmission path losses are the sum of many contributions from line sections, amplifiers, radio paths, etc. A fading channel is the result of multipath propagation. The same is true in many other fields. Broadly speaking, if some observable phenomenon has one dominant cause, we seek a determinate explanation for it. If it is due to a multitude of small causes, then for all practical purposes it is unpredictable in detail: so we seek a statistical theory to tell us what will happen, on the average. The material presented here is a natural, and powerful, subset of statistical theory.

† It is perhaps worth remarking that the proofs and derivations given here, some of them novel, are unusually brief and direct. For instance, compare the derivation of the Poisson distribution (1.35), and thereafter the gamma distribution (1.41), with those found in most textbooks on probability.

2

Spectral Characterisation of Cyclostationary Random Processes

J. J. O'Reilly

2.1 INTRODUCTION

Much of statistical communication theory is concerned with the representation and analysis of random or stochastic processes and perhaps especially with stationary processes, for which the related concepts of autocorrelation function and power spectrum provide particularly important statistics. In the present context, however, the processes of interest are not *stationary* but *cyclostationary* and some care is needed in deriving a valid spectral characterisation. In this contribution aspects of the correlation and spectral theory of stationary processes are reviewed and modifications to this theory necessary to accommodate cyclostationarity are introduced. In keeping with the present theme the development is illustrated in terms of digital signals but it should be appreciated that the ideas and techniques are more broadly applicable.

It is appropriate to commence with a brief resumé of some aspects of random process theory which will serve also to define notation.

2.2 STOCHASTIC PROCESSES—GENERAL CONCEPTS

From a physical viewpoint a stochastic process is a process in time governed to some degree by a random mechanism. More formally we define a stochastic process $X(t, \omega)$ as a function of two variables, t and ω, with $t \in T$, the index set and $\omega \in \Omega$ the event space. For all $t \in T$ there is a random variable $X(t, \omega)$, a sample of the random process, which takes values in a sample space S known in this context as the state space and assumed here to be real. There are thus four possible interpretations of $X(t, \omega)$ depending on the character of t and ω:

(1) t and ω fixed $\Rightarrow X(t_i, \omega_j)$ is a real number
(2) t and ω variable $\Rightarrow X(t, \omega)$ is a stochastic process
(3) t fixed, ω variable $\Rightarrow X(t_i, \omega)$ is a random variable
(4) t variable, ω fixed $\Rightarrow X(t, \omega_j)$ is a function of time, known as a *sample function* or *realisation* of the process.

With this view of a stochastic process, as an ensemble of time functions, the event space may be finite, countably infinite, or non-denumerable.

2.2.1 Classification of processes

Stochastic processes are usefully classified according to the characteristics of T and S. Considering first T, of particular importance to us are: (1) T a denumerably infinite sequence, in which case $X(t,\omega)$ is referred to as a *stochastic sequence* and (2) T the whole real line. Considering now S we may further classify processes or sequences as *continuous* or *discrete*, in accordance with the random variables $X(t, \omega)$ being able to assume any of a continuum of values or being restricted to certain discrete values.

To simplify the notation it is convenient to suppress specific reference to the event space, letting $X(t)$ represent the stochastic process and also the associated random variable at time t with $x(t)$ being a specific realisation of the process.

2.2.2 Statistics of a stochastic process

For a specific value of t the random variable $X(t)$ can be characterised by a first-order probability density function, which we denote as $f_X(x, t)$ to emphasise its possible time dependence. The first-order moments of $X(t)$ are thus:

$$M_X^{(n)}(t) \triangleq E\{X(t)^n\} = \int_{-\infty}^{\infty} x^n f_X(x, t)\, dx \tag{2.1}$$

where $E\{\ \}$ denotes *expectation* or *ensemble average*. Of particular interest is the mean value of the process $E\{X(t)\}$ which is in general time dependent.

For higher-order statistics we consider an arbitrary finite set of time instants $\{t_i\}$, the associated random vector $(X(t_1), X(t_2), \ldots, X(t_N))$ and Nth-order density function $f_X(x_1, \ldots, x_N; t_1, \ldots, t_N)$. In this way multivariate statistics can provide an appropriate vehicle for the description and analysis of a process, although a complete description, corresponding to a *specification* of the process, will not generally be possible since this requires f_X to be known for all $\{t_i\}$ and all N. In practice, we often rely on partial descriptions, or deal with processes which are fully specified once only a small set of statistics are known. In particular, it is often adequate to deal with statistics up to only second-order.

Considering specifically second-order statistics, we define an *autocorrelation function* for a stochastic process $X(t)$ by:

$$R_X(t_1, t_2) \triangleq E\{X(t_1)X(t_2)\} \tag{2.2}$$

a function of two variables, t_1 and t_2. Often it is convenient to rewrite eqn. (2.2) in terms of the difference between the sample times, as $R_X(t, t+\tau)$ where $t \equiv t_1$ and $\tau = t_2 - t_1$.

2.2.3 Stationarity

A process is said to be *stationary in the strict sense* if and only if all its statistics are time invariant, whence $X(t)$ and $X(t+\varepsilon)$ have the same statistics for all ε. A less demanding form of stationarity requires only that the mean and autocorrelation function be time invariant:

$$E\{X(t)\} = M_X = \bar{X} \text{ a constant}$$

$$E\{X(t)X(t+\tau)\} \equiv R_X(\tau),$$

$$\text{a function only of the time difference } \tau \tag{2.3}$$

Such a process is said to be *stationary in the wide sense* or *covariance stationary*. Clearly, a strict-sense stationary process is also wide-sense stationary but the converse is not guaranteed.

2.2.4 Time averages and ergodicity

If we now direct our attention to the individual member functions, the realisations of the stochastic process, it is appropriate to define a time averaging operator $A\{\ \}$ by

$$A\{\xi(t)\} \triangleq \lim_{T\to\infty} \frac{1}{T} \int_{-T/2}^{T/2} \xi(t)\, dt \tag{2.4}$$

for a continuous-time process and by

$$A\{\xi_r\} \triangleq \lim_{N\to\infty} \frac{1}{2N+1} \sum_{r=-N}^{N} \xi_r \tag{2.5}$$

for a stochastic sequence. Here ξ is a function of one or more realisations. Specific time averages of interest are: the mean of a sample function

$$\bar{x} = A\{x(t)\} \tag{2.6}$$

and the *time autocorrelation function*:

$$R_x(\tau) = A\{x(t)x(t+\tau)\} \tag{2.7}$$

For a given realisation $\bar{x}$ is a number and $R_x(\tau)$ is a function of τ, but both may depend on the particular realisation. Hence, when considering the ensemble of all possible realisations, $\bar{x}$ and $R_x(\tau)$ must be viewed as random variables. It is thus appropriate to evaluate expectations. If we assume that $X(t)$ is wide-sense stationary and that we can interchange the order of integration and of taking expectation then

$$\begin{aligned} E\{\bar{x}\} &= E\left\{\lim_{T\to\infty}\frac{1}{2T}\int_{-T}^{T} x(t)\,dt\right\} \\ &= \lim_{T\to\infty}\frac{1}{2T}\int_{-T}^{T} E\{X(t)\}\,dt \\ &= \bar{X} \end{aligned} \tag{2.8}$$

Similarly

$$E\{R_x(\tau)\} = R_X(\tau) \tag{2.9}$$

An *ergodic* process is a stationary process with the further property that ensemble averages and time averages are equal with probability one. There are degrees of ergodicity, which we define with respect to parameters of interest. For example, a wide-sense stationary process has $\bar{x} = \bar{X}$ if it is ergodic in the mean and $R_x(\tau) = R_X(\tau)$ if it is ergodic in autocorrelation.

2.3 POWER SPECTRUM FOR A STOCHASTIC PROCESS

For a wide-sense stationary process the autocorrelation function depends only on the time separation τ and has the properties:

$$\begin{aligned} R_X(\tau) &= R_X(-\tau) \\ |R_X(\tau)| &\leqslant R_X(0) = \overline{X^2} \end{aligned} \tag{2.10}$$

The Fourier transform of the autocorrelation function is the *power spectrum* or *power spectral density* of such a process:

$$S_X(f) \triangleq \int_{-\infty}^{\infty} R_X(\tau) e^{-j2\pi f t}\, d\tau \qquad * \tag{2.11}$$

which is real, even and non-negative.

It is instructive to investigate spectral characterisation in terms of operations on sample functions. For a realisation $x(t)$, define a truncated function $x_T(t)$:

$$x_T(t) \triangleq \begin{cases} x(t),\ |t| < T/2 \\ 0,\ \text{elsewhere} \end{cases}$$

such that

$$x(t) = \lim_{T\to\infty} x_T(t) \tag{2.12}$$

This truncation enables us to take the Fourier transform

$$X_T(f) = \int_{-\infty}^{\infty} x_T(t) e^{-j2\pi f t}\, dt \Leftrightarrow x_T(t) = \int_{-\infty}^{\infty} X_T(f) e^{j2\pi f t}\, df \tag{2.13}$$

and to write the energy as

$$\begin{aligned} E_T &= \int_{-\infty}^{\infty} |x_T(t)|^2\, dt = \int_{-T/2}^{T/2} |x(t)|^2\, dt \\ &= \int_{-\infty}^{\infty} |X_T(f)|^2\, df, \quad \text{by Parseval's theorem} \end{aligned} \tag{2.14}$$

With a power signal $x(t)$, we associate a power spectrum with area equal to the average power; it is appropriate to consider the power in $x(t)$ in $(-T/2, T/2)$:

$$P_T = \frac{E_T}{T} = \int_{-\infty}^{\infty} \frac{1}{T} |X_T(f)|^2\, df \tag{2.15}$$

The function under the integral describes the distribution of power with frequency and leads to the power spectrum for a realisation $x(t)$ as

$$S_x(f) \triangleq \lim_{T\to\infty} \frac{1}{T} |X_T(f)|^2 \tag{2.16}$$

Since eqn. (2.16) applies to an individual realisation, it constitutes a random variable with respect to the process. To obtain a characterisation for the process as a whole we must average over all realisations, performing the limiting operation last [1]:

$$S_X(f) = \lim_{T\to\infty} \frac{1}{T} E\{|X_T(f)|^2\} \tag{2.17}$$

We can use eqn. (2.17) to write $S_X(f)$ in terms of the autocorrelation function as follows:

$$S_X(f) = \lim_{T\to\infty} E\left\{\frac{1}{T}\int_{-T2}^{T/2} X(t_1)e^{j2\lambda f t_1}\,dt_1 \int_{-T/2}^{T/2} X(t_2)e^{-j2\pi f t_2}\,dt_2\right\}$$

$$= \lim_{T\to\infty}\frac{1}{T}\int_{-T/2}^{T/2}\int_{-T/2}^{T/2} E\{X(t_1)X(t_2)\}e^{-j2\pi f(t_2-t_1)}\,dt_2\,dt_1$$

$$= \lim_{T\to\infty}\frac{1}{T}\int_{-T/2}^{T/2}\int_{-T/2}^{T/2} R_X(t_1,t_2)e^{-j2\pi f(t_1-t_1)}\,dt_1\,dt_2$$

$$= \lim_{T\to\infty}\frac{1}{T}\int_{-T/2}^{T/2}\int_{-T/2}^{T/2} R_X(t,t+\tau)\,dt\,.\,e^{-j2\pi f\tau}\,d\tau$$

$$= \lim_{T\to\infty}\frac{1}{T}\int_{-T/2}^{T/2} A\{R_X(t,t+\tau)\}\,e^{-j2\pi f\tau}\,d\tau \tag{2.18}$$

In this interpretation, the power spectrum and the time average of the autocorrelation function form a Fourier transform pair

$$S_X(f) \Leftrightarrow A\{R_X(t,t+\tau)\} \tag{2.19}$$

If $X(t)$ is wide-sense stationary then $A\{R_X(t,t+\tau)\} = R_X(t,t+\tau) = R_X(\tau)$ for all t and we have the result corresponding to (2.11)

$$S_X(f) \Leftrightarrow R_X(\tau) \tag{2.20}$$

known as the Wiener–Kintchine theorem.

Notice that while eqn. (2.19) provides a power spectrum characterisation for a non-stationary process (provided always that the limit in (2.18) exists) it contains less information than the autocorrelation function $R_X(t, t+\tau)$ as a consequence of the time-averaging operation. A spectral characterisation which avoids this information loss and provides an equivalent of the Wiener–Kintchine theorem for non-stationary processes is available by way of a double Fourier transform

$$\Gamma_X(f_1,f_2) \triangleq \int_{-\infty}^{\infty}\int_{-\infty}^{\infty} R_X(t_1,t_2)e^{-j2\pi(f_1t_1-f_2t_2)}\,dt_1\,dt_2 \tag{2.21}$$

with inverse

$$R_X(t_1 t_2) = \int_{-\infty}^{\infty} \int_{-\infty}^{\infty} \Gamma_X(f_1 f_2) e^{j2\pi(f_1 t_1 - f_2 t_2)} \, df_1 \, df_2 \tag{2.22}$$

For our purposes, however, the time-averaged autocorrelation function and average power spectrum of (2.19) are sufficient. It can be shown [2] that

$$S_X(f) = \Gamma_X(f,f) \geqslant 0 \tag{2.23}$$

and that

$$S_Y(f) = S_X(f)|H(f)|^2 \tag{2.24}$$

for $Y(t)$ the stochastic process at the output of a linear system, with transfer function $H(f)$, driven by the process $X(t)$. The average power spectrum has the same properties as the power spectrum of a stationary process.

2.4 CYCLOSTATIONARY PROCESSES

Many stochastic processes encountered in practice are stationary. Important exceptions, common in telecommunications systems, are processes for which the statistics are periodic rather than time invariant. This can arise, for example, when a stationary process is subjected to a repetitive processing operation such as sampling, scanning, modulation or multiplexing. Such processes are said to be *cyclostationary* [3, 4].†

There are degrees of cyclostationarity, much as for stationarity and ergodicity. We need concern ourselves here only with wide-sense cyclostationarity, defined by:

$$\begin{aligned} M_X(t) \triangleq E\{X(t)\} &= M_X(t+T) \quad \text{for all } t \\ R_X(t, t+\tau) &= R_X(t+T, t+T+\tau) \quad \text{for all } t, \tau \end{aligned} \tag{2.25}$$

where T, the smallest time interval giving equality, is the period. A spectral characterisation is provided by our previous development in terms of the time-averaged autocorrelation function. However, since $R_X(t, t+\tau)$ is periodic in t it now suffices to average over just one period, obviating the limiting process:

$$\tilde{R}_X(\tau) \triangleq \frac{1}{T} \int_{-T/2}^{T/2} R_X(t, t+\tau) \, dt \equiv A\{R_X(t, t+\tau)\} \tag{2.26}$$

† The terms 'periodically stationary', 'periodically correlated' and 'periodic nonstationary' are also used [5, 6, 7].

2.4.1 Phase randomising

As an alternative to time averaging the statistics of a cyclostationary process, we can attempt to form a related stationary process from which time-invariant statistics can be obtained directly [4, 8]. This procedure is both appropriate and illuminating for a cyclostationary process. As an illustration, consider a synchronous pulse amplitude modulated (PAM) signal

$$X(t) = \sum_i a_i p(t - iT) \tag{2.27}$$

where $p(t)$ is the signal element and $\{a_i\colon i \in Z\}$ is a stochastic sequence defining the pulse amplitudes. Let $\{a_i\}$ be wide sense stationary, such tyat:

$$\left.\begin{aligned} E\{a_i\} &= \bar{a} \\ E\{a_i a_{i+k}\} &= R_a(k) = R_a(-k) \end{aligned}\right\} \quad \text{for all } i \tag{2.28}$$

Then

$$\begin{aligned} E\{X(t)\} &= \sum_i E\{a_i\} p(t - iT) = \bar{a} \sum_i p(t - T) \\ &= E\{X(t + T)\} \end{aligned} \tag{2.29}$$

Also:

$$\begin{aligned} R_X(t, t + \tau) &= E\{X(t)X(t + \tau)\} \\ &= \sum_i \sum_j E\{a_i a_j\} p(t - iT) p(t + \tau - jT) \end{aligned}$$

and putting $j = i + k$ gives

$$\begin{aligned} R_X(t, t + \tau) &= \sum_k R_a(k) \sum_i p(t - iT) p(t + \tau - iT - kT) \\ &= R_X(t + T, t + T + \tau) \end{aligned} \tag{2.30}$$

The PAM signal is thus wide-sense cyclostationary. We can obtain a related stationary process, however, by randomising the time reference for the individual realisations of eqn. (2.27). This involves the introduction of a phase-randomising variable Δ uniformly distributed over one period, say $[0, T)$. The process is then

$$X_\Delta(t) = \sum_i a_i p(t - iT - \Delta) \tag{2.31}$$

The mean is given by

$$
\begin{aligned}
E\{X_\Delta(t)\} &= \sum_i E\{a_i\} \frac{1}{T} \cdot \int_0^T p(t - iT - \Delta)\, d\Delta \\
&= \frac{\bar{a}}{T} \int_{-\infty}^{\infty} p(t)\, dt \\
&= E\{X_\Delta(t + \varepsilon)\} \quad \text{for all } \varepsilon
\end{aligned}
\tag{2.32}
$$

where we have used the fact that Δ uniformly distributed on $[0, T)$ has probability density $1/T$ on this interval. The autocorrelation function is obtained as follows

$$
\begin{aligned}
R_{X_\Delta}(t, t + \tau) &= \frac{1}{T} \sum_k \sum_i E\{a_i a_{i+k}\} \\
&\quad \cdot \int_0^T p(t - iT - \Delta) p(t + \tau - iT - kT - \Delta)\, d\Delta \\
&= \frac{1}{T} \sum_k R_a(k) \sum_i \int_0^T \sum_i \int_0^T p(t - iT - \Delta) p(t + \tau - iT - kT - \Delta)\, d\Delta \\
&= \frac{1}{T} \sum_k R_a(k) \int_{-\infty}^{\infty} p(t - \Delta) p(t + \tau - kT - \Delta)\, d\Delta \\
&= \frac{1}{T} \sum_k R_a(k) R_p(\tau - kT) \\
&= R_{X_\Delta}(t + \varepsilon, t + \varepsilon + \tau) \quad \text{for all } \varepsilon
\end{aligned}
\tag{2.33}
$$

Here

$$
R_p(\tau) \triangleq \int_{-\infty}^{\infty} p(t) p(t + \tau)\, dt \tag{2.34}
$$

is the autocorrelation function for the deterministic, finite energy signal element, $p(t)$.

From eqns. (2.32) and (2.33) we observe that the phase-randomised process $X_\Delta(t)$ has time-invariant mean and autocorrelation function and is thus wide-sense stationary, as desired. Notice, however, that by time averaging (2.29) and (2.30) the same results are obtained

$$
\begin{aligned}
\tilde{M}_X &\triangleq \frac{1}{T} \int_0^T E\{X(t)\}\, dt \\
&= \frac{1}{T} \int_0^T \bar{a} \sum_i p(t - iT)\, dt = \frac{\bar{a}}{T} \int_{-\infty}^{\infty} p(t)\, dt \\
&= E\{X_\Delta(t)\}
\end{aligned}
\tag{2.35}
$$

and

$$\begin{aligned}\tilde{R}_X(\tau) &\triangleq \frac{1}{T}\int_0^T R_X(t, t+\tau)\,dt \\ &= \frac{1}{T}\int_0^T \sum_k R_a(k) \sum_i p(t-iT)p(t+\tau-iT-kT)\,dt \\ &= \frac{1}{T}\sum_k R_a(k)\int_{-\infty}^{\infty} p(t)p(t+\tau-kT)\,dt \\ &= \frac{1}{T}\sum_k R_a(k)R_p(\tau-kT) \\ &= R_{X_\Delta}(t, t+\tau) \qquad (2.36)\end{aligned}$$

We have considered only a special case, but the equivalence of the two procedures can be established quite generally [8]: time averaging the statistics over one cycle is equivalent to modelling the time reference or phase of the process as a random variable uniformly distributed over one period.

In some circumstances this phase randomising is quite appropriate and we can replace a cyclostationary process by a stationary equivalent but in others the cyclostationarity is crucial. For example, carrier and timing recovery systems rely for their operation on the cyclostationary character of the signal process, which must be preserved in any analysis. This is usually accomplished by working directly with the time-dependent mean and autocorrelation function [9], although in some instances a spectral analysis supplemented by knowledge of the structural character of cyclostationarity may suffice (see Chapter 10). On the other hand, when considering mutual interference the wanted and unwanted signals may not be synchronised and phase randomising is appropriate. In these circumstances the average power spectrum is a useful aid to system performance assessment.

2.4.2 Power spectrum of a cyclostationary process

We cannot invoke the Wiener–Kitchine theorem directly to obtain a power spectral density characterisation of a cyclostationary process since the autocorrelation function is not time-invariant. However, eqn. (2.19) provides such a characterisation in terms of the Fourier transform of the time-average autocorrelation function.† It is clear from eqn. (2.36) that this corresponds also to the power spectrum of the phase-randomised process. That this procedure provides a proper spectral density for the cyclostationary process, in the sense that the area under $S_X(f)$ in the frequency interval $|f| \in \{f_1, f_2)$ represents the average power in that interval, is implicit in the analysis leading to eqns. (2.17) and (2.19). This can be shown explicitly for the PAM signal of eqn. (2.27) by representing the cyclostationary process as a sum of wide-sense stationary subprocesses [11]. The subprocesses are disjoint in frequency space so their individual power spectra can be summed to yield the power spectrum for the cyclostationary process; the result accords with that obtained by phase randomising or by time averaging the autocorrelation function. A rather more general treatment is also available [5, 12] which makes use of a harmonic series representation for the process and yields similar results.

2.4.3 Conclusion

The purpose of this chapter has been to provide a summary and tutorial exposition of those aspects of stochastic process theory relating to autocorrelation function and power spectral density characterisation. Some emphasis has been placed on the question of stationarity and on the respective roles of ensemble and time averaging, with cyclostationary processes receiving special attention. The treatment is by no means encyclopaedic, but should provide a reasonable basis for detailed examination of digital signal processes.

REFERENCES

1 Middleton, D., *An Introduction to Statistical Communication Theory*, McGraw-Hill, 152 (1960)
2 Papoulis, A., *Probability, Random Variables and Stochastic Processes*, McGraw-Hill, 448–451 (1965)
3 Bennett, W. R., 'Statistics of regenerative digital transmission', *Bell Syst. Tech. J.*, **37**, 1501–1542 (1958)
4 Franks, L. E., *Signal Theory*, Prentice-Hall, 204–225 (1969)
5 Ogura, H., 'Spectral representation of periodic nonstationary random processes', *IEEE Trans. Inf. Th.*, **IT-17**, 143–149 (1971)
6 Hurd, H. L., 'An introduction to periodically correlated stochastic processes', Ph.D. Thesis, Duke Univ. Durham, N.C. (1971)

† This can reasonably be taken as a uniform definition, accommodating wide-sense stationary and cyclostationary processes together with deterministic power signals by the simple expedient of modelling a deterministic signal as a degenerate stochastic process [10].

7 Stratanovich, R. L., *Topics in the Theory of Random Noise*, Gordon and Breach, Vol. 1, 139–141 (1963)
8 Hurd, H. L., 'Stationarising properties of random shifts', *SIAM J. Appl. Math.*, **26**, 203–212 (1974)
9 Franks, L. E. and Bubrouski, J. R., 'Statistical properties of timing jitter in a PAM timing recovery scheme', *IEEE Trans. Commun.*, **COM-22**, 913–920 (1974)
10 O'Reilly, J. J., 'Unified Treatment of Autocorrelation Function and Power Spectrum for Deterministic and Stochastic Signal Processes', University of Essex, Telecommunication Systems Group Report TGS-172 (1982)
11 Van der Wurf, P., 'On the spectral density of a cyclostationary process', *IEEE Trans. Commun.*, **COM-22**, 1727–1730 (1974)
12 Gardner, W. A. and Franks, L. E., 'Characterisation of cyclostationary random processes', *IEEE Trans. Inf. Th.*, **IT-21**, 4–14 (1975)

3

Markov Chains

J. J. O'Reilly

3.1 INTRODUCTION

Many probability structures encountered in communication systems studies can usefully be modelled in terms of a Markov process. To introduce the concept we consider a random sequence $\{X_n\} \triangleq \{X(t_n)\}$, the $\{t_n \in T\}$ corresponding to observation times. In general, the index set T may be discrete or continuous, as may the state space. We will concentrate here on discrete state-space systems known as Markov chains and will place most emphasis on discrete-time processes.

The simplest possible model for a random sequence would be to assume that the X_n are independent random variables, in which case the process possesses no memory whatsoever. The essential feature of the Markov model is that it provides for dependence between successive random variables in a sequence, under the minimal assumption that the future of the process depends only on the present state and not on the past history. We can state this more formally as follows: A stochastic process $\{X(t), t \in T\}$ is a Markov process if for any set of $n+1$ values $t_1 < t_2, \ldots, t_n < t_{n+1} \in T$ and any set of $n+1$ states $\{x_1, \ldots, x_{n+1}\}$ we have

$$P[X(t_{n+1}) = x_{n+1} \mid X(t_1) = x_1, X(t_2) = x_2, \ldots, X(t_n) = x_n] = P[X(t_{n+1}) = x_{n+1} \mid X(t_n) = x_n] \quad (3.1)$$

3.2 DISCRETE-TIME MARKOV CHAINS

We now restrict attention to a discrete index set which we denote as $T = \{0, 1, 2, \ldots\}$. Also, we can conveniently label the states in the discrete state-space by the positive integers $\{1, 2, \ldots\}$ and we will often assume the state-space to be finite dimensional.

The conditional probabilities $P[X_{n+1} = j \mid X_n = i]$ are called transition probabilities and will be denoted by $q_{ij}(n)$, allowing that in general they may depend on the index value, n. The next-state probabilities $p_j(n + 1)$ may be expressed in terms of the transition probabilities and the present state probabilities $p_i(n)$ as follows. The probability that the system will occupy state j at time $n + 1$ given that the system is in state i at time n is simply equal to the transition probability $q_{ij}(n)$. This is a conditional probability

$$q_{ij}(n) = P[X_{n+1} = j \mid X_n = i] \tag{3.2}$$

The outcome occurs with this probability conditional on the system being in state i at time n and the probability of this latter event is $p_i(n)$.

To remove the conditioning we average over all states at time n to obtain

$$p_j(n + 1) = \sum_i p_i(n) q_{ij}(n) \tag{3.3}$$

Equation (3.3) applies for each j and thus represents a whole set of equations which may be expressed conveniently in matrix form as

$$\mathbf{p}(n + 1) = \mathbf{p}(n)\mathbf{Q}(n) \tag{3.4}$$

If the state-space is N-dimensional then $\mathbf{p}(n)$, $\mathbf{p}(n + 1)$ are N-dimensional row vectors and $\mathbf{Q}(n)$ is an $N \times N$ (square) matrix known as the transition probability matrix.† If the $q_{ij}(n)$ are independent of time the Markov chain possesses stationary transition probabilities and is described as *stationary* or *time-homogeneous*. We then write eqn. (3.4) more simply as

$$\mathbf{p}(n + 1) = \mathbf{p}(n)\mathbf{Q} \tag{3.5}$$

3.2.1 Graphical representations

A useful graphical representation for a Markov chain is provided by the state-transition diagram (std). This is a graph in which the vertices correspond to states and the directed edges denote possible transitions. In the present context the edges are labeled with the transition probabilities q_{ij}. An std for a three-state process is presented by way of illustration in Fig. 3.1 together with the corresponding transition probability matrix and a lattice section representation. The lattice representation provides for easy visualization but is hardly practicable for large n. On the other hand,

† It is not uncommon for these equations to be written in transposed form with the state probabilities being represented by column vectors and the next state probability vector being obtained by premultiplying by the transpose of $\mathbf{Q}$. The two representations are, of course, equivalent. That adopted here is more common in the recent literature.

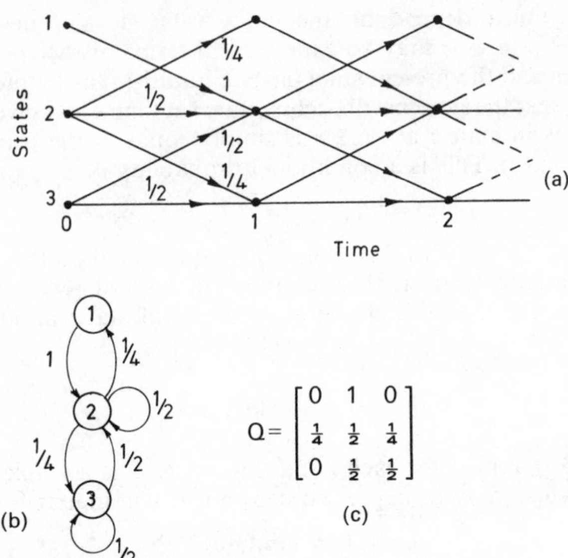

Fig. 3.1 Alternative representations for a three-state Markov chain. (a) Lattice representation, (b) state–transition diagram, (c) transition probability matrix

the std achieves a conveniently compact representation while, as we shall see shortly, providing a useful aid to calculation.

3.2.2 Stochastic matrices

The elements of the transition probability matrix **Q** are probabilities and are thus necessarily non-negative while the rows are conditional probability distributions. The $(i + 1)$th row gives the probability distribution of X_{n+1} given that $X_n = i$. The elements of each row must thus sum to one since some transition—possibly back to the same state—must occur. The elements of **Q** must thus satisfy

$$q_{ij}(n) \geqslant 0 \quad \text{for all } i, j \quad \text{and} \quad n = 0, 1, \ldots$$

$$\sum_j q_{ij}(n) = 1 \quad \text{for all } i \quad \text{and} \quad n = 0, 1, \ldots \tag{3.6}$$

Matrices satisfying eqn. (3.6) are termed *stochastic* and any stochastic matrix constitutes a valid transition probability matrix. Stochastic matrices have some special properties which we shall exploit in due course.

3.2.3 *n*-step transition probabilities

Repeated application of eqn. (3.4), given an initial state probability distribution $\mathbf{p}(0)$, gives

$$\begin{aligned} \mathbf{p}(1) &= \mathbf{p}(0)\mathbf{Q}(0) \\ \mathbf{p}(2) &= \mathbf{p}(1)\mathbf{Q}(1) = \mathbf{p}(0)\mathbf{Q}(0)\mathbf{Q}(1) \\ &\vdots \\ \mathbf{p}(n+1) &= \mathbf{p}(n)\mathbf{Q}(n) \\ &= \mathbf{p}(0)\mathbf{Q}(0)\mathbf{Q}(1), \ldots, \mathbf{Q}(n) \end{aligned} \tag{3.7}$$

For a time-homogeneous process this simplifies to

$$\mathbf{p}(n) = \sigma(0)\mathbf{Q}^n \tag{3.8}$$

and the state probabilities are completely determined for all $n > 0$ if we know $\mathbf{Q}$ and the initial state probability vector $\mathbf{p}(0)$. Defining the elements of $\mathbf{Q}^n$ as $q_{ij}^{(n)}$ we have

$$p_j(n) = \sum_i p_i(0) q_{ij}^{(n)} \tag{3.9}$$

or more generally

$$p_j(r+n) = \sum_i p_i(r) q_{ij}^{(n)} \tag{3.10}$$

Here, $q_{ij}^{(n)}$ is the conditional probability that a process $X(t)$ in state i at a given time r will be in state j after precisely n steps. We thus refer to $\mathbf{Q}^n$ as the n-step transition probability matrix with elements

$$q_{ij}^{(n)} = P[X_{r+n} = j \mid X_r = i] \tag{3.11}$$

for all $r \geqslant 0$ and $n > 0$.

In this notation $q_{ij}^{(1)} = q_{ij}$ and we can define $q_{ij}^{(0)}$ in terms of the Kronecker delta

$$q_{ij}^{(0)} = \delta_{ij} = \begin{Bmatrix} 1, \text{ if } i = j \\ 0, \text{ otherwise} \end{Bmatrix} \tag{3.12}$$

We can write the n-step transition probabilities as

$$q_{ij}^{(r+n)} = \sum_k q_{ik}^{(r)} q_{kj}^{(n)} \quad \text{for all } r, n, i, j \geqslant 0 \tag{3.13}$$

corresponding to a restricted form of the Chapman–Kolmogorov equations. The $q_{ij}^{(n)}$ satisfy the conditions of eqn. (3.6) for all $n \geqslant 0$, whence $\mathbf{Q}^n$ is a stochastic matrix.

3.2.4 Observed behaviour of Markov chains

We will concentrate now on homogeneous processes. We have just seen that given $\mathbf{Q}$ and an initial state probability vector $\mathbf{p}(0)$ we can determine $\mathbf{p}(n)$ for all later times; essentially all that is required is post-multiplication by $\mathbf{Q}^n$. Determination of $\mathbf{Q}^n$ can in principle be achieved by direct matrix multiplication but this is not computationally efficient for anything other than small n. We will examine shortly some alternative approaches but it is instructive first to proceed by way of direct successive multiplication by $\mathbf{Q}$ to draw attention to some possible behaviour patterns of Markov chains.

Consider the matrix

$$\mathbf{Q} = \begin{bmatrix} \frac{1}{4} & \frac{3}{4} \\ \frac{1}{4} & \frac{3}{4} \end{bmatrix} \tag{3.14}$$

and initial state probability vector $\mathbf{p}(0) = (0, 1)$. Successive state probability vectors are given by

$$\begin{aligned} \mathbf{p}(1) &= (0, 1)\begin{bmatrix} \frac{1}{4} & \frac{3}{4} \\ \frac{1}{4} & \frac{3}{4} \end{bmatrix} = (\tfrac{1}{4}, \tfrac{3}{4}) \\ \mathbf{p}(2) &= (\tfrac{1}{4}, \tfrac{3}{4})\begin{bmatrix} \frac{1}{4} & \frac{3}{4} \\ \frac{1}{4} & \frac{3}{4} \end{bmatrix} = (\tfrac{1}{16} + \tfrac{3}{16}, \tfrac{3}{16} + \tfrac{9}{16}) \\ &= (\tfrac{1}{4}, \tfrac{3}{4}) \end{aligned} \tag{3.15}$$

Hence $\mathbf{p}(n) = (\frac{1}{4}, \frac{3}{4})$ for all $n > 0$ if $\mathbf{p}(0) = (0, 1)$. We say that the chain has reached an equilibrium condition (rather rapidly in this judiciously selected example!) and refer to $(\frac{1}{4}, \frac{3}{4})$ as a stationary state probability distribution. In this particular case, as in many cases of practical importance, there is only one such equilibrium distribution and the chain converges to this distribution from any valid initial state probability assignment. A Markov chain of this type is referred to as irreducible; we will study irreducible chains in some detail in a later section. We note here, though, that since we are concerned with the behaviour as $n \to \infty$ for the system

$$\mathbf{p}(n) = \mathbf{p}(0)\mathbf{Q}^n$$

it is appropriate to consider

$$\mathbf{Q}^\infty \triangleq \lim_{n \to \infty} \mathbf{Q}^n \tag{3.16}$$

The limiting distribution for an irreducible chain, which we denote as π, is then obtained in a single step from any valid initial state probability vector by

$$\boldsymbol{\pi} = \mathbf{p}(0)\mathbf{Q}^{\infty} \tag{3.17}$$

Notice that our matrix (3.14) is a rather singular example in that it is already in this form, so that $\mathbf{Q} = \mathbf{Q}^{\infty}$ in this case.

Not all chains are irreducible; to illustrate this we introduce the following example

$$\mathbf{Q} = \begin{bmatrix} 0 & 1 \\ 1 & 0 \end{bmatrix} \tag{3.18}$$

which is a stochastic matrix since the elements satisfy eqn. (3.6). If the initial state probability vector is $\mathbf{p}(0) = (1, 0)$ successive state vectors are

$$\mathbf{p}(1) = (1, 0)\begin{bmatrix} 0 & 1 \\ 1 & 0 \end{bmatrix} = (0, 1)$$

$$\mathbf{p}(2) = (0, 1)\begin{bmatrix} 0 & 1 \\ 1 & 0 \end{bmatrix} = (1, 0) = \mathbf{p}(0) \tag{3.19}$$

and the system simply alternates between the two states such that $\mathbf{p}(n+2) = \mathbf{p}(n)$ for all $n \geqslant 0$ if $\mathbf{p}(0) = (1, 0)$. However, if the initial state probability is $(\frac{1}{2}, \frac{1}{2})$ we obtain

$$\mathbf{p}(1) = (\tfrac{1}{2}, \tfrac{1}{2})\begin{bmatrix} 0 & 1 \\ 1 & 0 \end{bmatrix} = (\tfrac{1}{2}, \tfrac{1}{2}) \tag{3.20}$$

whence $\mathbf{p}(n+1) = \mathbf{p}(n)$ for all $n \geqslant 0$ if $\mathbf{p}(0) = (\frac{1}{2}, \frac{1}{2})$. The system clearly possesses an equilibrium distribution but convergence to this distribution is not guaranteed.

An example of a chain with more than one stationary distribution is provided by

$$\mathbf{Q} = \begin{bmatrix} 1 & 0 & 0 \\ \frac{1}{3} & \frac{1}{3} & \frac{1}{3} \\ 0 & 0 & 1 \end{bmatrix} \tag{3.21}$$

Stable distributions are $(1, 0, 0)$ and $(0, 0, 1)$ as are linear combinations of the form $((1-\alpha), 0, \alpha)$ with $0 < \alpha < 1$. States 1 and 3 are examples of *absorbing states*; a state i is so designated if $q_{ii} = 1$. To appreciate this nomenclature it is necessary to shift our perspective away from the sequence of state probability vectors $\mathbf{p}(n)$, which relate to the expected behaviour, and consider individually the possible realizations of the process. For this purpose it is convenient to denote the states by E_1, E_2, E_3. If the system is initially in state E_1 then it remains in state E_1 and the only possible realization is

$$E_1, E_1, E_1, \ldots$$

Similar behaviour is observed for E_3. If, however, the initial state is E_2 then there are three possibilities, equally likely to occur, for the next state:

$$E_2, E_1; \quad E_2, E_2; \quad E_2, E_3$$

If the first or third occur then we are thereafter back to our earlier situation in which the system stays in a fixed state. While it is in principle possible for the system to remain in state E_2 the probability is vanishingly small for large n. For $n \to \infty$ the system enters E_1 or E_3 almost surely (with probability one). E_2 is referred to as a *transient state.*

A matrix which gives rise to an *absorbing set of states* rather than isolated absorbing states may be obtained by interchanging columns 1 and 3 of (21)

$$\mathbf{Q} = \begin{bmatrix} 0 & 0 & 1 \\ \frac{1}{3} & \frac{1}{3} & \frac{1}{3} \\ 1 & 0 & 0 \end{bmatrix} \tag{3.22}$$

The behaviour in relation to E_2 is unchanged but not so E_1, E_3. If the system enters E_1, E_3 at any stage it will thereafter alternate between these two states. We can group E_1, E_3 together and refer to them as an absorbing set of states.

The above example also provides a simple illustration of a *periodic* state. More generally a state E_j is said to be periodic with period c if $q_{ij}^{(n)} > 0$ only for $n = c$, $2c$, etc. with $c > 1$ the largest such integer. (The state E_j is *aperiodic* if no such $c > 1$ exists.) It is impossible for the system to return to E_j except at every cth place but it is not necessary that it should do so at every such place in individual realizations. Our example is thus a rather special case.

A noticeable feature of the matrices (3.21) and (3.22) is that if the system is in E_1 or E_3 then it cannot thereafter enter E_2. That is, E_2 is not *reachable* from E_1 or E_3. In general, state j of a Markov chain is said to be reachable from state i if it is possible for the chain to proceed from state i to state j in a finite number of steps. If i is reachable from j and j from i then i and j are said to *communicate* and if all the states communicate with one another the chain is said to be *irreducible.*

Some further important classifications of states may be noted briefly. Letting $f_{jk}^{(n)}$ denote the probability that starting from j the first entry to k occurs at the nth step, then

$$f_{jk} = \sum_{n=1}^{\infty} f_{jk}^{(n)} \tag{3.23}$$

Here $f_{jk} \leqslant 1$ is the probability that starting from j the system will at some time enter k. If $f_{jk} = 1$ $\{f_{jk}^{(n)}\}$ is the *first passage* probability distribution for state k and $\{f_{jj}^{(n)}\}$ is the distribution of *recurrence times* for state j. If $f_{jj} = 1$, such that return to state j is assured, then we can define the *mean recurrence time* for state j as

$$\mu_j = \sum_{n=1}^{\infty} n f_{jj}^{(n)} \tag{3.24}$$

We refer to state j as *persistent* if $f_{jj} = 1$ and transient if $f_{jj} < 1$. A *persistent* state is said to be *null* if μ is infinite; a recurrent state that is neither null nor periodic is referred to as an *ergodic* state.

Concerning irreducible Markov chains we note that all states belong to the same class: they are all transient, all null recurrent or all positive recurrent and they are either aperiodic or all periodic with the same period. For an aperiodic chain if the states are transient or null recurrent then no stable distribution exists and $q_{ij}^{(n)} \to 0$ as $n \to \infty$. On the other hand, if the states are ergodic then there is a unique stationary distribution, $q_{ij}^{(n)} \to \pi_j$ as $n \to \infty$, and $\pi_j = 1/\mu_j$ is the reciprocal of the mean recurrence time for state j. Such ergodic Markov chains are particularly important in practical applications and deserve special attention.

3.2.5 Analytic study of Markov chain dynamics

We observed previously that the behaviour of a homogeneous Markov chain is determined from the initial state probability vector $\mathbf{p}(0)$ by post multiplication by $\mathbf{Q}^n$. We now proceed to examine in more detail the behaviour of $\mathbf{p}(n)$ as n increases, paying particular attention to ergodic chains.

A direct but wholely impracticable approach is to compute $\mathbf{Q}^n$ for all n. Nevertheless direct matrix multiplication for small n can prove instructive as illustrated by the following numerical example.

(*1*) *A sequence of matrix powers*

$$\mathbf{Q} = \begin{bmatrix} 0.9 & 0.1 \\ 0.3 & 0.7 \end{bmatrix}, \qquad \mathbf{Q}^2 = \begin{bmatrix} 0.84 & 0.16 \\ 0.48 & 0.52 \end{bmatrix}$$

$$\mathbf{Q}^3 = \begin{bmatrix} 0.804 & 0.196 \\ 0.588 & 0.412 \end{bmatrix}, \qquad \mathbf{Q}^4 = \begin{bmatrix} 0.7824 & 0.2176 \\ 0.6528 & 0.3472 \end{bmatrix}$$

$$\mathbf{Q}^9 = \begin{bmatrix} 0.753 & 0.247 \\ 0.742 & 0.258 \end{bmatrix}, \qquad \mathbf{Q}^{10} = \begin{bmatrix} 0.752 & 0.248 \\ 0.745 & 0.255 \end{bmatrix}$$

(2) *Analytic expression for* $\mathbf{Q}^n$

$$\mathbf{Q}^n = \begin{bmatrix} \frac{3}{4} & \frac{1}{4} \\ \frac{3}{4} & \frac{1}{4} \end{bmatrix} + (0.6)^n \begin{bmatrix} \frac{1}{4} & -\frac{1}{4} \\ -\frac{3}{4} & \frac{3}{4} \end{bmatrix}$$

(3) *The limiting matrix* $\mathbf{Q}^\infty$

$$\mathbf{Q}^\infty = \begin{bmatrix} \frac{3}{4} & \frac{1}{4} \\ \frac{3}{4} & \frac{1}{4} \end{bmatrix}$$

Example 1: Illustrative matrix powers

Covergence towards the limiting matrix is apparent; the analytic expressions for $\mathbf{Q}^n$ and $\mathbf{Q}^\infty$ will be discussed shortly.

Decomposition of the initial state vector

If our interest is in $\mathbf{p}(n)$ for a given $\mathbf{p}(0)$ then a possible approach is to express $\mathbf{p}(0)$ in terms of the eigenvectors $\{\mathbf{x}_i\}$ of $\mathbf{Q}$. The eigenvectors are defined by

$$\lambda_i \mathbf{x}_i = \mathbf{x}_i \mathbf{Q} \tag{3.25}$$

where $\{\lambda_i\}$ are the corresponding eigenvalues. If the λ_i are distinct then there are N independent eigenvectors. Hence $\{\mathbf{x}_i\}$ is a basis and any initial state vector may be written as

$$\mathbf{p}(0) = \sum_{i=1}^{N} a_i \mathbf{x}_i \tag{3.26}$$

The state probability vector at time n is thus given by

$$\mathbf{p}(n) = \mathbf{p}(0)\mathbf{Q}^n = \sum_i a_i \mathbf{x}_i \mathbf{Q}^n$$

$$= \sum_i a_i \lambda_i^n \mathbf{x}_i \tag{3.27}$$

by way of eqn. (3.25).

Consider now the eigenvalues ordered such that $|\lambda_1| > |\lambda_2| > |\lambda_3|$, etc., with $\lambda_1 = 1$, corresponding to an ergodic process. We then have

$$\mathbf{p}(n) = a_1 \mathbf{x}_1 + \sum_{i=2}^{N} a_i \lambda_i^n \mathbf{x}_i \tag{3.28}$$

and convergence to $a_1 \mathbf{x}_1 = \boldsymbol{\pi}$ is assured since $|\lambda_i| < 1$ for $i \neq 1$ and the terms for $i > 1$ will die away to zero for sufficiently large n. This expansion is useful mainly in that it demonstrates the factors influencing the rate of convergence to the equilibrium distribution. Convergence will be slow if the initial vector $\mathbf{p}(0)$ contains a substantial component of x_i, $i \neq 1$, and λ_i is close to one or if $\mathbf{p}(0)$ contains only a small component of $\mathbf{x}_1 \equiv \boldsymbol{\pi}$. However, if a unique stationary distribution exists then the constraints on $\mathbf{p}(0)$ of a state probability vector ensure that it contains some component of $\boldsymbol{\pi}$ and this is sufficient to ensure ultimate convergence to $\boldsymbol{\pi}$.

Diagonalisation of **Q**

If a matrix **Q** is normal it can be written as

$$\mathbf{Q} = \mathbf{MDM}^{-1} \tag{3.29}$$

Where **D** is a diagonal matrix whose non-zero elements are the eigenvalues of **Q**, **M** is the modal matrix with columns the corresponding eigenvectors and $\mathbf{M}^{-1}$ is its inverse. This decomposition provides for ready determination of $\mathbf{Q}^n$ since

$$\begin{aligned} \mathbf{Q} &= \mathbf{MDM}^{-1} \\ \Rightarrow \mathbf{Q}^2 &= (\mathbf{MDM}^{-1})(\mathbf{MDM}^{-1}) \\ &= (\mathbf{MD})(\mathbf{M}^{-1}\mathbf{M})(\mathbf{DM}^{-1}) = \mathbf{MD}^2\mathbf{M}^{-1} \end{aligned}$$

and by induction

$$\mathbf{Q}^n = \mathbf{MD}^n\mathbf{M}^{-1} \tag{3.30}$$

For a diagonal matrix **D** we have

$$\mathbf{D}^n = \begin{bmatrix} \lambda_1 & 0 & 0 & \cdots \\ 0 & \lambda_2 & 0 & \cdots \\ 0 & 0 & \lambda_3 & \cdots \\ \vdots & \vdots & \vdots & \cdots \\ \vdots & \vdots & \vdots & \cdots \end{bmatrix}^n = \begin{bmatrix} \lambda_1^n & 0 & 0 & \cdots \\ 0 & \lambda_2^n & 0 & \cdots \\ 0 & 0 & \lambda_3^n & \cdots \\ \vdots & \vdots & \vdots & \cdots \\ \vdots & \vdots & \vdots & \cdots \end{bmatrix} \tag{3.13}$$

and computation of $\mathbf{Q}^n$ is greatly simplified. We may thus determine $\mathbf{p}(n)$ for any n and given $\mathbf{p}(0)$ from

$$\mathbf{p}(n) = \mathbf{p}(0)\mathbf{MD}^n\mathbf{M}^{-1} \tag{3.32}$$

The generating function technique

Let $\mathbf{P}(z)$ denote the vector generating function for the vector sequence $\{\mathbf{p}(n), n = 0, 1, \ldots\}$

$$\mathbf{P}(z) = \sum_{n=0}^{\infty} \mathbf{p}(n)z^n \tag{3.33}$$

Postmultiplication of $\mathbf{p}(n)$ by $\mathbf{Q}$ produces $p(n+1)$, hence

$$\begin{aligned}\mathbf{P}(z)\mathbf{Q} &= \sum_{n=0}^{\infty} p(n+1)z^n \\ &= z^{-1}\sum_{n=0}^{\infty} p(n+1)z^{n+1} \\ &= z^{-1}\sum_{n=1}^{\infty} p(n)z^n \\ &= z^{-1}\{\mathbf{P}(z) - \mathbf{p}(0)\}\end{aligned}$$

$$\Rightarrow \mathbf{P}(z)[\mathbf{Q} - z^{-1}\mathbf{I}] = -z^{-1}\mathbf{p}(0)$$

whence

$$\mathbf{P}(z) = p(0)[\mathbf{I} - \mathbf{Q}z]^{-1} \tag{3.34}$$

where $\mathbf{I}$ is the identity matrix.

The sequence of probability vectors $\{\mathbf{p}(n), n = 0, 1, 2, \ldots\}$ may thus be obtained by inverse transformation of $\mathbf{P}(z)$. For the limiting distribution, however, we can employ the final value theorem, whence

$$\boldsymbol{\pi} = \lim_{n\to\infty} \mathbf{p}(n) = \lim_{z\to 1} (1-z)\mathbf{P}(z) \tag{3.35}$$

and, from eqn. (3.34)

$$\mathbf{Q}^{\infty} = \lim_{z\to 1} (1-z)[\mathbf{I} - \mathbf{Q}z]^{-1} \tag{3.36}$$

or more generally

$$\mathbf{Q}^n \Leftrightarrow [\mathbf{I} - \mathbf{Q}z]^{-1} \tag{3.37}$$

where $\Leftrightarrow$ denotes correspondence between distributions and generating functions (i.e., a z-transform pair).

As an illustration of this method, consider the two-state process with transition probability matrix given by

$$\mathbf{Q} = \begin{bmatrix} 1-a & a \\ b & 1-b \end{bmatrix} \tag{3.38}$$

Here

$$\mathbf{I} - \mathbf{Q}z = \begin{bmatrix} 1-(1-a)z & -az \\ -bz & 1-(1-b)z \end{bmatrix} \tag{3.39}$$

and the inverse may be written as

$$[\mathbf{I} - \mathbf{Q}z]^{-1} = \frac{1}{(a+b)(1-z)}\begin{bmatrix} b & a \\ b & a \end{bmatrix} + \frac{1}{(a+b)^2 z}\begin{bmatrix} a & -a \\ -b & b \end{bmatrix} \tag{3.40}$$

whence

$$\mathbf{Q}^n = \frac{1}{(a+b)}\begin{bmatrix} b & a \\ b & a \end{bmatrix} + \frac{(1-a-b)^n}{a+b}\begin{bmatrix} a & -a \\ -b & b \end{bmatrix} \tag{3.41}$$

and

$$\mathbf{Q}^\infty = \frac{1}{(a+b)}\begin{bmatrix} b & a \\ b & a \end{bmatrix} \tag{3.42}$$

Note that eqn. (3.42) may be obtained as the limit for $z \to 1$ in (3.36) or equivalently as the limit for $n \to \infty$ in (3.41). Also, (3.41), (3.42) are the general forms of the expressions for $\mathbf{Q}^n, \mathbf{Q}^\infty$ noted previously in Example 1.

The balance equations

If we think in terms of probability flows then in the steady-state the total flow into and out of any state must be balanced. Hence for each vertex i we have

$$\sum_j p_j q_{ji} = \sum_j p_i q_{ij} \tag{3.43}$$

(total flow into i) = (total flow out of i)

This provides a set of N equations which may be solved for $\boldsymbol{\pi}$ assuming a unique stationary distribution. A state transition diagram aids visualisation; we will illustrate the procedure with reference to the system of Fig. 3.2.

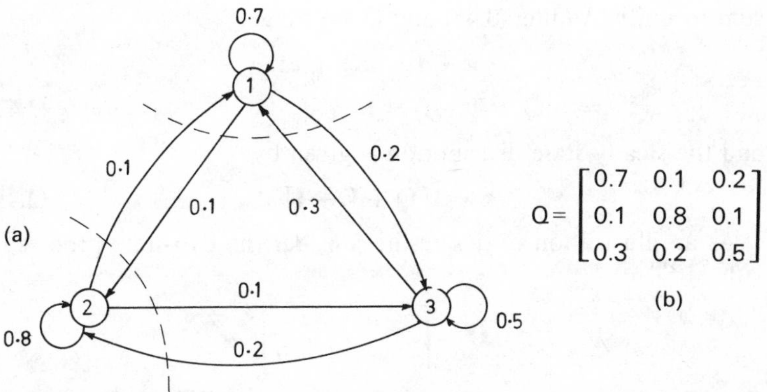

Fig. 3.2 An ergodic three-state Markov chain. (a) State-transition diagram, (b) transition probability matrix

Letting π_i denote the steady-state probability for state i we have:

For state 1

$$0.1\pi_2 + 0.3\pi_3 = (0.1 + 0.2)\pi_1 \tag{3.44a}$$

For state 2

$$0.1\pi_1 + 0.2\pi_3 = (0.1 + 0.1)\pi_2 \tag{3.44b}$$

If we now consider state 3 we obtain linearly dependent equations. Instead we make use of the observation that $\boldsymbol{\pi}$ is a probability distribution

$$\pi_1 + \pi_2 + \pi_3 = 1 \tag{3.44c}$$

Equations (3.44) may now be solved by standard methods to yield

$$\boldsymbol{\pi} = (\pi_1, \pi_2, \pi_3) = (0.364, 0.409, 0.227)$$

as the equilibrium distribution.

Essentially the same procedure may be developed and expressed in a matrix form, which can prove useful for numerical evaluation, as follows:

The stationary distribution satisfies

$$\boldsymbol{\pi} = \boldsymbol{\pi}\mathbf{Q} \tag{3.45}$$

and

$$\mathbf{1} = \boldsymbol{\pi}\mathbf{E} \tag{3.46}$$

where **1** is an N-dimensional row vector with all elements unity and **E** is a $N \times N$ matrix with all elements unity. Equation (3.46) follows from the normalisation of $\boldsymbol{\pi}$, it is a probability vector so its elements sum to unity. Adding (3.45) and (3.46) we get

$$\boldsymbol{\pi} + \mathbf{1} = \boldsymbol{\pi}\mathbf{Q} + \boldsymbol{\pi}\mathbf{E}$$

$$\Rightarrow \boldsymbol{\pi}(\mathbf{Q} + \mathbf{E} - \mathbf{I}) = \mathbf{1} \tag{3.47}$$

and the steady-state distribution is given by

$$\boldsymbol{\pi} = \mathbf{1}[\mathbf{Q} + \mathbf{E} - \mathbf{I}]^{-1} \tag{3.48}$$

As an illustration of this result consider the two-state process of eqn. (3.38)

$$\mathbf{Q} = \begin{bmatrix} 1-a & a \\ b & 1-b \end{bmatrix}$$

$$\Rightarrow \mathbf{Q} + \mathbf{E} - \mathbf{I} = \begin{bmatrix} 1-a & 1+a \\ 1+b & 1-b \end{bmatrix}$$

$$\Rightarrow [\mathbf{Q} + \mathbf{E} - \mathbf{I}]^{-1} = \frac{1}{2(a+b)} \begin{bmatrix} b-1 & a+1 \\ b+1 & a-1 \end{bmatrix}$$

$$\pi = (1, 1) \begin{matrix} \dfrac{b-1}{2(a+b)} & \dfrac{a+1}{2(a+b)} \\ \dfrac{b+1}{2(a+b)} & \dfrac{a-1}{2(a+b)} \end{matrix}$$

$$= \left(\frac{a}{a+b}, \frac{b}{a+b} \right) \tag{3.49}$$

The implications of the various foregoing results concerning Markov chain dynamics are perhaps most readily appreciated with reference to specific systems. By way of example, graphs are presented in Fig. 3.3 showing the transient behaviour and the emergence of equilibria for two and three state systems of Fig. 3.2 and Example 1. In each case the process has been initiated with the system in a single well-defined state. It should perhaps be emphasised that the results

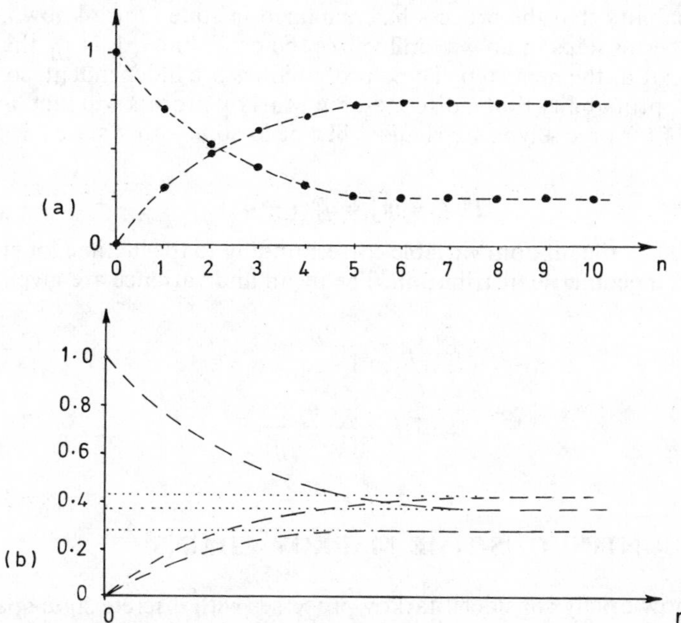

Fig. 3.3 Transient behaviour of illustrative two and three state discrete time Markov chains. (a) Two state chain of Example 1 with P(0) = (0,1), (b) three state chain of Fig. 3.2 with P(0) = (1,0,0)

presented relate to behaviour on the average rather than to specific realisations. A system occupies only one state at any time; the graphs represent probability vectors $\mathbf{p}(n)$ as functions of time and for any given n each component of the vector gives the probability that the system will occupy the corresponding state.

3.2.6 The memoryless property in discrete time

The Markov property of eqn. (3.1) corresponds to dependency in a random sequence extending backwards in time by just one step, so that the present state summarises the entire past history. This is often referred to as the *memoryless* property and it constrains the distribution for the time T_j that a process may remain in a given state, j. We can determine this distribution as follows: let the system be in state j at some time $n = k$. The probability that it will remain in this state at $n = k + 1$ is q_{jj} while $1 - q_{jj}$ is the probability that it will exit from j at the next step. If the system does remain in state j then the probability that it will survive a further step is q_{jj} and we can extend this over m steps to $n = k + m$. Now by the Markov property the probability that the process has remained in state j for a known, m, number of steps in no way influences the probability, $1 - q_{jj}$, that it will exit at the next step. These probabilities are independent, so we have: probability that a discrete-time Markov process will remain in state j for precisely m steps and will exit to some other state $i \neq j$ at step $m + 1$

$$P[T_j = m] = q_{jj}^m(1 - q_{jj}) \tag{3.50}$$

Hence T_j, the random variable corresponding to the lifetime for state j, has a geometric distribution. The mean and variance are given by

$$E\{T_j\} = \frac{q_{jj}}{1 - q_{jj}} \tag{3.51a}$$

$$V\{T_j\} = \frac{q_{jj}}{(1 - q_{jj})^2} \tag{3.51b}$$

3.3. CONTINUOUS-TIME MARKOV CHAINS

We now briefly consider Markov processes with discrete state-space but continuous index set: continuous-time Markov chains. Much of what has been discussed previously in the context of discrete-time processes carries over with only minor modifications and, indeed, it is often convenient to obtain approximate solutions to continuous-time problems by way of an appropriate discrete approximation, although we will not examine this explicitly here. We will restrict attention to ergodic processes.

It is now appropriate to consider transition rates (or intensities) r_{ij} defined by

$$P[X(t + \Delta t) = j \,|\, X(t) = i] \simeq r_{ij}\,\Delta t; \quad i \neq j \tag{3.52}$$

$$\sum_j r_{ij} = 0 \qquad \text{for all } i \tag{3.53}$$

From eqn. (3.53) we note that

$$r_{ii} = -\sum_{\substack{j \\ (j \neq i)}} r_{ij} \tag{3.54}$$

We thus consider a transition rate matrix $\mathbf{R}$ with elements r_{ij} denoting the rate at which the process moves from state i to state j, defined by

$$\mathbf{R} = \lim_{\Delta t \to 0} \frac{\mathbf{Q}(\Delta t) - \mathbf{I}}{\Delta t} \tag{3.55}$$

where $\mathbf{Q}(\Delta t)$ is the state transition probability matrix for a small time increment Δt. The evolution of the state probabilities is thus given by

$$\mathbf{p}(t + \Delta t) \simeq \mathbf{p}(t)\mathbf{Q}(\Delta t)$$
$$\Rightarrow \mathbf{p}(t + \Delta t) - \mathbf{p}(t) \simeq \mathbf{p}(t)[\mathbf{Q}(\Delta t) - \mathbf{I}]$$

which, on proceeding to the limit, becomes

$$\dot{\mathbf{p}}(t) = \mathbf{p}(t)\mathbf{R} \tag{3.56}$$

This matrix differential equation has the solution:

$$\mathbf{p}(t) = \mathbf{p}(0) \exp\{\mathbf{R}t\} \tag{3.57}$$

where $\mathbf{p}(0)$ is the initial condition state vector and

$$\exp\{\mathbf{R}t\} \triangleq \mathbf{I} + \mathbf{R}t + \mathbf{R}^2 \frac{t^2}{2!} + \tag{3.58}$$

From eqn. (3.57) it is clear that $\exp\{\mathbf{R}t\}$ for a continuous-time Markov chain corresponds to $\mathbf{Q}^n$ for the discrete-time case. For the equilibrium distribution, since $\dot{\mathbf{p}}(t) \to 0$ as $t \to \infty$, we have from (3.56)

$$\lim_{t \to \infty} \mathbf{p}(t)\mathbf{R} = \mathbf{0} = \boldsymbol{\pi}\mathbf{R} \tag{3.59}$$

with the usual constraint

$$\sum_i \pi_i = 1 \tag{3.60}$$

since $\boldsymbol{\pi}$ is a probability distribution.

3.3.1 State transition rate diagrams

Useful visualisation for a continuous-time process is provided by the state transition rate diagram; an example is shown in Fig. 3.4. The

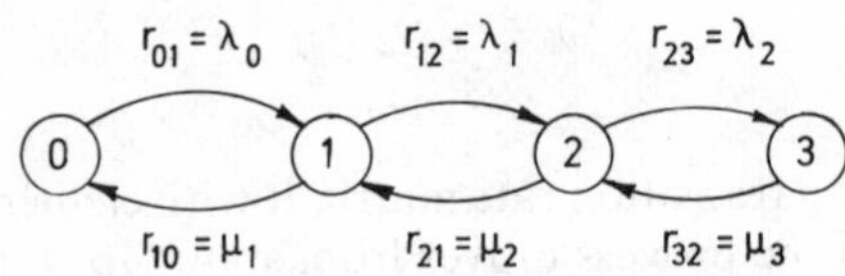

$$R = \begin{bmatrix} r_{00} & r_{01} & 0 & 0 \\ r_{10} & r_{11} & r_{12} & 0 \\ 0 & r_{21} & r_{22} & r_{23} \\ 0 & 0 & r_{32} & r_{33} \end{bmatrix} = \begin{bmatrix} -\lambda_0 & \lambda_0 & 0 & 0 \\ \mu_1 & -(\lambda_1 + \mu_1) & \lambda_1 & 0 \\ 0 & \mu_2 & -(\lambda_2 + \mu_2) & \lambda_2 \\ 0 & 0 & \mu_3 & -\mu_3 \end{bmatrix}$$

Fig. 3.4 State-transition rate diagram and corresponding transition rate matrix for a continuous-time process

arcs on the diagram are labelled with the transition *rates* rather than *probabilities.* Here, as for the std employed with discrete-time processes, the equilibrium probability distribution is readily determined by setting up and solving a set of *balance equations.* For the example of Fig. 3.4, considering each state in turn, this gives

$$-\lambda_0 p_0 + \mu_1 p_1 = 0 \tag{3.61a}$$

$$\lambda_0 p_0 - (\lambda_1 + \mu_1)p_1 + \mu_2 p_2 = 0 \tag{3.61b}$$

$$\lambda_1 p_1 - (\lambda_2 + \mu_2)p_2 + \mu_3 p_3 = 0 \tag{3.61c}$$

$$\lambda_2 p_2 - \mu_3 p_3 = 0 \tag{3.61d}$$

These equations may then be solved to determine the equilibrium state probabilities. Notice that the equation set (3.61), deduced from balance equation arguments, agrees with the general result of (3.59).

3.3.2 Some important special processes

Birth–death processes

If we consider Markov chains for which transitions occur between *nearest neighbour* states only then the matrix **R** (and indeed **Q** for a discrete-time chain) is of tri-diagonal form having all elements zero except for the leading diagonal and the elements immediately above and below this diagonal. Such a chain is often referred to as a *birth–death* process. A transition from state j to $j + 1$ corresponds to a 'birth' and from j to $j - 1$ to a 'death' in a notional population with the size of the population being denoted by the state number. It is appropriate in these circumstances to denote the lowest state as '0' rather than '1' as formerly. Birth–death processes are frequently encountered in communications; for example in switching, traffic and queuing systems studies. In such applications the state may correspond, for example, to the number of calls in progress, links occupied, jobs being processed or in a queue, or to the number of processes active on a computer system.

We will consider here the relatively general case for which arrival rates λ and departure rates μ may be arbitrarily state dependent, as shown in Fig. 3.5.

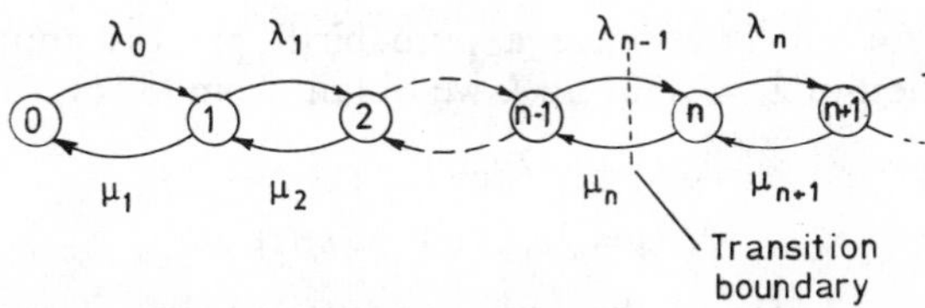

Fig. 3.5 State-transition rate diagram for a general birth-death process

Since only nearest-neighbour states are coupled, we can identify transition boundaries between states and in equilibrium the flow across each boundary must be balanced. Hence:

$$\lambda_{n-1}\pi_{n-1} = \mu_n\pi_n$$

$$\Rightarrow \pi_n = \frac{\lambda_{n-1}}{\mu_n}\pi_{n-1} \tag{3.62}$$

Considering the states in turn, we have

$$\pi_1 = \frac{\lambda_0}{\mu_1}\pi_0$$

$$\pi_2 = \frac{\lambda_1}{\mu_2}\pi_1 = \frac{\lambda_0\lambda_1}{\mu_1\mu_2}\pi_0$$

$$\pi_n = \frac{\lambda_0\lambda_1, \ldots, \lambda_{n-1}}{\mu_1\mu_2, \ldots, \mu_n}\pi_0 \tag{3.63}$$

Also

$$\sum_i \pi_i = \pi_0 + \pi_1 + \pi_2 + \cdots = 1$$

$$= \pi_0 + \frac{\lambda_0}{\mu_1}\pi_0 + \cdots \frac{\lambda_0\lambda_1, \ldots, \lambda_{n-1}}{\mu_1\mu_2, \ldots, \mu_n}\pi_0 + \cdots$$

$$\Rightarrow \frac{1}{\pi_0} = 1 + \frac{\lambda_0}{\mu_1} + \frac{\lambda_0\lambda_1}{\mu_1\mu_2} + \cdots \tag{3.64}$$

The equilibrium distribution is thus obtained by solving eqns. (3.62) and (3.63).

3.3.3 The Poisson process

If $\mu_k = 0$ for all k we have a pure birth process and with the simplification of $\lambda_k = \lambda$ for all k we obtain from (3.56)

$$\dot{p}_0(t) = -\lambda p_0(t) \tag{3.65a}$$

$$\dot{p}_k(t) = \lambda p_{k-1}(t) - \lambda p_k(t) \tag{3.65b}$$

Solving eqns. (3.65) with the initial conditions $p_0(0) = 1$; $p_k(0) = 0$ for all $k > 0$, we obtain

$$p_k(t) = \frac{(\lambda t)^k}{k!}\exp(-\lambda t), \quad k = 0, 1, 2, \ldots \tag{3.66}$$

This is the Poisson distribution giving, in this instance, the probability distribution for the size of the population as a function of time. The associated sequence of birth epochs define a random point process—a Poisson process. The essential features of the (time-homogeneous) Poisson process for our purposes are:

(1) event occurrences are purly random with constant intensity λ
(2) $P[\text{one event in } \Delta t] \simeq \lambda\Delta t$

(3) P[more than one event in Δt] is negligible compared with P[one event in Δt] for sufficiently small Δt.

These features can be used as the basis for a direct derivation of eqn. (3.66).

An important characteristic for the process is the inter-event time distribution. Denoting the inter-event times as τ and assuming that an event occurs at time $t = 0$, we have

$$\begin{aligned} P[t < \tau \leqslant t + \Delta t] &= p_0(t). \quad P[\text{event in } \Delta t] \\ &\simeq \frac{(\lambda t)^0}{0!} \exp(-\lambda t). \quad \lambda \Delta t \\ &= \lambda \exp(-\lambda t). \quad \Delta t \end{aligned}$$

whence the probability density function (pdf) for inter-event times is given by

$$f_\tau(t) = \lambda \exp(-\lambda t) \tag{3.67}$$

If we are concerned with the waiting time to the nth event, W_n, we may note that this is the sum of n independent, identically distributed (iid) random variables each with distribution (3.67). The pdf for W_n is thus an n-fold convolution of (3.67)

$$f_{W_n}(t) = \lambda^n \frac{t^{n-1}}{(n-1)!} \exp(-\lambda t) \tag{3.68}$$

a gamma distribution. This result is perhaps most readily derived with the aid of generating functions, the n-fold convolution of the inter-event time distribution corresponding to $(F_\tau(Z))^n$, the nth power of the probability generating function $F_\tau(Z)$ for the inter-event time. Inverse transformation then gives (3.68).

3.4 ILLUSTRATIVE PRACTICAL EXAMPLES

Random binary signals in discrete time

We consider here a random binary Markov sequence $\{X_n\}$ with state space $\{0, 1\}$. We will determine the equilibrium distribution and the autocorrelation function for such a sequence in terms of the transition probabilities. Equation (3.38) provides the transition probability matrix for such a system

$$\mathbf{Q} = \begin{bmatrix} 1-a & a \\ b & 1-b \end{bmatrix} \tag{3.69}$$

and the nth power, given in eqn. (3.41), is

$$\mathbf{Q}^n = \frac{1}{a+b}\begin{bmatrix} b & a \\ b & a \end{bmatrix} + \frac{(1-a-b)^n}{a+b}\begin{bmatrix} a & -a \\ -b & b \end{bmatrix} \quad (3.70)$$

The equilibrium distribution is thus

$$\pi = \left(\frac{b}{a+b}, \frac{a}{a+b}\right) \quad (3.71)$$

The autocorrelation function for a stationary stochastic sequence is defined by:

$$R_X(k) \triangleq E\{X_n X_{n+k}\} = \sum_{i,j} x_i P[X_n = x_i] x_j P[X_{n+k} = x_j \mid X_n = x_i] \quad (3.72)$$

Here, then, we have

$$\begin{aligned} R_X(k) = {} & 0 \cdot P[X_n = 0] \cdot 0 \cdot P[X_{n+k} = 0 \mid X_n = 0] \\ & + 0 \cdot P[X_n = 0] \cdot 1 \cdot P[X_{n+k} = 1 \mid X_n = 0] \\ & + 1 \cdot P[X_n = 1] \cdot 0 \cdot P[X_{n+k} = 0 \mid X_n = 1] \\ & + 1 \cdot P[X_n = 1] \cdot 1 \cdot P[X_{n+k} = 1 \mid X_n = 1] \end{aligned}$$

$= P[X_n = 1] \cdot P[X_{n+k} = 1 \mid X_n = 1]$ (all other terms being zero)

$$= \pi_1 \cdot q_{11}^{(k)} \quad (3.73)$$

where π_1 is given by eqn. (3.71) and $q_{11}^{(k)}$, *the* k-step transition probability, is available from (3.70). Hence

$$\begin{aligned} R_X(k) &= \frac{a}{a+b}\left\{\frac{a}{a+b} + \frac{b}{a+b}(1-a-b)^k\right\} \\ &= \left(\frac{a}{a+b}\right)^2 \left\{1 + \frac{b}{a}(1-a-b)^k\right\} \end{aligned} \quad (3.74)$$

Note that if $a + b = 1$ we have $\pi_1 = a$; $R_X(k) = a^2$ independent of k, and the elements in the sequence are uncorrelated. In particular, if $a = b = \frac{1}{2}$ this reduces to

$$\pi_0 = \pi_1 = \tfrac{1}{2}; \quad R_X(k) = \tfrac{1}{4}$$

corresponding to a purely random binary sequence with equiprobable uncorrelated elements.

Random telegraph wave—a binary signal in continuous time

We consider now a random binary signal $X(t)$ with state-space $\{0, 1\}$ for which transitions can occur at any time, t. The state transition rate

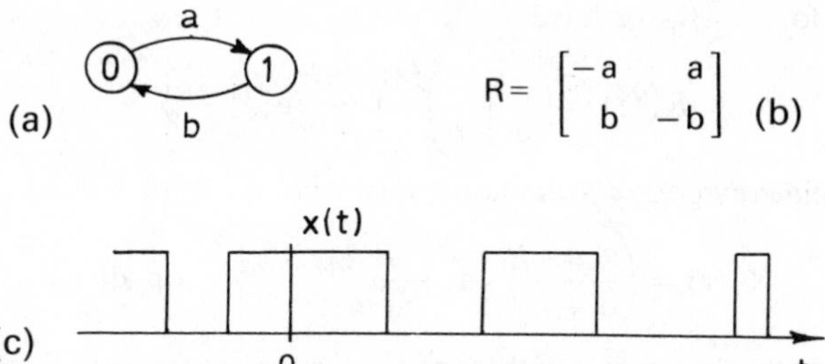

Fig. 3.6 Representation of a random telegraph wave. (a) State-transition rate diagram, (b) transition rate matrix, (c) illustrative realization of X(t)

diagram and matrix are shown in Fig. 3.6, together with an illustrative segment of a realisation, or sample function, of the process. This is a simple birth–death process and the balance equations have the form

$$\pi_0 a = \pi_1 b$$

$$\pi_0 + \pi_1 = 1 \tag{3.75}$$

The equilibrium distribution is thus

$$\pi = \left(\frac{b}{a+b}, \frac{a}{a+b}\right) \tag{3.76}$$

The autocorrelation function is given by

$$\begin{aligned} R_X(t, \tau) &= E\{X(t)X(t+\tau)\} \\ &= R_X(\tau) \quad \text{(for a stationary process as here)} \\ &= \sum_{i,j} x_i P[X(t) = x_i] x_j P[X(t+\tau) = x_j \mid X(t) = x_i] \\ &= P[X(t) = 1] P[X(t+\tau) = 1 \mid X(t) = 1] \\ &= \pi_1 q_{11}(\tau) \end{aligned} \tag{3.77}$$

where $q_{11}(\tau)$, the probability that the system will be in state 1 at time $t + \tau$ given that it is in state 1 at time t is, from eqn. (3.57), given by the corresponding element of the matrix $\exp\{\mathbf{R}\tau\}$ defined in (3.58).

Spectral decomposition yields

$$\mathbf{Q}(\tau) = \exp\{\mathbf{R}\tau\} = \frac{1}{a+b}\begin{bmatrix} b & a \\ b & a \end{bmatrix} + \frac{e^{-(a+b)\tau}}{a+b}\begin{bmatrix} a & -a \\ -b & b \end{bmatrix} \tag{3.78}$$

whence, for $\tau > 0$, we have

$$R_X(\tau) = \left(\frac{a}{a+b}\right)^2 \left\{1 + \frac{b}{a} e^{-(a+b)\tau}\right\}$$

From symmetry considerations we then obtain

$$R_X(\tau) = \left(\frac{a}{a+b}\right)^2 \left\{1 + \frac{b}{a} e^{-(a+b)|\tau|}\right\} \quad \text{for all } \tau \tag{3.79}$$

For the special case of equal transition rates, $a = b = 1/2T$, we have

$$\pi_0 = \pi_1 = \tfrac{1}{2}; \quad R_X(\tau) = \tfrac{1}{4}\{1 + e^{-|\tau|/T}\} \tag{3.80}$$

Limited capacity work pool without storage

We will consider the following scenario: A computer installation is equipped with three matrix printers. Users send print requests to the printer pool at random with mean rate λ and print tasks are of varying length with mean print time $1/\mu$. There is no spooler on the system to allow tasks to be queued so requests occurring when all printers are busy are not accepted, the message 'printers busy' being displayed on the users' terminal. We will assume that users wait for a sufficiently long random time interval before re-trying so that these may be modelled as new requests already accommodated in λ. We wish to determine the printer utilization and the probability that a user will find a printer free to service his request. The state transition rate diagram is shown in Fig. 3.7.

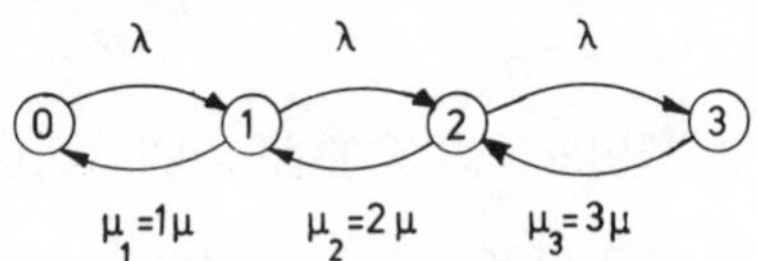

Fig. 3.7 State-transition diagram for a pool of printers

Notice that $\mu_n = n\mu$, since the rate at which tasks are completed is proportional to the number of printers in operation. The balance equations for this system give

$$\begin{aligned} \frac{1}{\pi_0} &= \left\{1 + \frac{\lambda}{\mu_1} + \frac{\lambda^2}{\mu_1\mu_2} + \frac{\lambda^3}{\mu_1\mu_2\mu_3}\right\} \\ &= \left\{1 + \frac{\lambda}{\mu} + \frac{1}{2}\left(\frac{\lambda}{\mu}\right)^2 + \frac{1}{2.3}\left(\frac{\lambda}{\mu}\right)^3\right\} \\ &= \{1 + \rho + \tfrac{1}{2}\rho^2 + \tfrac{1}{6}\rho^3\} \end{aligned} \tag{3.81}$$

where $\rho \triangleq \lambda/\mu$ and

$$\lambda\pi_{n-1} = \mu_n\pi_n = n\mu\pi_n$$

$$\pi_n = \frac{1}{n}\rho\pi_{n-1} \tag{3.82}$$

For example, if $\rho = 2$ we have

$$\pi = (\tfrac{3}{19}, \tfrac{6}{19}, \tfrac{6}{19}, \tfrac{4}{19})$$

and the probability of finding all printers busy is $\frac{4}{19}$, whence the probability of finding a printer free is $\frac{15}{19}$. The printer utilisation is given by

$$\text{Utilisation} = \frac{\sum n \cdot P[n \text{ printers are busy}]}{\text{Total number of printers available}}$$

$$= \tfrac{1}{3}(P_1 + 2P_2 + 3P_3) = \tfrac{10}{19}$$

A single-server queue

This situation could apply, for example, to a computer with a single printer equipped with a spooler to allow print tasks to be queued. We will assume unlimited queueing capacity is available; the state transition rate diagram has the form shown in Fig. 3.8.

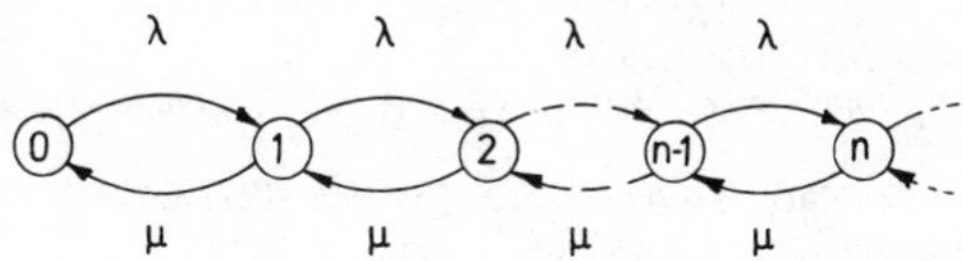

Fig. 3.8 State-transition rate diagram for a single sequence

We now have

$$\frac{1}{\pi_0} = \{1 + \rho + \rho^2 + \cdots\} = 1/(1 - \rho)$$

and require $\rho = \lambda/\mu < 1$ for stability. Also

$$\lambda\pi_{n-1} = \mu\pi_n$$

$$\pi_n = \rho\pi_{n-1} = \rho^n\pi_0 = \rho/(1 - \rho)$$

For example, if we replace the three printers of the previous example by a single printer with three times the print rate, so that $\rho = 2/3$, and add a spooler to allow queueing, we find

$$\pi_0 = P[\text{printer free for immediate access}] = \tfrac{1}{3}$$

$$\text{utilisation} = 1 - \pi_0 = \tfrac{2}{3}$$

3.5 CONCLUDING REMARKS

The aim of this chapter has been to provide a tutorial review of those aspects of Markov chain theory most commonly encountered in engineering applications. Emphasis has been placed on discrete-time processes since, generally speaking, these are easier to visualise, but the relationship, and extension, to continuous-time processes has been outlined. Applications have been considered only briefly since several examples are provided in other chapters of this book. While the treatment provided is introductory, and is most certainly far from comprehensive, it should provide a sound basis for further study of the literature.

BIBLIOGRAPHY

Chung, K. L., *Elementary Probability Theory with Stochastic Processes*, 3rd ed., Springer-Verlag (1979)

Feller, W., *An Introduction to Probability Theory and its Applications*, Vol. 1, 3rd ed., Wiley (1968)

Whittle, P., *Probability*, Wiley (1970)

Kleinrock, L., *Queuing Systems, Vol. 1: Theory, Vol. 2: Computer Applications*, Wiley (1975, 1976)

Allen, A. O., *Probability, Statistics and Queuing Theory with Computer Science Applications*, Academic Press (1978)

Kobayashi, H., *Modelling and Analysis: An Introduction to System Performance Evaluation Methodology*, Addison-Wesley (1978)

4

Bounds and Approximations in Probability Theory

J. J. O'Reilly

4.1 INTRODUCTION

It is frequently the case that statistics of interest for practical applications cannot be readily determined, even when the probability distribution has a known closed-form analytic expression. In yet other instances the distribution itself may not be known but only partly characterised by a limited set of statistics. In such circumstances it is useful to establish relationships between known or readily calculated statistics and desired statistics. Of particular value are relationships which are insensitive to the details of the underlying distribution, these are termed *distribution free*.

Broadly speaking there are two approaches open to us. We can seek an approximation or—and this is particularly appropriate if we are concerned with potential hazard conditions—upper and/or lower bounds. In the latter instance the statistic of interest often taken the form of a tail probability, tail properties are generally very much more difficult to determine than central properties.

The two approaches, bounding and approximating, are not unrelated. An adequately tight bound may serve very well as an approximation. If both upper and lower bounds are available and agree closely then either can be used as an approximation and the difference provides a measure of uncertainty. There are bounds, too, which derive from approximations. For example, we can select a convenient comparison function which is reasonably close to, but does not lie uniformly above or below, a target function. We can then obtain both upper and lower bounds by considering the convex hull of the approximating curve.

It is both convenient and appropriate to begin our discussion of this subject by considering the Gaussian distribution. It is central to much of statistical communication theory, plays a part in certain approximation techniques and provides an elementary illustration of the premise that desired statistics may not be readily calculable even for a known distribution.

4.2 GAUSSIAN APPROXIMATIONS AND BOUNDS

From the central limit theorem we know that the sum of a large number of random variables tends in the limit, assuming finite variance, towards Gaussian. Hence, when dealing with such a sum it is reasonable to examine the possibility of employing an approximation based on a Gaussian distribution with appropriate mean and variance. While this is often appropriate, some words of caution are necessary:

(1) First, the central portion of the distribution of a sum may converge quite rapidly as more terms are added, but the tails are another matter. This is illustrated in Fig. 4.1 where the distribution of the sum of n identical uniformly distributed random variables is presented for $1 \leqslant n \leqslant 4$. From casual observation, concentrating on the central region, the $n = 4$ case could be thought close to Gaussian. But to estimate a tail probability such as $P(x > 4)$ on these grounds is clearly inappropriate.

(2) A second caution relates to skew in the contributors; asymmetry tends to reduce only slowly under convolution.

(3) Finally, we should not overlook the possibility that a random variable may derive from many contributions in some manner other than a sum. For example, if multiplicative rather than additive effects dominate it is likely that the log of the composite random variable will be close to Gaussian, rather than the variate itself. The appropriate limit distribution for approximation in these circumstances is the log-normal.

Before relying on a Gaussian approximation, then, we should check that there are sufficient contributors and that they are of a suitable form to ensure convergence of the statistic of interest. If this is not so then a series approximation may be more useful.

4.2.1 Bounds on the Gaussian tail probability

From the above comments it is clear that the estimation of tail probabilities in particular must be done with care. Yet even if the random variable of interest is known to be Gaussian, or sufficiently close thereto, a closed form analytic expression for the tail probability

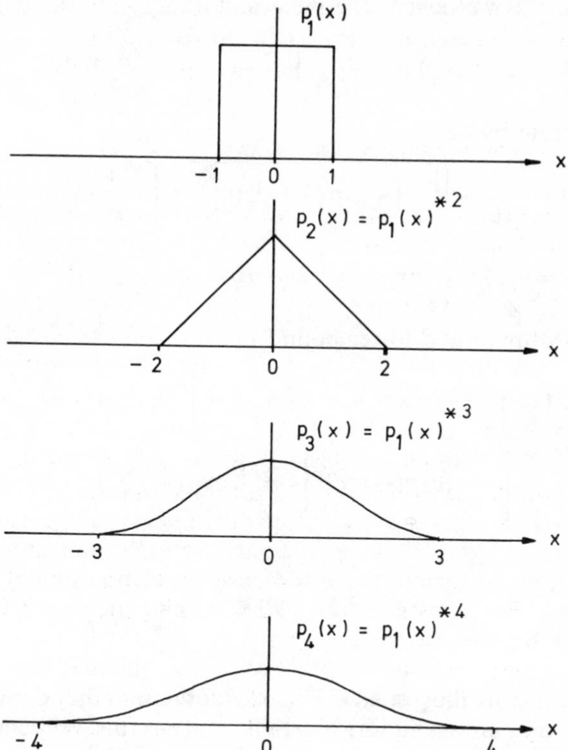

Fig. 4.1 Probability density functions for successive sums of uniform iid random variables

is not available. We consider here, therefore, various specific upper and lower bounds on the Gaussian tail.

For convenience and without loss of generality we will restrict attention to the zero mean unit variance case, transformation of variables readily provides the generalization. The probability density is given by

$$f(x) = \frac{1}{\sqrt{(2\pi)}} \exp\{-x^2/2\} \tag{4.1}$$

and the probability of X exceeding some value $\gamma > 0$ is given by

$$P(X > \gamma) = \int_\gamma^\infty \frac{1}{\sqrt{(2\pi)}} \exp(-x^2/2)\, dx \tag{4.2}$$

To bound this we rewrite the integrand

$$P(X > \gamma) = \frac{1}{\sqrt{(2\pi)}} \int_{\gamma}^{\infty} \frac{1}{x} x \exp(-x^2/2)\, dx$$

and integrate by parts:

$$P(X > \gamma) = \frac{1}{\sqrt{(2\pi)}} \left\{ \left[\frac{1}{x}(-\exp(-x^2/2)) \right]_{\gamma}^{\infty} - \int_{\gamma}^{\infty} \frac{1}{x^2} \exp(-x^2/2)\, dx \right\}$$

$$= \frac{1}{\sqrt{(2\pi)}} \left\{ \frac{1}{\gamma} \exp(-\gamma^2/2) - I \right\}$$

We now upper and lower bound I:

$$0 < I = \int_{\gamma}^{\infty} \frac{1}{x^3} x \exp(-x^2/2) dx < \frac{1}{\gamma^3} \int_{\gamma}^{\infty} x \exp(-x^2/2) dx$$

$$= \left[-\frac{1}{\gamma^3} \exp(-x^2/2) \right]_{\gamma}^{\infty} = \frac{1}{\gamma^3} \exp(-\gamma^2/2)$$

Hence

$$\left(1 - \frac{1}{\gamma^2}\right) \frac{1}{\sqrt{(2\pi)}\gamma} \exp(-\gamma^2/2) < P(X > \gamma) < \frac{1}{\sqrt{(2\pi)}\gamma} \exp(-\gamma^2/2) \tag{4.3}$$

These bounds are illustrated in Fig. 4.2. Notice that they converge as γ increases, they are asymptotically tight, and it is thus very common to

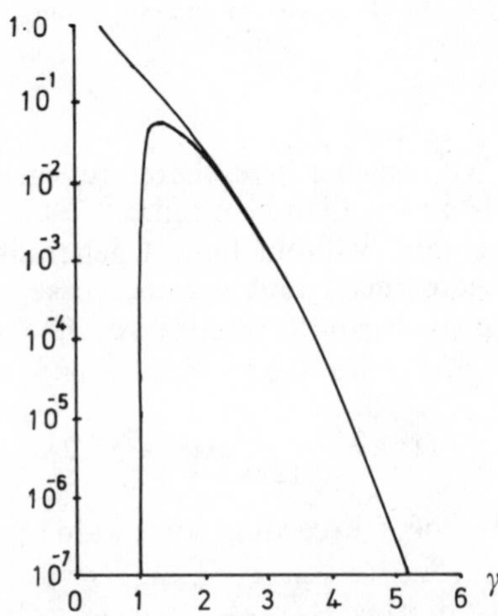

Fig. 4.2 Upper and lower bounds on Gaussian tail function

adopt the upper bound as a tail approximation for analytic purposes.

Another useful approximation is provided by the asymptotic expression [1]:

$$P(X > \gamma) \simeq \frac{1}{\sqrt{(2\pi)}\gamma} \exp(-\gamma^2/2)\left\{1 - \frac{1}{\gamma^2} + \cdots \frac{(-1)^n(2n-1)!!}{\gamma^{2n}}\right\} \tag{4.4}$$

where the notation ( )!! admits only alternate terms of the factorial sequence. Notice that the first term and the first two terms taken together correspond to our upper and lower bounds respectively. A further lower bound which is also a very good approximation is:

$$P(X > \gamma) \geqslant \frac{1}{\sqrt{(2\pi)}} \exp(-\gamma^2/2)\left\{\frac{(\gamma^2 + 4)^{1/2} - \gamma}{2}\right\} \tag{4.5}$$

4.3 BOUNDS AS EXPECTATION FUNCTIONALS

A number of useful bounds can be obtained by applying the expectation operator to bounds on an indicator function. Let I_A be the indicator for a set A:

$$I_A(X) = \begin{cases} 1, & X \in A \\ 0, & \text{otherwise} \end{cases}$$

Then

$$P(X \in A) = \int_{-\infty}^{\infty} I_A(x) f(x)\, dx = E\{I_A(X)\} \tag{4.6}$$

If now we have, $B_1(X) \leqslant I_A(X) \leqslant B_2(X)$ for all X, then

$$E\{B_1(X)\} \leqslant P(X \in A) \leqslant E\{B_2(X)\} \tag{4.7}$$

Provided we can find suitable bounding functions B_1, B_2 lying uniformly below and above I_A we can obtain upper and lower bounds by this means. Much depends, though, on the choice of bounding function. Three examples leading to upper bounds are examined here:

4.3.1 The Markov bound

Consider a non-negative random variable $X \geqslant 0$. Then

$$P(X \geqslant \gamma) = E\{I_{[\gamma,\infty)}(X)\} = E\{U(X - \gamma)\}$$

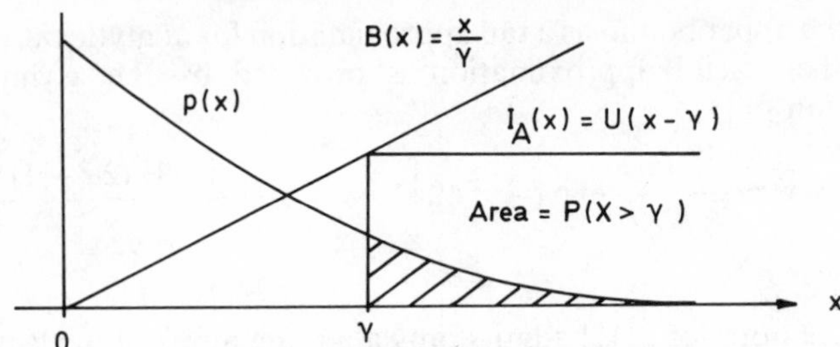

Fig. 4.3 Markov bound

where $U(x)$ denotes the unit step function. Now

$$U(x-\gamma) \leqslant \frac{x}{\gamma}, \quad x \geqslant 0$$

so this provides a suitable bounding function as shown in Fig. 4.3. Hence

$$P(X \geqslant \gamma) = E\{U(X-\gamma)\} \leqslant E\left\{\frac{X}{\gamma}\right\} = \frac{1}{\gamma}E\{X\} \tag{4.8}$$

and we can thus upper bound the tail in terms of the mean value only, although the bound obtained is far from tight.

4.3.2 The Chebyshev inequality

For a random variable X with mean $m = E\{X\}$ and variance $\sigma^2 = E\{(X-m)^2\}$ this bound considers $P(|X-m| > \gamma)$. Here $A = \{x: |x-m| > \gamma\}$, so

$$I_A(X) = \begin{cases} 1, & |X-m| > \gamma \\ 0, & \text{otherwise} \end{cases}$$

We choose a bounding function illustrated in Fig. 4.4, given by

$$B(x) = \left(\frac{x-m}{\gamma}\right)^2 \geqslant I_A(x)$$

Hence

$$\begin{aligned} P(|X-m| > \gamma) &= E\{I_A(X)\} \leqslant E\{B(X)\} \\ &= E\left\{\left(\frac{X-m}{\gamma}\right)^2\right\} = \frac{1}{\gamma^2}E\{(X-m)^2\} \\ &= \frac{\sigma^2}{\gamma^2} \end{aligned} \tag{4.9}$$

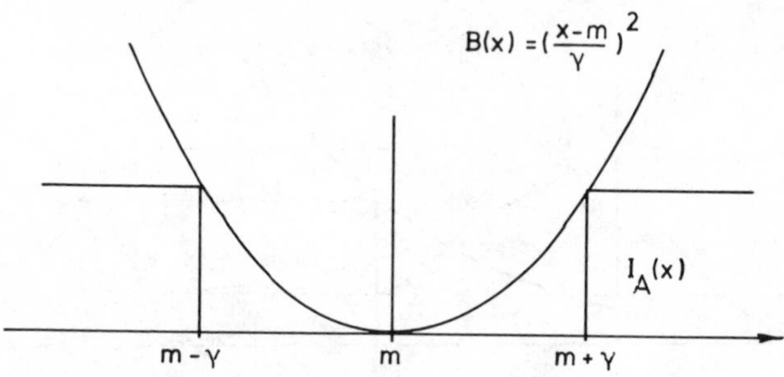

Fig. 4.4 Chebyshev inequality

Here we have been able to upper bound the tails in terms of the mean and the variance.

The two foregoing bounds are often encountered in theoretical developments, but they are of little value for direct application to practical problems since they are far from tight. This should not surprise us since they make use of very little information concerning the random variable of interest. This is not, though, a limitation of the general method, much depends on the form of the bounding function as the next example illustrates.

4.3.3 The Chernoff bound

Here we employ an exponential bounding function, using

$$e^{s(x-\gamma)} \geqslant U(x-\gamma), \quad s>0$$

when considering the positive tail,† as illustrated in Fig. 4.5, so that

$$\begin{aligned} P(X>\gamma) &= E\{U(X-\gamma)\} \\ &\leqslant E\{\exp[s(X-\gamma)]\} = E\{\exp(sX)\exp(-s\gamma)\} \\ &= \exp(-s\gamma)E\{\exp(sX)\} \\ &= \exp(-s\gamma)M_X(s) \end{aligned} \tag{4.10}$$

where $M_X(s)$ is the moment generating function (mgf) for the random variable X. Here s is a strictly positive real number and *any* such s provides a bound. If we want the tightest possible bound we must choose s appropriately to minimise the bound. This is most readily accomplished by considering the log of the bound, which we define by

$$\begin{aligned} LB(s) &= \ln\{\exp(-s\gamma)M_X(s)\} \\ &= -s\gamma+\psi_X(s) \end{aligned}$$

† The same idea is, of course, readily adapted to a negative tail.

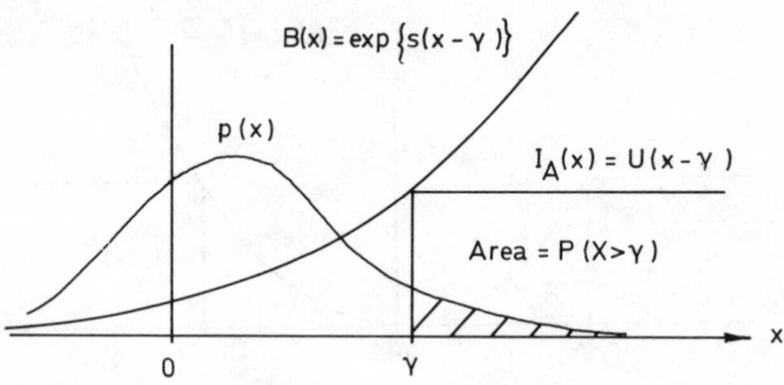

Fig. 4.5 Chernoff bound

where ψ is the log mgf, sometimes called the cumulant generating function. It suffices now to find s_{opt} to minimise $LB(s)$; since ln is a monotonic increasing function this same value of s minimises the Chernoff bound also. Differentiating with respect to s and setting the result to zero to find a stationary point gives

$$\frac{d}{ds} LB(s) = -\gamma + \psi'_X(s) = 0$$

The solution of this equation provides s_{opt}. On substituting s_{opt} into eqn. (4.10) we obtain the tightest form of the Chernoff bound for a given problem. We need also to be sure that s_{opt} minimises the bound, rather than providing a maximum or a stationary point of inflexion. There are no major difficulties here, the procedure guarantees a minimum if the mgf exists for all s. Unfortunately the mgf may not always be so well behaved. For example, one can encounter problems for which $M_X(s)$ is defined only for $(0 \leqslant s < c)$ and for which the bound decreases with increasing $s \to c$. The Chernoff bound is then only defined for these s values and the tightest version is found by letting $s \to c$ rather than by finding a stationary value. In these circumstances it can be desirable to consider complex values of s and to employ a saddlepoint approximation [2].

Various modifications and extensions of the Chernoff bound are possible. For example, if the mgf is not known but an upper bound $BM_X(s)$ for the unknown mgf is available then we can simply replace $M_X(s)$ by $BM_X(s)$ in (4.10) and obtain

$$P(X > \gamma) \leqslant \exp(-s\gamma)M_X(s) \leqslant \exp(-s\gamma)BM_X(s) \qquad (4.11)$$

This holds for all s for which $BM_X(s)$ is well defined and bounds $M_X(s)$. Once again an optimum value of s may be determined to tighten the bound.

Modification to the Chernoff bound relating specifically to error probability in digital and optical communications have recently been reported [3, 4]. Here the principles of analytic continuation and contour integration are employed to obtain upper and lower bounds. The bounds are straightforward to apply while being tight and useful for cases of practical interest.

4.3.4 General formulation

We have concentrated so far on tail probabilities but the idea is in principle applicable more generally. Consider a random variable X and two further ordered random variables r_1, r_2 defined as functions of X by

$$r_1(X) \leqslant r_2(X) \qquad \text{for all } X$$

then

$$E\{r_1(X)\} \leqslant E\{r_2(X)\}$$

An upper bound for the statistic of X defined by $E\{r_1(X)\}$ is obtained as the expectation of a bounding function $r_2(X)$ as before. In our earlier examples $r_1(X)$ was simply an indicator function, but this need not be the case. Also, while we have concentrated on upper bounds it is clear that the method can in principle be used to establish upper and lower bounds to any desired statistic. The difficulty comes in finding suitable bounding functions lying uniformly above and below the function defining the statistic of interest.

4.4 MOMENT SPACE METHODS

Moment space methods avoid the problem noted above of identifying bounding functions. Here bounds are obtained with reference to an *approximating* function. These methods have proved particularly useful in obtaining bounds on error probability for digital transmission [5]. We must begin, though, with some definitions relating to moments.

4.4.1 Generalised moments

Consider a random variable X with probability density $f(x)$ defined on $a \leqslant x \leqslant b$. If a function $r(x)$ is single valued on this interval then a generalised moment is defined by

$$E\{r(X)\} = \int_a^b r(x)f(x)\,dx \tag{4.12}$$

4.4.2 Moment space

Consider a set of generalised moments

$$m_i = E\{r_i(X)\}, \quad i = 1, 2, \ldots, n$$

For a given distribution the coordinates $m = (m_1, m_2, \ldots, m_n)$ can be considered as specifying a point in E_n, n dimensional Euclidean space. The subset of that space comprising all points m corresponding to a valid probability distribution is termed a moment space. It can be shown [6] that the set is closed, bounded and convex; the points corresponding to all possible distributions are clustered into a limited region. Stating this formally, if the coordinates

$$u_i = r_i(x), \quad i = 1, 2, \ldots, n; \quad a \leqslant x \leqslant b$$

define a curve in E_n, where x is a parameter, then all points $m = (m_1, m_2, \ldots, m_n)$ corresponding to a valid probability distribution lie in the convex hull† of the curve.

This statement requires some clarification. To this end we will restrict attention to a two-dimensional example and will provide a supportive plausibility argument for the case of discrete random variables only. More rigorous and comprehensive developments are available elsewhere [5, 6].

Consider a discrete random variable X with probability density function given by

$$f(x) = \sum_i f_i \delta(x - x_i), \quad a < x_i < b \tag{4.13}$$

Taking each possible outcome individually we can state formally the trivial conditional density functions

$$f(x \mid X = x_i) = \delta(x - x_i) \tag{4.14}$$

Each of these 'single impulse' conditional density functions corresponds to a point on the curve in moment space with coordinates $(r_1(x_i), r_2(x_i))$. Considering the ensemble of all possible x_i we generate the curve $(r_1(x), r_2(x))$ in E_2 over the range defined by $a < x < b$.

If we now consider pairs of possible values for X we obtain conditional density functions

$$f(x \mid X \in \{x_j, x_k\}) = f_j' \delta(x - x_j) + f_k' \delta(x - x_k); \ \forall x_j, x_k \in \{x_i\}, \quad j \neq k \tag{4.15}$$

where f_j', f_k' are normalised by $f_j' = f_j/(f_j + f_k)$, etc. Any such conditional density function corresponds to a point in the moment

† Convex hull H of a set $G \triangleq$ smallest convex set $\supset G$.

space on a straight line between $(r_1(x_j), r_2(x_j))$ and $(r_1(x_k), r_2(x_k))$: the first coordinate pair corresponds to $f_j' = 1, f_k' = 0$ such that x_j occurs almost surely, the second to $f_j' = 0, f_k' = 1$ such that x_k occurs almost surely and points in between to distributions for which x_j, x_k may both occur with non-zero probability. All points in E_2 generated by this process lie within the convex hull of the curve, that is, within the closed region defined by the curve and all possible straight lines between points on the curve.

If we relax the conditioning still further we obtain

$$f(x|X\in\{x_j, x_k, x_l\}) = f_j'\delta(x - x_j) + f_k'\delta(x - x_k) + f_l'\delta(x - x_l);$$

$$\forall x_j, x_k, x_l \in \{x_i\}, \quad i \neq j \neq k \qquad (4.16)$$

where the f_j', etc., are now normalised relative to $(f_j + f_k + f_l)$. Any such conditional density function corresponds to a point between p_α and p_β. Considering all possible, we cover points within the closed triangular region defined by p_1, p_2, p_3; considering all possible p_1, p_2, p_3 leads us once more to the convex hull of the curve.

We may further relax conditioning and obtain, in general, polygonal regions of any degree. However, by considering conditional distributions we can always reduce the problem to a union of triangular regions of the form defined above and we are thereby led inexorably to the convex hull of the curve defining a region in E_2 (and in general in E_n) containing all valid discrete distributions.

Illustrative curves in two dimensions and the associated convex hulls are shown in Fig. 4.6. Since the point in E_n corresponding to *any* valid distribution lies in the convex hull of the curve, then for any given value of $m_1 = E\{r_1(X)\}$ the extremes of the convex hull provide upper and lower bounds on $E\{r_2(X)\}$. The bounds obtained in this way are distribution free.

As an example, we consider quantising noise in pulse code modulation. The normalised quantising noise power can be expressed as [14]

$$\frac{1}{3N^2}\int r_2(x) f(x)\,dx \equiv \frac{1}{3N^2}\,E[r_2(x)]$$

where N is the number of encoded quantum intervals, $f(x)$ the probability distribution of signal amplitude, and $r_2(x)$ a function depending on the companding characteristic of the pcm system (specifically, the square of the inverse slope of the compression law). With the logarithmic companding law designed for pcm telephony

$$\begin{aligned} r_2(x) &= \left(\frac{1 + \log A}{A}\right)^2 \qquad 0 \leqslant x \leqslant \frac{1}{A} \\ &= x^2(1 + \log A)^2 \qquad \frac{1}{A} \leqslant x \leqslant 1 \end{aligned}$$

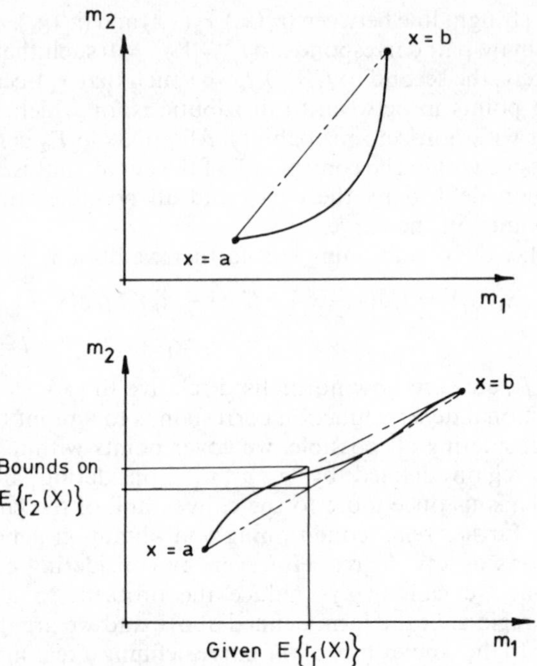

Fig. 4.6 Convex hulls and moment space bounds

where A is the companding parameter. The signal power is $S^2 = E(X^2)$. Taking $r_1(x) = x^2$, we can find moment-space bounds to the quantising noise power. The moment space is shown in Fig. 4.7. The lower bound, and the tighter upper bound A, are derived from the convex hull as described above. The line B, drawn parallel to the lower bound, is a looser but obviously valid upper bound. It is convenient to express the bounds to noise/signal ratio as a function of signal level

$$\frac{\text{Noise power}}{\text{Signal power}} \leqslant \frac{(1+\log A)^2}{3N^2}\left[1+\frac{1-S^2}{A^2S^2}\right] \qquad \text{(Bound } A\text{)}$$

$$\leqslant \frac{(1+\log A)^2}{3N^2}\left[1+\frac{1}{A^2S^2}\right] \qquad \text{(Bound } B\text{)}$$

$$\geqslant \frac{(1+\log A)^2}{3N^2}\left[\max\left\{1, \frac{1}{A^2S^2}\right\}\right] \qquad (4.17)$$

Upper and lower bounds are nowhere more than 3 dB apart (at $S = 1/A$): they converge at both low and high signal levels.

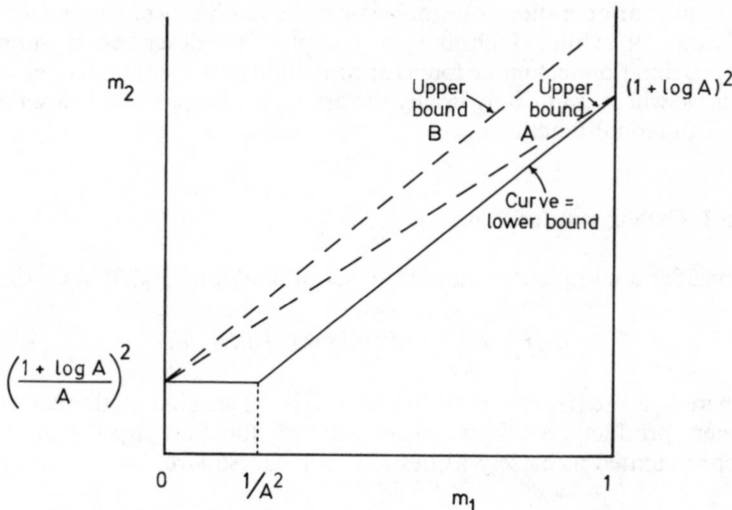

Fig. 4.7 Moment space bounds for quantising noise

The looser bound *B*, originally derived by a different argument [14], is a tight moment-space bound if *x* be allowed to go to infinity. It shows that, although the moment space is usually developed within a finite range, we may be able to allow a semi-infinite range if the relevant curve has an asymptote.

The treatment of moment space bounds presented here is necessarily brief, more details are available in the literature. For example, Dresher [6] provides a more formal, general treatment of the mathematical basis while Yao [7] gives a useful summary with illustrative applications in the field of digital communications. Also, it should be noted that, while we have concentrated here on two-dimensional moment spaces, multi-dimensional extensions are practicable [8, 9] and often provide very tight bounds.

Finally, we observe that the method has recently been extended to accommodate uncertainty in the moments from which the bounds are derived [15].

4.5 SERIES APPROXIMATIONS

Useful approximations can often be obtained by considering an expansion of the desired variable as the sum of a series of mutually orthogonal terms. For example, the use of power series and orthogonal function expansions to assess and quantify the influence of nonlinear operations on stochastic processes has been discussed in Chapter 9 while Cochrane in Chapter 17 describes a series approximation technique for error probability assessment. In view of this we will examine only briefly the principle of expansion in a series of orthogonal functions.

4.5.1 Orthogonal functions

Consider a complete orthonormal set of functions $\{r_i(x)\}$ such that

$$(r_i, r_j) \triangleq \int_{-\infty}^{\infty} r_i(x) r_j(x) w(x)\, dx = \delta_{ij} \tag{4.18}$$

where δ_{ij} is the Kronecker delta and $w(x)$ is the weight function for the inner product. A finite mean square function $q(x)$ can be approximated as closely as desired (in mean square) by

$$q(x) \simeq \sum_i a_i r_i(x) \tag{4.19}$$

where $a_i = (r_i, q)$.

It follows that if X is a random variable with probability density function $w(x)$, then

$$E\{(q(x))^2\} = \sum_i a_i^2 \tag{4.20}$$

Similarly, if $q(x)$, $p(x)$ are two such functions with series coefficients a_i, b_i then

$$E\{q(X)p(X)\} = \sum_i a_i b_i \tag{4.21}$$

we can thus compute a wide range of statistics by using the appropriate orthogonal function expansions. Orthogonal polynomials corresponding to many useful probability distributions are available: a fairly full list, with references to further information, is given in Table 9.1 of Chapter 9.

4.5.2 Hermite polynomials and Gram–Charlier series

Perhaps the most important continuous distribution is the Gaussian. The appropriate orthogonal functions are the Hermite polynomials, which are defined and discussed in Chapter 9, Section 9.7.2. They are readily used for statistical estimates of the form (4.19), and two examples of statistics deriving from Gaussian random processes are given in Chapter 7.

Another, related, approach is to express a non-Gaussian probability distribution in terms of the Gaussian and its derivatives, by means of the *Gram–Charlier series.* Consider a normalised (zero mean, unit variance) random variable X whose characteristic function can be expanded in a series of the form

$$\phi_x(j\theta) = e^{-(1/2)\theta^2}\left\{1 + \sum_{n=3}^{\infty} c_n(j\theta)^n\right\} \tag{4.22†}$$

The Fourier transform relationship, eqn. (9.79) of Chapter 9, shows the probability density to be

$$p(x) = (2\pi)^{-1/2}e^{-(1/2)x^2}\left\{1 + \sum_{n=3}^{\infty} c_n H_n(x)\right\} \tag{4.23}$$

Integrating term by term and using eqn. (9.80) of Chapter 9 gives the tail probability

$$p(x > \gamma) = \int_\gamma^\infty p(x)\,dx = Q(\gamma) + (2\pi)^{-1/2}e^{-(1/2)\gamma^2}\sum_{n=3}^{\infty} c_n H_{n-1}(\gamma) \tag{4.24}$$

where $Q(\gamma)$ is the Gaussian tail probability. The series (4.24) converges if

$$\int_\gamma^\infty \exp(x^2/4)p(x)\,dx$$

converges [13]: conditions for convergence of the distribution (4.22) are more exacting.

The coefficients c_n are related to the cumulants k_r of Chapter 1, Section 1.5. For the normalised variable, the cumulant generating function is

$$\phi(s) = \tfrac{1}{2}s^2 + \sum_{r=3}^{\infty} \frac{k_r}{r!} s^r$$

whence the moment generating function is

$$\exp(\tfrac{1}{2}s^2)\exp\left\{\sum_{r=3}^{\infty} \frac{k_r}{r!} s^r\right\}$$

† The index of summation has the lower limit 3 because with the normalisation used c_1 and c_2 are identically zero: see [12], Section 17.6.

equating this to eqn. (4.22) with the substitution $s = j\theta$ gives

$$\sum_{n=3}^{\infty} c_n S^n = \sum_{m=1}^{\infty} \frac{1}{m!}\left\{\sum_{r=3}^{\infty} \frac{k_r}{r!} S^r\right\}^m \tag{4.25}$$

which allows the c_n to be determined from the k_r. The first few are given by

$$3!\,c_3 = k_3$$
$$4!\,c_4 = k_4$$
$$5!\,c_5 = k_5$$
$$6!\,c_6 = k_6 + 10k_3^2$$
$$7!\,c_7 = k_7 + 35k_3k_4$$
$$8!\,c_8 = k_8 + 35k_4^2 - 56k_3k_4$$

and higher-order terms may be found with increasing labour.

A series of this kind is most useful if (1) it converges rapidly, (2) the terms may be calculated easily. This may be true under the following, non-exclusive, conditions:

(i) The distribution is nearly Gaussian
(ii) The random variable is the sum of many independent random components, its cumulants being therefore the sum of component cumulants
(iii) The random variable is the sum of independent components, one or more of which is Gaussian and thereby contributes nothing to the cumulants of third and higher order (see Chapter 7 for an example).

If the random variable is the sum of N independent identically distributed random components, then the coefficients c_n include factors which are powers of $N^{1/2}$. Unfortunately, the power of $N^{1/2}$ does not increase uniformly with n. The *Edgeworth series* is a true asymptotic series in powers of $N^{1/2}$: it may be considered as a rearrangement of the terms of a Gram–Charlier series which in some circumstances improves the convergence [12]. The difference is often fairly minor, and in the communication theory literature the basic Gram–Charlier series appears to be more widely used.

Acknowledgement

I have had the benefit of seeing an early draft of a forthcoming book on communication engineering by Professor Ken Cattermole. This deals with several of the topics discussed in this contribution and his treatment has undoubtedly influenced my approach. In particular, I am grateful to him for providing the pcm example of moment space methods.

REFERENCES

1. Abramowitz, M. and Stegun, I. A., *Handbook of Mathematical Functions*, Dover Publications Inc. (1972)
2. Yue, O. C., 'Saddle point approximation for error probability in PAM systems with intersymbol interference', *IEEE Trans. Commun.*, **COM-27**, 1604–1609 (1979)
3. Prabhu, V. K., 'Modified Chernoff bounds for PAM systems with noise and interference', *IEEE Trans. Inf. Th.*, **IT-22**, 95–100 (1982)
4. Rocha, J. R. F. and O'Reilly, J. J., 'A modified Chernoff bound for binary optical communication', *Electron. Lett.*, **18**, No. 16, 108–110 (1982)
5. Yao, K. and Tobin, R. M., 'Moment space upper and lower bounds for digital systems with intersymbol interference', *IEEE Trans. Inf. Th.*, **IT-22**, 65–74 (1976)
6. Dresher, M., 'Moment spaces and inequalities', *Duke Math. J.*, **20**, 261–271 (1953).
7. Yao, K., 'Moment space error bounds in digital communication systems', in Skwirzynski, J. K. (ed.), *Communication Systems and Random Process Theory*', Sijthoff and Noordhoff (1978)
8. Yao, K. and Biglieri, E., 'Moment inequalities for error probabilities in digital communication systems', *Proc. N.T.C.*, December (1977)
9. King, M. A., 'Multi-Dimensional Moment Bounds with Applications to Communication Theory', UCLA Ph.D. Thesis (1977)
10. Queen, N. M., *Methods of Applied Mathematics*, 212–218, Nelson (1980)
11. Eisberg, R. M., *Fundamentals of Modern Physics*, 256–266, Wiley (1961)
12. Cramer, H., *Mathematical Methods of Statistics*, Princeton University Press (1946)
13. Whalen, A. D., *Detection of Signals in Noise*, Academic Press (1971)
14. Cattermole, K. W., *Principles of Pulse Code Modulation*, Iliffe, London (1969)
15. Biglieri, E., 'Probability of error for digital systems with inaccurately known interference', *IEEE International Symposium on Information Theory*, Les Arcs, France (1982)

5

Statistical Modelling of Traffic in Networks

K. W. Cattermole

5.1 INTRODUCTION

The characteristic property of communication networks as treated in this chapter is that they serve to interconnect many users who make independent demands for service. The aggregate demand, however quantified, is a random process: in consequence, the behaviour of the network has a component of randomness and the design must be based in fact upon statistical inference. There are two respects in which the user is faced with uncertainty.

(1) Availability of a message channel. Connection may be blocked or transmission delayed in response to overload.
(2) Quality of message transmission. The loss, noise, distortion, error rate and other signal impairments may vary.

Most networks are designed in such a way that these problems are to a large extent separable. We shall concentrate here on the first aspect, that of availability. Some of the signal problems in multiple-access systems have been discussed previously [1].

The availability problem arises primarily because the network is designed to serve the users' needs economically: usually, improvements could be made by more lavish provision. Reasonable compromise requires a knowledge of the trade-off between provision and performance.

The analysis and estimation of availability depends on three interacting factors:

(1) The topology of the network which interconnects the users.
(2) The statistical properties of the traffic as deriving from individual sources and as aggregated into groups.
(3) The protocols governing the operation of the network.

This chapter concentrates on statistical modelling and its interaction with network topology: while postulating only simple and classical protocols. More details on networks, protocols and their interaction may be found in the literature [e.g. 2–7].

5.2 STATISTICAL MODELLING OF TRAFFIC GROUPS

5.2.1 General ideas

This section deals with the statistical theory of individual traffic groups such as might occur at a source or a node, or in a group of trunks. It is an essential preliminary to the statistical treatment of networks in Section 5.3. Useful references throughout are standard books on teletraffic [10, 11], queueing theory [8, 9, 12], probability [13], and stochastic processes [14, 15].

We shall frequently use statistics *conditional* on some event. The concept of conditional probability is well known, but other statistics such as conditional generating functions are also very useful. Consider an exhaustive set $Y = \{Y_1, Y_2, \ldots\}$ of mutually exclusive events to which a probability measure Prob (Y_i) can be assigned. Let $S(X)$ be a statistic of the random variable X, and let $S(X \mid Y_i)$ be the value assumed by S under the condition $Y = Y_i$. Then

$$S(X) = \sum_i S(X \mid Y_i) \operatorname{Prob}(Y_i) \tag{5.1}$$

The conditions Y_i may be numerical (e.g. i trunks busy) or otherwise (e.g. a specific configuration of links in a network). This will become clear by example in this and the succeeding sections.

We are concerned mainly with *stationary* random processes, whose statistics are unchanging with time. More precisely, we shall invoke the concept of *statistical equilibrium*, introduced by A. K. Erlang. The discrete random process of interest may be defined in terms of two sets of probabilities: the *state probabilities* Prob $(X = i) = p_i$, and the *transition probability densities* p_{ij} such that

$$\operatorname{Prob}[X(t + dt) = j \mid X(t) = i] = p_{ij}\,dt$$

In equilibrium, the transition probabilities have been stationary for long enough to ensure that the state probability has attained a steady distribution.

5.2.2 The Poisson process

Consider a point process whose events occur randomly but equiprobably along the time scale: the probability of an event in the interval $(t, t + dt)$ being $\rho\, dt$, where ρ is an unvarying parameter (the *rate* of the process). This is the basic definition of a *Poisson process.* Empirically, it is a good model for many practical point processes. The fundamental properties which would seem to be the basis for its theoretical and practical utility are that it is a *stable* and a *limiting* process. Consider a point process formed by the superposition of several contributory point processes. The following statements about the composite process can be proved:

(1) the sum of Poisson processes is a Poisson process (whose rate is the sum of the rates)
(2) the sum of a large number of point processes each with stationary but otherwise arbitrary interval distribution tends to a Poisson process in the limit.

In the light of these properties, it seems natural that the Poisson has the same dominant role among point processes that the Gaussian has among probability distributions.† Again, it seems a natural model for processes arising from the superposition of many independent contributions: and this is obviously true of traffic streams in a large multi-user system. We shall usually take the Poisson process as a model for the point process containing a point-event for the beginning of each demand to establish a connection or send a message. There are two random variables whose distributions are of interest:

(1) the number of point-events in a given time interval of length t
(2) the time interval between adjacent point-events and, more generally, the waiting time to the rth successive event.

Distribution of numbers of events

Let the probability of r events in the interval t be $p_r(t)$, with generating function $P(z, t)$. It is shown in Chapter 1 that

$$P(z, t) = e^{\rho t(z-1)} \tag{5.2}$$

whence

$$p_r(t) = \frac{(\rho t)^r e^{-\rho t}}{r!} \tag{5.3}$$

This is the *Poisson distribution.* It may be shown that for any Poisson variate X with parameter λ $(=\rho t$ in this case)

$$E(X) = V(X) = \lambda \tag{5.4}$$

† The second property is subject to conditions which (like those of the Gaussian central limit theorem) amount to saying that no one contribution is dominant.

Distribution of intervals

Let the probability density of the time from initial event to the *rth* event thereafter be $f_r(t)$. The probability that the rth event occurs in $(t, t + dt)$ if $f_r(t)\,dt$. But this is the composite event that r point-events occur in the main interval $(0, t)$ and 1 in the increment $(t, t + dt)$, whose probability is $p_{r-1}(t)\cdot\rho\,dt$. Consequently

$$f_r(t) = \rho p_{r-1}(t) = \frac{\rho^r t^{r-1} e^{-\rho t}}{(r-1)!} \tag{5.5}$$

which is known as the *gamma distribution.*

The most important special case is the interval between adjacent events, with the *negative-exponential distribution*

$$f_1(t) = \rho e^{-\rho t} \tag{5.6}$$

Since the time to the rth event is the sum of r inter-event times, it follows that

$$f_r(t) = [f_1(t)]^{*r} \tag{5.7}$$

(the notation implies the convolution of r similar terms) and this may be verified by direct calculation. The Laplace transform of $f_1(t)$ is easily found to be $\rho/(\rho + s)$, so that of the gamma distribution is

$$F_r(s) = \left(\frac{\rho}{s+\rho}\right)^r \tag{5.8}$$

Note that $f_1(t)$ has dual roles: it is the distribution both of inter-event intervals and of waiting time from an arbitrary origin to the next event.†

The mean of a negative-exponential variate is $1/\rho$ and its variance $1/\rho^2$: correspondingly, the gamma distribution has mean r/ρ and variance r/ρ^2. The negative-exponential distribution is also widely used as a model for the distribution of holding or service times. This implies, by the converse to our derivation, that the clear-down of a connection, or completion of service, can be regarded as a random event occurring with uniform probability density. As we shall see, this assumption simplifies a great deal of traffic theory. It is empirically justified for some, but not all, applications; we shall commonly assume its use, with some stated exceptions.

5.2.3 A model for random traffic in equilibrium

Consider a communication link in systems subject to random demands from many sources. The number X of users in the system

† The paradoxical nature of this dual role is fully discussed by Feller [13].

(e.g. telephone calls in progress or messages waiting in a queue for transmission) is a random variable. Its sequence of values as a function of time is a discrete random process. In an infinitesimal increment of time, X may increase by 1 (new call arrives): or diminish by 1 (call in progress is cleared): or remain the same. Multiple events have probabilities which are infinitesimal of higher order, and in the limit may be ignored. Single events are taken as random in the Poisson sense, i.e. defined by probability-density parameters. These parameters may, in the most general case considered, depend on the *state* of the system (characterised by the value of X) but are otherwise time-invariant. The set of states, and the transitions between them, are represented by Fig. 5.1.

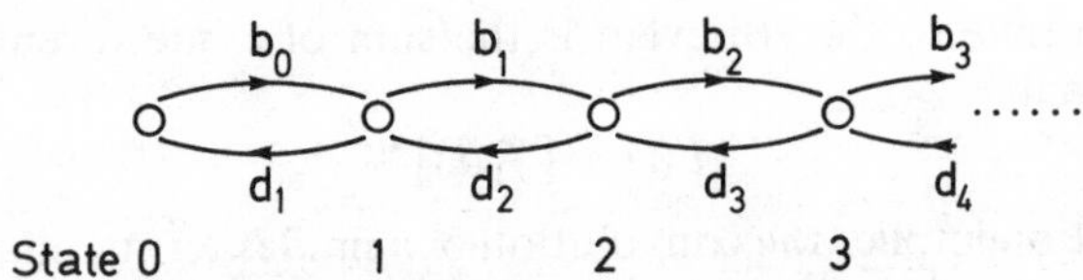

Fig. 5.1 Stochastic process state diagram

In a variety of practical problems, the transition probability densities can be defined by simple *a priori* arguments and the problem is to deduce the distribution of state probabilities: it is the latter which determine the significant dimensions of the system (how many trunk circuits do we need, or how many message locations in a buffer store). Now the process as defined comes under the general class of Markov processes, and may be analysed by the general methods for that class (see Chapter 3). However, there is a much simpler procedure, depending on the concept of statistical equilibrium.

In equilibrium, the upward movement from state i to state $i + 1$ must be balanced in probability by a downward movement from state $i + 1$ to state i. That is,

$$p_i b_i \, dt = p_{i+1} d_{i+1} \, dt \tag{5.9}$$

If we use the notation

$$c_i = \frac{b_0 b_1, \ldots, b_{i-1}}{d_1 d_2, \ldots, d_i}, \quad c_0 = 1 \tag{5.10}$$

then this balance implies that

$$p_i = c_i p_0 \tag{5.11}$$

Since the p_i must sum to unity it follows that

$$p_i = \frac{c_i}{\sum_j c_j} \tag{5.12}$$

As a first example of this process, consider an ideal situation in which the capacity of the system is unlimited. Let demands arise with constant probability density ρ so that $b_i = \rho$ for all i. Let each user drop out of the system on completion of his task, say by clearing down after a telephone call, the holding times being exponentially distributed with mean h: then the probability density of clear-down is $1/h$ for each active user, so that $d_i = i/h$.

Define a traffic intensity parameter $A = \rho h$. Then

$$p_r = \frac{A^r/r!}{\sum_{i=0}^{\infty} A^i/i!} = \frac{A^r e^{-A}}{r!} \tag{5.13}$$

which is a Poisson distribution with mean A. Note that this derivation is quite different in principle from our original derivation of the Poisson distribution. The latter would yield the same distribution for the number of call arrivals in a time h, and hence (very obviously) for the number of simultaneous calls in progress if the holding times were all equal to the constant h. It is a non-trivial result that a completely different holding time distribution gives the same distribution for calls in progress.

5.2.4 Loss systems

The typical telephone system either makes a connection on demand, or rejects it if all available trunks are busy. We will suppose, as a fairly general case, that there are S sources each of rate ρ, and N trunks. Then the upward probability density due to calls arising from free sources is $b_i = \rho(S - i)$: the downward probability density is $d_i = i/h$. Then

$$p_r = \frac{(\rho h)^r \binom{S}{r}}{\sum_{i=0}^{\min(S,N)} (\rho h)^i \binom{S}{i}} \tag{5.14}$$

We shall consider four special cases:

(1) Very large number of sources each with low rate, and relatively few trunks ($S \gg N$).

Take a traffic parameter $A = \rho Sh$ and let $S \to \infty$. Then

$$p_r = \frac{A^r/r!}{\sum_{i=0}^{N} A^i/i!} \tag{5.15}$$

which is Erlang's distribution. The probability of blocking is the probability that all trunks are busy, that is p_N (Erlang's *B*-formula, of which tables and charts are widely available: see References [10, 11, 16]. This applies quite well to the concentration stages of a telephone exchange, in which many subscribers are connected to a relatively small group of trunks.

(2) Small number of sources, and enough trunks for blocking to be small or zero.

Take a traffic parameter $p = \rho h/(1 + \rho h)$ and let $N = S$. Then

$$p_r = \frac{(\rho h)^r \binom{S}{r}}{(1 + \rho h)^s} = \binom{S}{r} p^r (1 - p)^{S - r} \tag{5.16}$$

which is the binomial distribution, corresponding to S trunks each with occupancy probability p independently. This applies quite well to the central stages of a telephone exchange, handling the smoothed traffic coming from the concentrating stages.

(3) Finite number of sources S, and limited trunks $N < S$ so that there is significant blocking. Then the upper limit of summation in eqn. (5.14) is N; there is no analytic simplification, but numerical calculation is fairly easy. This is the *Engset distribution* of telephony [10].

(4) The foregoing examples are standard situations in telecommunications, and can be expressed in fairly general form: but the same principles can be applied to a great variety of non-standard problems. The essential requirement is that the terms of the problem give a means of finding the coefficients b_i and d_i of Fig. 5.1: then the solution follows from eqns. (5.10) to (5.12).

As an illustration, we take a modern pabx with facilities for automatic transfer of incoming calls to another extension if the called extension is busy. Let extension A be transferable to B, B to C and C to A: an incoming caller gets no answer only if all three extensions are busy. Let each extension originate calls at a rate ρ, and let there be incoming calls at rate ρ directed to each extension. Mean holding times are the same for all types of call, and for simplicity we normalise to unit holding time. We ignore calls among A, B and C. What are the probabilities that an incoming caller

(a) is answered by the party he is calling
(b) is answered by someone else
(c) receives busy tone?

There are four states, with the numbers of busy extensions among A, B, C being $\{0, 1, 2, 3\}$. The transition probability densities are

$$b_i = \rho(6 - i), \quad i = 0, 1, 2$$
$$d_i = i, \qquad\quad i = 1, 2, 3$$

It readily follows from eqn (5.12) that

$$\begin{bmatrix} p_0 \\ p_1 \\ p_2 \\ p_3 \end{bmatrix} = \frac{1}{1 + 6\rho + 15\rho^2 + 20\rho^3} \begin{bmatrix} 1 \\ 6\rho \\ 15\rho^2 \\ 20\rho^3 \end{bmatrix} \tag{5.17}$$

which is an Engset distribution for $S = 6$, $N = 3$. The probabilities required are

$$\begin{bmatrix} p_a \\ p_b \\ p_c \end{bmatrix} = \begin{bmatrix} 1 & \frac{2}{3} & \frac{1}{3} & 0 \\ 0 & \frac{1}{3} & \frac{2}{3} & 0 \\ 0 & 0 & 0 & 1 \end{bmatrix} \begin{bmatrix} p_0 \\ p_1 \\ p_2 \\ p_3 \end{bmatrix}$$

$$= \frac{1}{1 + 6\rho + 15\rho^2 + 20\rho^3} \begin{bmatrix} 1 + 4\rho + 5\rho^2 \\ 2\rho + 10\rho^2 \\ 20\rho^3 \end{bmatrix} \tag{5.18}$$

For example, if $\rho h = 0.1$, then an incoming call has an 82% chance of being answered by the called party, a 17% chance of being answered by someone else, and a 1% chance of not being answered at all.

The reader is advised to verify every step in this example, and to study some other traffic problem in the same general category. For some suggestions see Section 5.2.6. below.

We shall return to loss systems in Section 5.3, and there is much more information to be found in the literature [10, 11]. The formulae (5.15) and (5.16) are important in themselves, and are also a foundation for further results. Moreover, a few exercises of the kind suggested above leave one with the clear impression that this general method is much more versatile than would appear from study confined to the usual textbook cases.

5.2.5 Delay systems

The typical data network includes switching centres of a store-and-forward type: that is, incoming messages are placed in a queue for transmission when an appropriate channel is vacant. Similar queueing operations may take place within the signalling and control systems of telephone networks, notably in common-channel signalling techniques. In the archetypal case, messages are never rejected but are subject to a random delay. Such a system can be modelled as in Fig. 5.1 if arrivals are Poisson and holding-times (or message lengths) negative-exponential.

(1) Let there be a source of constant mean traffic intensity, a finite number N of service channels, and unlimited provision for queueing. Then in state 0 the system is idle, in states $1, 2, \ldots, N$ some or all service channels are busy but there is no queue, and in states $N + j$ $(j = 1, 2, \ldots)$ all service channels are busy and there are j users in the queue. The transition probability densities are $b_i = \rho$, and $d_i = h^{-1} \min(i, N)$.

With a traffic parameter $A = \rho h$ we have

$$p_r = \frac{\dfrac{A^r}{r!}}{\sum_{i=0}^{N-1} \dfrac{A^i}{i!} + \dfrac{A^N}{N!}\left(\dfrac{N}{N-A}\right)}, \quad r \leqslant N$$

$$= \frac{\dfrac{A^r}{N! N^{r-n}}}{\sum_{i=0}^{N-1} \dfrac{A^i}{i!} + \dfrac{A^N}{N!}\left(\dfrac{N}{N-A}\right)}, \quad r \geqslant N \tag{5.19}$$

The probability that a message is delayed is the probability that all service channels are busy, namely

$$p_D = \sum_{j=0}^{\infty} p_{N+j} = p_N \sum_{j=0}^{\infty} \left(\frac{A}{N}\right)^j = p_N\left(\frac{N}{N-A}\right) \tag{5.20}$$

which (substituting for p_N) is Erlang's C-formula [8, 11].

The waiting time W_D of a delayed message is a random variable whose statistics can be calculated by summing statistics conditional on the state r. Assuming that messages are sent in order of arrival, the conditional waiting time is the time required to clear the $j + 1$ prior messages. For simplicity, we will normalise to unit waiting time. Then the probability density of clear-down is N, and the interval between successive clear-downs has negative-exponential distribution with mean $1/N$: its Laplace transform, from eqn. (5.10), is $N/(s + N)$, and so the conditional Laplace transform of the waiting time is

$$E(e^{-sW} \mid r = N + j) = \left(\frac{N}{s + N}\right)^{j+1} \tag{5.21}$$

The Laplace transform of the waiting time distribution for all delayed calls is therefore

$$E(e^{-sW_D}) = \frac{\sum_{j=0}^{\infty} p_{N+j} E(e^{-sW} \mid r = N + j)}{\sum_{j=0}^{\infty} p_{N+j}}$$

$$= \frac{N - A}{s + N - A} \tag{5.22}$$

the second form following easily on substitution from eqns. (5.19) and (5.21). This defines a negative-exponential distribution with mean

$$E(W_D) = \frac{1}{N - A} \tag{5.23}$$

in units of mean holding time. Note that this refers to the waiting time of delayed messages. The waiting time W of *all* messages is different, since it takes into account the fact that a proportion $1 - p_D$ have zero waiting time. Consequently

$$E(W) = \frac{p_D}{N - A} \tag{5.24}$$

(2) The special case of a single server is sufficiently important for us to distinguish it. The formulae reduce to:

$$p_r = (1 - A)A^r \tag{5.25}$$

$$p_D = A \tag{5.26}$$

$$E(W_D) = \frac{1}{1 - A} \tag{5.27}$$

$$E(W) = \frac{A}{1 - A} \tag{5.28}$$

The total time that a message spends in the system is the sum of its waiting time and its service time, which has expectation (in units of h)

$$\frac{A}{1 - A} + 1 = \frac{1}{1 - A} \tag{5.29}$$

The coincidence of expressions (5.27) and (5.29) sometimes causes confusion, and they must be conceptually distinguished.

Physical reasoning is perhaps little easier with the single server. For example, all the traffic passes through the one server (assuming that an equilibrium condition is reached, which we comment on below) so the proportion of time that the server is busy is $\rho h = A$. Delay occurs if the server is busy, so $p_D = A$: if there is no delay the system is idle, so $p_0 = 1 - A$. If delay occurs, the mean number of messages ahead of the newcomer is

$$1 + \frac{\sum_{j=0}^{\infty} jA^j}{\sum_{j=0}^{\infty} A^j} = \frac{1}{1 - A} \tag{5.30}$$

and so for delayed messages

(Mean delay) = (Mean service time)

× (Mean number of messages ahead) (5.31)

an intuitively plausible result which, although here derived on the assumption of Poisson input and exponential service time, is more generally applicable.

(3) We conclude this subsection with a remark on the conditions for equilibrium. Several significant statistics for delay systems—mean delay, mean queue length, etc.—assume small values for $A < N$ but diverge as $A \to N$: thus the theory fails if $A > N$. Intuitively, it is clear that the mean traffic A (the product of mean arrival rate and mean holding time) is the mean occupancy of servers and so cannot exceed the number of servers. If too much traffic is offered, the queue length grows, equilibrium is lost, and eventually the system fails to operate. An ideal loss system, on the other hand, always attains an equilibrium. In a practical system, even lost calls place some load on the system and one cannot postulate ideal behaviour: but very large overloads are needed to cause system failure. A well-designed system should always have some provision for shedding excessive load.

As an example of a queueing system with provision for loss in the event of overload, we consider a single-server queue in which there is a limit on queue length. Let there be $M - 1$ waiting places, and therefore a finite set of states $\{0, 1, 2, \ldots, M\}$: any demand made when all places are full is rejected, conceptually without load on the system. The transition probability densities are

$$b_i = \rho, \quad i = 0, 1, \ldots, M - 1$$

$$d_i = 1/h, \quad i = 1, 2, \ldots, M$$

Then

$$p_r = A^r \left(\frac{1 - A}{1 - A^{M+1}} \right) = p_0 A^r \tag{5.31}$$

There are three possible outcomes of a new demand. It is served immediately with probability

$$p_0 = \frac{1 - A}{1 - A^{M+1}} \tag{5.32}$$

It is rejected with probability

$$p_B = p = \frac{A^M - A^{M+1}}{1 - A^{M+1}} \tag{5.33}$$

It is delayed with probability

$$p_D = 1 - p_0 - p_M = \frac{A - A^M}{1 - A^{M+1}} \tag{5.34}$$

The mean delay conditional on there being j calls ahead of the new demand is

$$E(W \mid r = j) = j, \quad j = 0, 1, \ldots, M - 1$$

in units of mean holding time. The mean delay for delayed calls is

$$E(W_D) = \frac{\sum_{j=1}^{M-1} jA^j}{\sum_{j=1}^{M-1} A^j} = \frac{1 - MA^{M-1} + (M-1)A^M}{(1 - A)(1 - A^{M-1})} \tag{5.35}$$

The mean delay for all calls served is

$$E(W) = \frac{\sum_{j=0}^{M-1} jA^j}{\sum_{j=0}^{M-1} A^j} = \frac{A - MA^M + (M-1)A^{M+1}}{(1 - A)(1 - A^M)} \tag{5.36}$$

For large M and small A, these expressions approach (5.27) and (5.28), respectively: but, unlike these latter, they do not diverge as $A \rightarrow 1$. In fact

$$E(W) < E(W_D) < \tfrac{1}{2}M \qquad A < 1 \tag{5.37a}$$

$$< M - 1 \quad \text{all } A \tag{5.37b}$$

Figure 5.2 shows the mean delay and loss probability as a function of the traffic offered, for several values of M. The loss probability curves lie between the curve for a pure loss system (corresponding to $M = 1$) and a lower bound min $\{0, 1 - 1/A\}$.

It is possible to operate a composite system of this kind at high efficiency (A a little less than 1) with low loss and constrained delay. We may view this as either (1) a means of dealing with overload in a delay system, or (2) a means of improving efficiency in a loss system. The latter aspect applies to usage in telephony for example to access a bank of registers. The former aspect applies to flow control in store-and-forward networks, which may be more complex than our example but use the same basic principle of combined loss and delay.

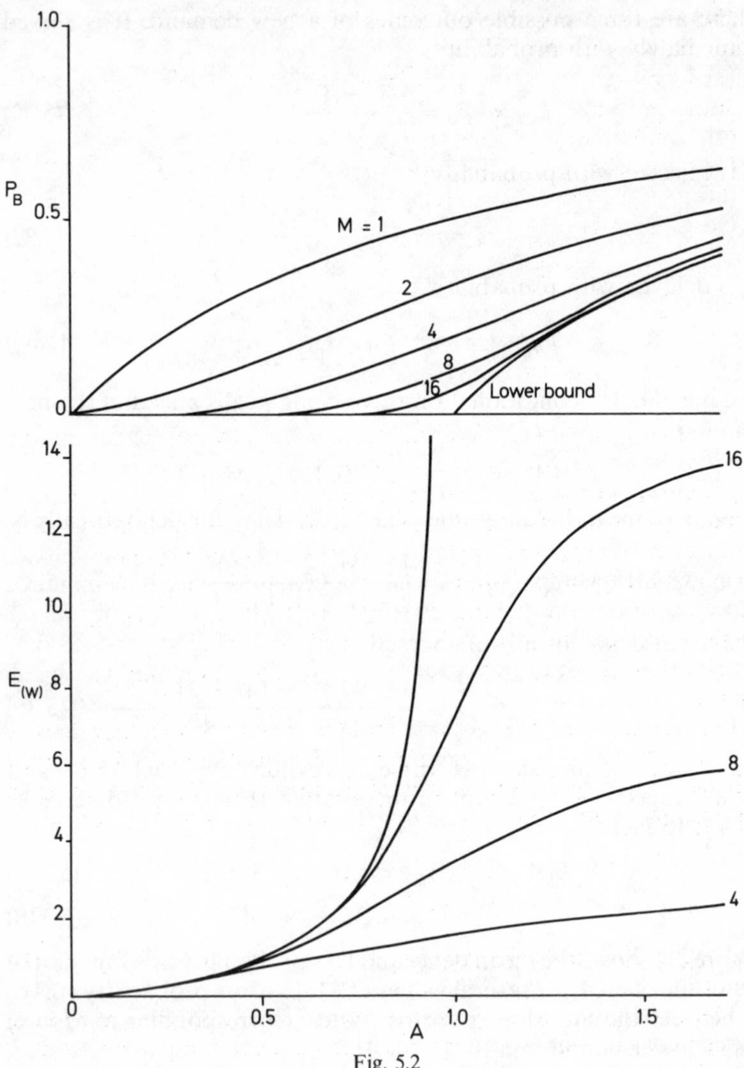

Fig. 5.2

5.2.6 Other applications of the simple model

There are many other cases which can be fitted to the model of Fig. 5.1 by suitable choice of transition probability densities, including

(1) delay systems with finite sources

(2) delay systems in which arrival rate or service time depend on queue length
(3) multi-server systems with finite bounds to queue length, which suffer both loss and delay
(4) combinations of the foregoing
(5) telephone links with two-way traffic
(6) special telephonic situations involving party lines, pabxs, closed user groups, differential calling rates, etc.

We cannot discuss these in detail, but the general approach should be obvious.†

5.2.7 Effect of service-time distribution

Of the many possible generalisations, perhaps the most natural and obvious is to service-time distributions other than the negative-exponential. It is clear that in some situations, such as control signalling, messages or processes of standard format and therefore uniform duration may arise; also that messages and processes may be an aggregate of simpler units, whose aggregate service time might be expected to have a gamma distribution. In face, the theory of the single-server queue with Poisson arrival and arbitrary service time distribution is well-established. We begin by deriving the mean waiting time.

Let service time be a random variable X with probability density $b(x)$. We will consider the waiting time W_D conditional on the delayed call having arrived while a call of duration $x < X < x + dx$ is in service. The proportion of busy time occupied by calls of this duration is

$$\frac{xb(x)\,dx}{\int_0^\infty xb(x)\,dx} \tag{5.38}$$

and with Poisson arrivals this must be the probability of the condition.

The conditional waiting time comprises the unexpired portion of the current call, whose average duration (given random arrivals) is $\frac{1}{2}x$, plus the holding time of calls ahead in the queue (assuming service in order of arrival). Now while a queue persists, its average length is the average number of calls which arrive while one call moves from the tail to the head of the queue, namely $\rho E(W_D)$: so the average total holding time is $AE(W_D)$. Consequently

$$E(W_D \mid x) = \tfrac{1}{2}x + AE(W_D) \tag{5.39}$$

† For further discussion of several such delay systems see Kleinrock [8].

and on making the weighted summation in accordance with the conditional probability density (5.38)

$$E(W_D) = \frac{E(X^2)}{2E(X)} + AE(W_D) \tag{5.40}$$

whence

$$E(W_D) = \frac{E(X^2)}{2(1-A)E(X)} = \frac{E(X)}{2(1-A)}\left\{1 + \frac{V(X)}{E^2(X)}\right\} \tag{5.41}$$

and also $E(W) = AE(W_D)$.

A more complex approach enables us to derive a relationship between the holding time distribution and the waiting time distribution. This is most easily expressed in terms of their Laplace transforms $B(s)$ and $W(s)$, respectively: it is

$$W(s) = \frac{(1-A)sB(s)}{s - A + AB(s)} \tag{5.42}$$

It is easily established (1) that this is consistent with the expectation derived above, (2) that on substituting the exponential holding time distribution $B(s) = (1+s)^{-1}$ the result is

$$W(s) = \frac{1-A}{1-A+s} \tag{5.43}$$

which defines another exponential distribution with mean $(1-A)^{-1}$, consistently with (5.22). For a full discussion of the general-holding-time problem, see Kleinrock [8], or Cox and Smith [12]. For dimensioning data with negative-exponential and with constant holding times see the SEL tables [16].

The main point is that mean waiting time increases with the variance of service time. For the negative-exponential distribution $V/E^2 = 1$, and so the mean delay is twice that for constant service time: most practical cases fall somewhere between these extremes.

Thus queueing systems need to be dimensioned more generously if the service time is variable. This is not, in general, true of loss systems. We have seen that with Poisson arrivals and no loss or delay, the number of calls in service has the same distribution whether holding times are constant or negative-exponential: in fact this is independent of holding-time distribution. The same is true for a loss system with Poisson arrivals of constant mean rate: the Erlang distribution of busy trunks is independent of holding-time distribution. When we come to deal with overflow traffic in Section 5.3.3 we shall see that constant holding time actually requires somewhat more generous dimensioning, which for most people is counter-intuitive.

5.3 STATISTICAL THEORY OF TRAFFIC IN NETWORKS

5.3.1 Scope of the theory

The analytical theory of traffic in networks is much less complete than the theory of simple traffic groups, and a good deal of practical traffic engineering now depends on computer simulations. It is nevertheless well worth studying the theory, for several reasons. Firstly, it gives a good deal of insight into qualitative behaviour of networks. Secondly, the cases known to be soluble are often quite good approximations to practical problems. Thirdly, simulation is often expensive, and one does not want to resort to it unnecessarily.

We shall discuss four specific cases, each of which gives both insight and useful numerical results.

(1) Telephone networks with independently-controlled circuit switching centres.
(2) Overflow and alternate routing in telephone networks.
(3) Connection networks with coordinated control.
(4) Data networks with independent store-and-forward control.

One important limitation of all these studies (and indeed of almost all traffic theory) is that they relate to stationary conditions. In practice, traffic demand shows both cyclical variation and erratic peaks, which on the whole have not been adequately taken into account [10]. The normal practice is to design for some tolerable small impairment (loss or delay) in an "average busy hour"; the latter is, when one stops to analyse it, a very subtle concept, and is in practice defined rather arbitrarily.

5.3.2 Networks of independent circuit switches

A telephone connection may commonly pass through anything from two to eight switching centres, and even larger numbers can occur. At any one of these, blocking is possible. The blocking is due to traffic peaks arising from so many other sources that, at least on a short time scale, it seems plausible to consider the successive stages as independent. Let the blocking probabilities be B_i at the ith switching centre in the chain. Then the probability of blocking, overall, should be

$$B = 1 - \prod_i (1 - B_i) \approx \sum_i B_i, \quad \text{if } B_i \text{ very small} \qquad (5.44)$$

It is usual in telephone networks to assign target values for the B_i such that, if blocking were additive as suggested by eqn. (5.44), the overall result would be tolerable in all the commoner classes of connection. Practical observation shows that overall blocking is often less than would appear from adding the target B_i. There are (at least) three reasons for this, all rather difficult to quantify:

(1) Switching centres are normally installed and extended with some reserve capacity for traffic growth. In a chain of several exchanges, it is likely that some will be operating near the planned maximum capacity but others will have excess capacity in hand.
(2) In a long connection, busy hours may not coincide at the various locations.
(3) There may be some positive correlation between traffic levels of successive links.

Taking all this into account, the simple estimate (5.44) can be considered as a useful upper bound.

5.3.3 Overflow traffic in loss systems

An important category of alternative routing uses a first choice group of circuits and an overflow group. It is intuitively clear that overflow traffic is very far from Poissonian. We will develop a simple model for overflow traffic, partly for its significance in the network, and partly as an illustration of another technique for calculating traffic distributions.

Traffic enters an overflow route only when the first-choice group is blocked. This occurs only for short intervals of time which, under the conditions of Section 5.2.4, are of random duration (with negative-exponential distribution) and random incidence (approximately a Poisson process). We will treat them as random bursts (Fig. 5.3).

How many calls are there in a burst ? If the burst duration be t, then the conditional generating function is given by eqn. (5.4). But the duration has a negative-exponential distribution of the form (5.8) but

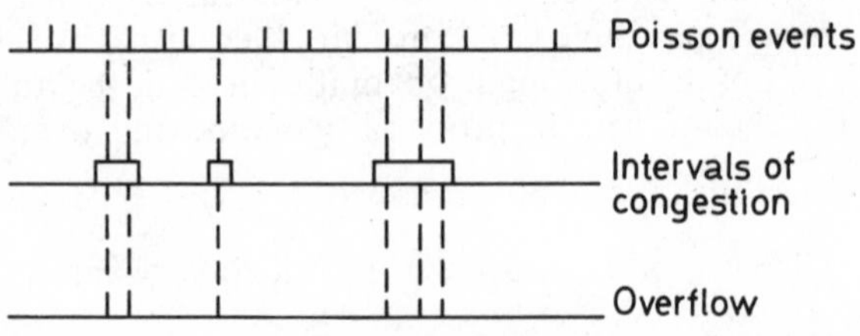

Fig. 5.3 Overflow traffic

with mean duration $1/\beta$ (say). So the unconditional generating function is

$$\int_0^\infty e^{\rho t(z-1)}\beta e^{-\beta t}\,dt = \frac{\beta}{\beta + \rho - \rho z} \tag{5.45}$$

This describes a geometric distribution with

$$p_r = (1-\lambda)\lambda^r, \quad \lambda = \frac{\rho}{\rho + \beta} \tag{5.46}$$

The overflow burst will normally be much shorter than a mean holding time, so we will suppose that to a first approximation the r calls of a burst begin together. If the holding time distribution is negative-exponential with mean h, then the expected duration of a state with $j \leqslant r$ calls, conditional on a burst of r calls, is h/j. The expected duration of states with j calls, randomised over all r, is

$$\frac{h}{j}\sum_{r=j}^{\infty} p_r = \frac{h}{j}(1-\lambda)\sum_{r=j}^{\infty}\lambda^r = \frac{h\lambda^j}{j} \tag{5.47}$$

Suppose that a burst may begin equiprobably anywhere in a long interval of duration T. The number of calls in progress at the end of the interval, due to this burst, is a random variable with generating function

$$F(z) = 1 - \frac{h}{T}\sum_{j=1}^{\infty}\frac{\lambda^j}{j}(1-z^j) = 1 + \frac{h}{T}\log\left(\frac{1-\lambda}{1-\lambda z}\right) \tag{5.48}$$

Now the number of bursts in the interval T is a Poisson variable, with mean γT where γ is the mean rate of bursts. So the generating function for the total number of calls in progress at the end of the interval (hence at an arbitrary time) is

$$e^{\gamma T\{F(z)-1\}} = \left(\frac{1-\lambda}{1-\lambda z}\right)^{\gamma h} \tag{5.49}$$

This is a negative-binomial distribution, with index γh equal to the mean number of bursts in a holding time. It is interesting to note that Wilkinson [17] has given a heuristic argument for the relevance of the negative-binomial distribution, together with empirical evidence. More exact theories of overflow traffic [11], [18] give numerically similar results.

The variance/mean ratio of overflow traffic is, from the distribution (5.49)

$$\frac{1}{1-\lambda} = 1 + \frac{\rho}{\beta} > 1 \tag{5.50}$$

whereas pure Poisson traffic has a ratio of exactly 1, and smoothed traffic (i.e. a traffic group from which the highest peaks have been removed by blocking) a ratio less than 1. Consequently, for given blocking probability, the ratio N/A must be higher for overflow traffic than for traffic elsewhere in the network.

We can apply a variant of this method to the case of constant holding time. All calls have duration h, and the number of calls per burst has the distribution (5.46). So the number of calls in progress at the end of a long interval T due to one burst has the generating function

$$F(z) = 1 - \frac{h}{T}(1-\lambda)\sum_{j=1}^{\infty}\lambda^j(1-z^j)$$

$$= 1 - \frac{h}{T}\left\{1 - \frac{1-\lambda}{1-\lambda z}\right\} \tag{5.51}$$

The generating function for the number of calls due to a Poisson distribution of bursts is

$$e^{\gamma T\{F(z)-1\}} = \exp\left\{\gamma h\left[\frac{1-\lambda}{1-\lambda z} - 1\right]\right\} \tag{5.52}$$

The mean and variance may be found by differentiation of the generating function, as explained in Chapter 1. The results show that, if the holding time is constant, the variance/mean ratio is $1 + 2\rho/\beta$, i.e. higher than for negative-exponential. This conclusion is consistent with the results of Burke [19], especially at low overflow traffic for which the present model should be most accurate. Compare the discussion of holding-time distributions in Section 5.2.7.

5.3.4 Connection networks with conditional selection

Within a single telephone exchange, it is possible for passage through many stages of switching to proceed in a step-by-step manner, so that overall blocking probability is (approximately) the product of loss probabilities in each stage, as in Section 5.3.2. In modern exchange systems, however, the control mechanism makes a search for free paths through a multi-stage network, and selects an initial link only if the latter has a free outlet through which the complete connection can be made; this is known as *conditional selection.* The topology of such networks is described in, for example, refs. [4], [10], [20], [21], [22], [24] and [28]. It is sufficient to remark here that for traffic calculations the network may usually be represented by its *channel graph*; this is a linear graph showing all paths between an inlet and an outlet which are to be connected, with switchblocks represented by nodes and the links joining them represented by edges.

There are many possible paths, but since all links are accessible to random traffic from many sources it may happen that all paths are blocked. Let this event be denoted by Y. The probability $P(Y)$ will, in general, be far from obvious: the number of states and transitions in the network is usually so great as to render the simple equilibrium method impracticable. However, we may use conditional probabilities as in Section 5.2.1. For example, let us fix some variable X, say the number of links busy in a particular group. The conditional probability $P(Y \mid X = x_i)$ may be much simpler, perhaps taking the form of a known result for a single stage. We are not confined to numerical variables (the outcomes of X might be topological configurations of busy links, a generalisation of the technique which is very useful with complex networks) but will first take a numerical example.

Consider a connection via two links in tandem. We designate the stages B and C, and suppose that there are m links in each group; the effect is the simple channel graph of Fig. 5.4. We define probability

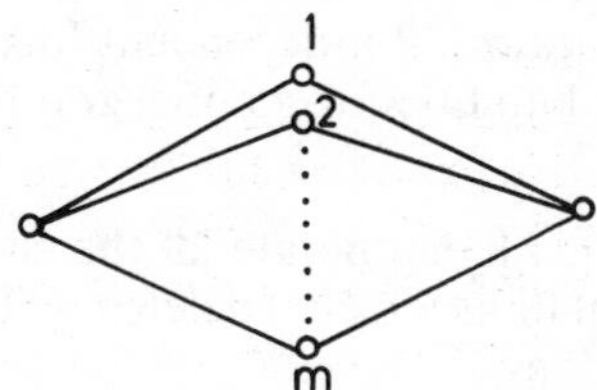

Fig. 5.4 Two-stage channel graph

distributions $G(x)$ that precisely x of the C-links are busy, and $H(y)$ that a nominated set of y B-links are busy (i.e. do not include a free link, so that $H(0) = 1$). Then the blocking probability over two stages is

$$p_B = \sum_{x=0}^{m} G(x)H(m-x) \tag{5.53}$$

This method was pioneered by Palm and Jacobeus [20]: it was extended and surveyed by Elldin [21], and many examples may be found in these sources and in Bear [10].

According to specific situations, the distributions G and H may be of Erlang, Engset or binomial form (eqns. (5.15, 5.14, 5.16), respectively): the final blocking probability sometimes emerges as a ratio of Erlang-*B* probabilities. We present here only one example, simple in this context but significant because it suggests the foundation of a more general theory of connection networks.

Suppose the *B* and *C* links have binomial distributions (often valid for well-smoothed traffic, as we remarked in Section 5.2.4) with link occupancy probabilities *b* and *c* respectively. Then

$$G(x) = \binom{m}{x} c^x (1-c)^{m-x}$$

$$H(m-x) = b^{m-x}$$

$$p_B = (b + c - bc)^m \tag{5.54}$$

This result can be derived in another way. Let us consider the channel graph in Fig. 5.3 to have independent occupancy of links. Any one of the *m* parallel paths is free if both its links are free, so it is blocked with probability $1 - (1 - b)(1 - c)$. The channel graph is blocked if all paths are blocked, and by multiplication of independent probabilities

$$p_B = \{1 - (1 - b)(1 - c)\}^m \tag{5.55}$$

Simple manipulation shows that expressions (5.54) and (5.55) are identical. The method may be extended to any combination of series-parallel paths.

On the same assumption of independent link occupancy, we can calculate blocking probabilities of channel graphs of any form. There are two methods:

(1) Following Lee [22], enumerate all the cut-sets of the channel graph, i.e. the sets of links which if deleted will cause blocking, and sum their probabilities.

(2) Define the simplest set of conditions *X* which reduce the channel graph to easily calculable special cases, and sum the conditional probabilities according to eqn. (5.1). This method may give tractable calculations even for such complex channel graphs as those of Cartesian product networks [23], or for Takagi's optimal multistage graph [24]: see also [4].

Conditions in the form of configurations are perhaps less familiar than those in the form of numbers, so we give an example of this very powerful technique. Let us verify Takagi's contention that the channel graph of Fig. 5.5b has uniform lower blocking probability than that of Fig. 5.5a, which uses the same number of links and switches but with a different pattern of interconnection. We suppose that each link is independently busy with probability *p*, and free with probability $q = 1 - p$. The blocking probability of channel graph (a) can be found by series/parallel calculations similar to those used

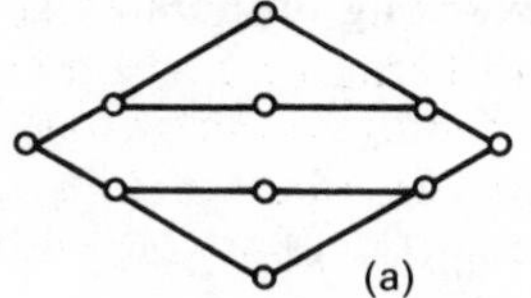

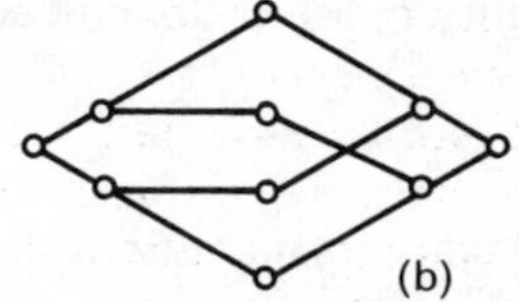

Fig. 5.5

above: but channel graph (b) is not amenable to such an approach. Lee's method, though valid in principle, calls for quite a lot of enumeration even for this rather small graph. The simplest method is to enumerate only the configurations of blocked and free links in the two stages adjacent to the terminal points: the channel graph conditional on each configuration is a simple series/parallel graph whose conditional blocking probability can be expressed in terms of

$$b = 1 - q^2$$

which is the blocking probability of two links in tandem.

Figure 5.6 shows the configurations C_i (their free links being drawn): their probabilities of occurrence p_i taking account of obvious permutations; their conditional channel graphs G_i; and the blocking

i	C_i	P_i	G_i	b_i
1		q^4		b^4
2		$4q^3p$		b^2
3		$2q^2p^2$		b
4		$2q^2p^2$		b
5		$2q^2p^2$		
6		$4q\,p^3$		
7		p^4		

Fig. 5.6

probability b_i for each G_i. The overall blocking probability is

$$p_B = \sum_{i=1}^{7} p_i b_i$$
$$= q^4 b^4 + 4q^3 pb^2 + 2q^2 p^2 (2b + 1) + 4qp^3 + p^4 \tag{5.56}$$

It is not difficult to find the blocking probability p_A of Fig. 5.5a, either by the conditional method or by series/parallel calculations. It is left to the reader to show that

$$p_A - p_B = 2q^4 p^2, \quad > 0 \text{ for } 0 < p < 1 \tag{5.57}$$

which proves Takagi's optimal construction for the special case of a 5-stage switched network. This result also has general implications for the interconnection of redundant units in the design of reliable equipment.

5.3.5 Store-and-forward networks with Poisson arrivals and departures

The Poisson process, and the associated negative-exponential distribution, constitute a special case of unique importance in the statistics of store-and-forward networks. Briefly, they are *stable* statistics: a network whose inputs are Poisson and whose service-times are exponential generates Poisson arrivals, and so the negative-exponential distribution of inter-arrival times, in each and every link, regardless of the topology.

The first step in our demonstration is to show that with these stipulations the departure process from a queue is Poisson. We prove a generalisation of Burke's theorem [25]. Let a queue be modelled as a stochastic process of the class described in Section 5.2.3, with upward probability densities $b_i = b$ (a constant, representing a constant rate of Poisson arrivals) and downward probability densities d_i (which may in general be different, representing service times which may be state-dependent). The equilibrium distribution is given by (5.12) with

$$c_i = \frac{b^i}{d_1 d_2, , , , , d_i}$$

We define $F_k(t)$ as the joint probability that (1) the system is in state k, (2) the current interdeparture interval is greater than t. We further define functions $G_k(t)$ by

$$F_k(t) = p_k G_k(t)$$

where the p_k are equilibrium state probabilities, and the $G_k(t)$ are conditional tail probabilities with $G_k(0) = 1$. The procedure is to set up differential equations for the $F_k(t)$ and thence for the $G_k(t)$. Over an infinitesimal interval dt, the only way for $F_0(t)$ to change is by the arrival of a new call. Consequently

$$F_0(t + dt) = F_0(t)(1 - b\,dt)$$

whence

$$F_0'(t) = -bF_0(t)$$

and also

$$G_0'(t) = -bG_0(t)$$

which with the boundary condition at the origin gives

$$G_0(t) = e^{-bt} \tag{5.58}$$

For $k > 0$, $F_k(t)$ can be reduced by either arrival or departure, and can also be augmented by arrival. The differential equation is

$$F_k'(t) = bF_{k-1}(t) - (b + d_k)F_k(t)$$

whence

$$p_k G_k'(t) = p_{k-1} b G_{k-1}(t) - p_k(b + d_k)G_k(t)$$

Since in equilibrium $p_{k-1}b = p_k d_k$, this is

$$G_k'(t) = -bG_k(t) + d_k\{G_{k-1}(t) - G_k(t)\}$$

with the obvious solution

$$G_{k-1}(t) = G_k(t) = e^{-bt} \tag{5.59}$$

the bracketed term in the differential equation being zero and the remaining terms constituting a simple differential equation for $G_k(t)$. Equations (5.58) and (5.59) show that the conditional interdeparture distribution is the same for all k, and is negative-exponential with the same mean as the interarrival distribution. If arrivals and service times are independent, then the departures are a Poisson process.

Next we invoke the property that the superposition of several Poisson processes of rates ρ_i is a Poisson process of rate $\rho = \sum_i p_i$ (Section 5.2.2). The stability of Poisson statistics in the network follows. For the input to any node is the sum of original inputs (Poisson by postulate) and/or outputs from other nodes: if each output is Poisson, then so are the aggregate inputs, which in turn guarantees the property in the outputs. Also, this is a stable balance. In Chapter 1, Section 1.4.4, we gave an alternative derivation for the departure process of a single server queue with Poisson arrivals and negative-exponential service times. Equation (1.51b) may be written as

$$D(s) = AB(s) + (1 - A)B(s)C(s) \tag{5.60}$$

where A is the mean traffic, and $B(s)$, $C(s)$, $D(s)$ are the Laplace transforms of service, next-arrival and interdeparture times, respectively. In the special case, we found that $D(s) = C(s)$. Let us suppose, however, that the arrival process is not strictly Poisson, but is a more general renewal process with independent interarrival times having a distribution departing somewhat from negative-exponential. It is clear from (5.60) that a small perturbation in $C(s)$ will produce a smaller perturbation in $D(s)$, the effect being most obviously diluted at high traffic levels. The effect is particularly small if each node has several inputs since, again invoking the cardinal properties from Section 5.2.2, the Poisson is a limiting process, approximated by the sum of many independent point processes even if the latter have arbitrary stationary distributions of interarrival times. Addition of even 3 or 4 independent sources may give a good approximation to the negative-exponential distribution of intervals; see, for example, the simulations reported by Kleinrock [7].

There are further key results on queueing networks whose proofs go beyond the scope of the present chapter. Jackson [26] proved that in a network under the present assumptions the joint occupancy distribution of the nodes equals the product of their marginal distributions: in other words, the node occupancy statistics behave as if the nodes were independent. Other product form solutions have been derived by Gordon and Newell [30] for closed networks with state-dependent service times; and by Baskett, Chandy and others [31, 32] for an extensive class of queueing networks including those with several types of customer. The customer-type approach may be used, among other things, to deal with non-random routing, by distinguishing at each node customers whose overall paths are different. Reiser and Lavenberg [33] show that many useful statistics derivable from a product-form joint distribution may also be attained by simpler methods. Kelly [34] has given an extensive treatment of stochastic networks based on the concept of reversibility, which is closely allied to the concept of local balance invoked in some of the previously cited works [31, 32].

Analysis of Poissonian queueing networks on the assumption that the several links are independent reveals some interesting properties, which are largely validated by simulation and practical experience. The following discussion is based mainly on the work of Kleinrock [7].

In the light of the foregoing discussion, we can estimate the delays in a store-and-forward network by means of a simple model. Each link is considered separately, together with the queue at its entry node. Let link number i have a capacity C_i (in appropriate units, e.g. bits/second) and a mean message rate of λ_i messages/second. Let the messages throughout the network have a negative-exponential length distribution (or more precisely a discrete approximation thereto) with mean length m bits. Then the mean holding time is m/C_i and the mean occupancy $A_i = \lambda_i m/C_i$. Traffic queueing for transmission on this link has a mean delay given by simple queueing theory as

$$T_i = \frac{1}{C_i/m - \lambda_i} \tag{5.61}$$

The total delay suffered by all messages entering in unit time is

$$S = \sum_i \lambda_i T_i = \sum_i \frac{\lambda i}{C_i/m - \lambda_i} \tag{5.62}$$

This can, of course, be reduced indefinitely by providing large capacity on all routes, but it is interesting to see what can be done within a capacity constraint $\sum_i C_i = C$. It can be shown [6, 7] that S is minimised by assigning link capacities

$$C_i = m\lambda_i + (C - m\lambda)\left\{\frac{\sqrt{\lambda_i}}{\sum_i \sqrt{\lambda_i}}\right\} \tag{5.63}$$

where $\lambda = \sum_i \lambda_i$. That is, each link has a capacity exceeding its average traffic, and the excess is shared out in proportion to the square root of traffic. The link delays are then inversely proportioned to $\sqrt{\lambda_i}$, and the total delay is

$$S = \frac{m\{\sum_i \sqrt{\lambda_i}\}^2}{C - m\lambda} \tag{5.64}$$

This is a lower bound to total delay: the assignments are not very critical so long as there is reasonable excess capacity on all links. The interesting point is that the total delay of a network is inversely proportional to total excess capacity, just as a simple queue has delay proportional to $1/(N - A)$.

To establish the mean delay of individual messages, we have to know something about their routing. In a network of moderate connectivity, many of them will traverse more than one link. Let the total mean rate at which messages enter the network be γ: then the mean path length, in links, is $L = \lambda/\gamma$. Taking the optimal assignment (eqn. (5.63)) as giving a lower bound, we have for mean message delay

$$T = \frac{S}{\gamma} \geqslant \frac{mL}{C - m\gamma L}\left\{\sum_i \sqrt{\frac{\lambda_i}{\lambda}}\right\}^2 \tag{5.65}$$

As we might expect, the reciprocal variation with excess capacity is still exhibited. This expression also enables us to draw two significant conclusions about routing:

(1) The path length L should be as short as reasonably possible. For given C, the delay T may increase rapidly with long routings.
(2) Large routes are more efficient than small ones: delay can be reduced if several small routes can be aggregated. This follows from the dependence on $\{\sum \sqrt{\lambda_i}\}^2$.

These factors may conflict, so that some compromise is necessary. The main qualitative conclusion is that fairly simple networks, without too many alternative routes, are most efficient in general.

In store-and-forward networks, departure from independence is often the result of deliberate design for efficient routing and flow control; these practices and their effects are described by Schwartz [3, 35]. There is a large and growing body of material on tandem queues and queueing networks in the literature of communications, computing systems and statistics: good surveys will be found in the books by Kleinrock [8, 9], Newell [27], Kelly [34] and Kobayashi [36].

REFERENCES

1. Cattermole, K. W., 'Stochastic problems of multiple communication', in Skwirzynski, J. K. (ed.), *Communication Systems and Random Process Theory*, Sijthoff and Noordhoff (1978)
2. Symons, F. J. W., 'Representation, analysis and verification of communication protocols', in ref. 29
3. Schwartz, M., 'Routing and flow control in data networks, in ref. 29
4. Cattermole, K. W., 'Graph theory and communication networks', in Wilson, R. J. and Beineke, L. W. (eds.), *Applications of Graph Theory*, Academic Press (1979)
5. Flood, J. E., *Telecommunication Networks*, Peter Peregrinus (1975)
6. Schwartz, M., *Computer-Communication Network Design and Analysis*, Prentice-Hall (1977)
7. Kleinrock, L., *Communication Networks*, McGraw-Hill (1964); reprinted Dover (1972)
8. Kleinrock, L., *Queueing Systems: Vol. I, Theory*, Wiley-Interscience (1975)
9. Kleinrock, L., *Queueing Systems: Vol. II, Computer Applications*, Wiley-Interscience (1976)
10. Bear, D., *Principles of Telecommunication Traffic Engineering*, Peter Peregrinus (1976)
11. Kosten, L., *Stochastic Theory of Service Systems*, Pergamon (1973)
12. Cox, D. R. and Smith, W. L., *Queues*, Methuen (1961)
13. Feller, W., *An Introduction to Probability Theory and Its Applications, Vol. II*, Wiley (1966)
14. Parzen, E., *Stochastic Processes*, Holden-Day (1962)
15. Snyder, D. L., *Random Point Processes*, Wiley-Interscience (1975)

16. Sarkowski, H. (ed.), *Teletraffic Engineering Manual*, Standard Elektrik Lorenz, Stuttgart (1966)
17. Wilkinson, R. I., 'Theories for toll traffic engineering in the U.S.A.', *Bell System Tech. J.*, **35**, 421–514 (1956)
18. Riordan, J., Appendix to ref. 17.
19. Burke, P. J., 'The overflow distribution for constant holding time', *Bell System Tech. J.*, **50**, 3195–3210 (1971)
20. Jacobeus, C., 'Blocking computations in link systems', *Ericsson Review*, **3**, 86–100 (1947)
21. Elldin, A., *Automatic Telephone Exchanges with Crossbar Switches: Switch Calculations: General Survey*, L. M. Ericsson, Stockholm, 3rd ed. (1967)
22. Lee, C. Y., 'Analysis of switching networks', *Bell System Tech. J.*, **34**, 1287–1315 (1955)
23. Cattermole, K. W. and Summer, J. P., 'Communication networks based on the product graph', *Proc. IEE*, **124**, 38–48 (1977)
24. Takagi, K., 'Optimum channel graph of link systems', *Electronics and Communications in Japan*, **54A**, No. 8, 1–10 (1971)
25. Burke, P. J., 'The output of a queueing system', *Operations Research*, **4**, 699–704 (1956)
26. Jackson, J. R., 'Networks of waiting lines', *Operations Research*, **5**, 518–521 (1957)
27. Newell, G. F., *Approximate Behaviour of Tandem Queues*, Springer-Verlag (1979)
28. Cattermole, K. W., 'Topological and statistical theory of communication networks', in ref. 29
29. Skwirzynski, J. K. (ed.), *New Concepts in Multi-user Communication*, Sijthoff and Noordhoff (1981)
30. Gordon, W. J. and Newell, G. F., 'Closed queueing systems with exponential servers', *Operations Research*, **15**, 254–265 (1967)
31. Baskett, F., Chandy, K. M., Muntz, R. R. and Palacios, F. G., 'Open, closed and mixed networks of queues with different classes of customers', *J.A.C.M.*, **22**, 248–260 (1975)
32. Chandy, K. M., Howard, J. H. and Towsley, D. F., 'Product form and local balance in queueing networks', *J.A.C.M.*, **24**, 250–263 (1977)
33. Reiser, M. and Lavenberg, S. S., 'Mean value analysis of closed multi-chain queueing networks', *J.A.C.M.*, **27**, 313–322 (1980)
34. Kelly, F. P., *Reversibility and Stochastic Networks*, Wiley (1979)
35. Schwartz, M., *Computer Communication Network Design and Analysis*, Prentice-Hall (1977)
36. Kobayashi, H., *Modelling and Analysis*, Addison-Wesley (1978)

6

Traffic Overflow and the Interrupted Poisson Process

D. J. Songhurst

6.1 INTRODUCTION

The interrupted Poisson process (IPP) is becoming widely used as a model for non-random telephone traffic. Its popularity derives from its ability to model non-random traffic (having three variable parameters) while retaining a near-Markovian form that allows it to be applied to more than just the simplest systems.

The IPP was first described fully by Kuczura [1], who derived its basic properties and showed that it could provide a very accurate model for overflow traffic.

In Section 6.2 we shall give a brief summary of the properties of the IPP. In Section 6.3 we describe the properties of overflow traffic and indicate how the IPP may be used to model this type of traffic. In Section 6.4 we present one method of solving the equations of state for a system whose offered traffic is represented by a combination of IPPs.

6.2 BASIC PROPERTIES OF THE IPP

6.2.1 Definition

We start with a Poisson process of intensity λ, which we shall assume to represent an arrival process of telephone calls. The process is interrupted by periods during which no arrivals occur. The on-times and off-times have independent negative exponential distributions with means $1/\gamma$ and $1/\omega$ respectively.

It is possible to define more general interrupted processes in which the on-times and off-times may have other distributions. Except where otherwise stated, we shall use the term 'IPP' to refer only to the simple process defined above.

6.2.2 Interarrival time distribution

Let $F(t) = P\{\text{interarrival time} \leqslant t\}$.

An arrival must occur within an on-period. The next arrival may occur within the same on-period, or within a subsequent on-period. We identify three different ways in which the next arrival may occur within time t:

A: the residual on-time (after the first arrival) is less than t, and the next arrival occurs within this time

B: the residual on-time is greater than t, and the next arrival occurs before t

C: the residual on-time plus the following off-time is less than t, and the next arrival occurs within a subsequent on-period before time t.

These three events are mutually exclusive, hence

$$F(t) = P(A) + P(B) + P(C) \tag{6.1}$$

Since the process is Markovian we note that all points within an on-period are renewal points both for the residual on-time and for the call interarrival time. Hence

$$P(A) = \int_0^t \gamma e^{-\gamma x}(1 - e^{-\lambda x})\, dx$$

$$P(B) = \int_t^\infty \gamma e^{-\gamma x}(1 - e^{-\lambda t})\, dx$$

$$P(C) = \int_0^t \gamma e^{-\gamma x} e^{-\lambda x}\left[\int_0^{t-x} \omega e^{-\omega y} F(t - x - y)\, dy\right] dx$$

Then by taking transforms in eqn. (6.1), the Laplace transform of the interarrival density function is

$$\frac{\lambda(\omega + s)}{s^2 + (\lambda + \gamma + \omega)s + \lambda\omega}$$

Kuczura [1] gives this distribution as a mixture of two exponential components

$$F(t) = k_1(1 - e^{-r_1 t}) + k_2(1 - e^{-r_2 t}) \tag{6.2}$$

where

$$r_1 = \tfrac{1}{2}[\lambda + \omega + \gamma + \{(\lambda + \omega + \gamma)^2 - 4\lambda\omega\}^{1/2}]$$

$$r_2 = \tfrac{1}{2}[\lambda + \omega + \gamma - \{(\lambda + \omega + \gamma)^2 - 4\lambda\omega\}^{1/2}]$$

$$k_1 = \frac{\lambda - r_2}{r_1 - r_2}, \quad k_2 = 1 - k_1$$

Milne [2] gives an alternative decomposition as a mixture of two distributions. The first component is exponential, representing a subsequent arrival within the same on-period. The second component is a convolution of exponentials, representing a subsequent arrival within a later on-period.

6.2.3 State equations

We assume that the IPP represents an arrival process of calls at an infinite circuit group (i.e. all offered calls are carried with no delay), and that calls have negative exponentially distributed holding times with mean $1/\mu$. Let $p(m, \varepsilon, t)$ be the probability that m calls are in progress at time t, and the process is then in state ε ($\varepsilon = 0$ will represent the off-state, 1 the on-state). The time-dependent state equations are as follows:

$$\frac{d}{dt}p(m,0,t) = \gamma p(m,1,t) + (m+1)\mu p(m+1,0,t) - (m\mu + \omega)p(m,0,t)$$

$$\frac{d}{dt}p(m,1,t) = \omega p(m,0,t) + (m+1)\mu p(m+1,1,t) + \lambda p(m-1,1,t)$$

$$- (m\mu + \gamma + \lambda)p(m,1,t) \qquad (6.3)$$

The probability generating functions are $G_\varepsilon(Z-1,t) = \sum_{m=0}^{\infty} p(m,\varepsilon,t)Z^m$, ($G_\varepsilon(Z,t)$ are then factorial moment generating functions). Equations (6.3) then give

$$\frac{\partial G_0}{\partial t} + \mu Z\frac{\partial G_0}{\partial Z} = \gamma G_1 - \omega G_0$$

$$\frac{\partial G_1}{\partial t} + \mu Z\frac{\partial G_1}{\partial Z} = \omega G_0 - \gamma G_1 + \lambda Z G_1 \qquad (6.4)$$

Milne derives the solution to the simultaneous partial differential equations (6.4). From these time-dependent generating functions it is then possible to obtain the covariance function

$$R(t) = E[N(0)N(t)] - E[N(0)]E[N(t)]$$

where $N(t)$ is the number of calls in progress at time t

$$R(t) = Ve^{-t} + (V-A)\frac{e^{-(\gamma+\omega)t} - e^{-t}}{1-(\gamma+\omega)}$$

Here A and V are the mean and variance of the number of calls in progress, which are derived below in Section 6.2.4.

The equilibrium state equations give

$$\mu Z \frac{\partial G_0}{\partial Z} = \gamma G_1 - \omega G_0$$

$$\mu Z \frac{\partial G_1}{\partial Z} = \omega G_0 - \gamma G_1 + \lambda Z G_1 \qquad (6.5)$$

6.2.4 Equilibrium state probabilities

Kuczura solved the eqn. (6.5) to give

$$G_0(Z) = \frac{\gamma}{\gamma+\omega} M\left(\frac{\omega}{\mu}, 1 + \frac{\gamma}{\mu} + \frac{\omega}{\mu}, \frac{\lambda}{\mu} Z\right)$$

$$G_1(Z) = \frac{\omega}{\gamma+\omega} M\left(1 + \frac{\omega}{\mu}, 1 + \frac{\gamma}{\mu} + \frac{\omega}{\mu}, \frac{\lambda}{\mu} Z\right)$$

where M is the confluent hypergeometric function defined by

$$M(a, b, c) = \sum_{n=0}^{\infty} \frac{(a)_n c^n}{(b)_n n!}$$

where

$$(a)_n = \begin{cases} 1, & n = 0 \\ (a+n-1)(a)_{n-1}, & n \geqslant 1 \end{cases}$$

The combined state probabilities have the generating function

$$G(Z-1) = G_0(Z-1) + G_1(Z-1) = M\left(\frac{\omega}{\mu}, \frac{\gamma}{\mu} + \frac{\omega}{\mu}, \frac{\lambda}{\mu}(Z-1)\right)$$

and are given by

$$p(m) = p(m, 0) + p(m, 1)$$

$$= \frac{\left(\frac{\lambda}{\mu}\right)^m}{m!} \cdot e^{-\lambda/\mu} \frac{\left(\frac{\omega}{\mu}\right)_m}{\left(\frac{\gamma+\omega}{\mu}\right)_m} \cdot M\left(\frac{\gamma}{\mu}, m + \frac{\gamma}{\mu} + \frac{\omega}{\mu}, \frac{\lambda}{\mu}\right)$$

Now $G(Z)$ is the generating function for the factorial moments of the number of calls in progress, and the factorial moments are

$$f_k = G^{(k)}(0) = \left(\frac{\lambda}{\mu}\right)^k \frac{\left(\frac{\omega}{\mu}\right)_k}{\left(\frac{\gamma+\omega}{\mu}\right)_k}$$

In particular the mean and variance are

$$A = f_1 = \frac{\lambda}{\mu}\left(\frac{\omega}{\gamma + \omega}\right)$$

$$V = f_2 + f_1 - f_1^2 = \frac{\lambda}{\mu}\left(\frac{\omega}{\gamma + \omega}\right)\left\{1 + \frac{\lambda\gamma}{(\gamma + \omega)(\gamma + \omega + \mu)}\right\}$$

6.3 OVERFLOW TRAFFIC

6.3.1 Alternative routing

The chief teletraffic application of the IPP is to model overflow traffic. Telephone networks commonly use alternative routing schemes, whereby traffic offered to a first-choice circuit group may overflow to other circuit groups whenever the first-choice group is busy. Usually the first-choice groups are direct high-usage routes to the destination exchanges (*AB* and *AC* in Fig. 6.1), and traffic from a number of such groups may overflow to a common final-choice group which will be a hierarchical route (*AD*).

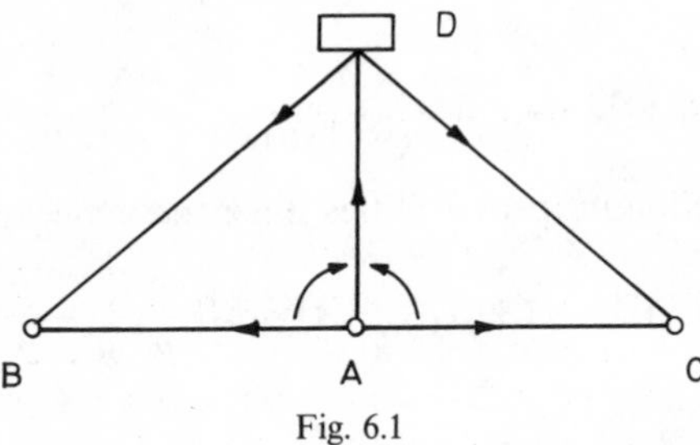

Fig. 6.1

In order to dimension such networks it is usually necessary to use decomposition methods in which each circuit group is calculated individually on the basis of the properties of the traffic offered to that group. Hence we need to know the properties of overflow traffic.

6.3.2 Overflow from a single circuit

Assume that a traffic process with mean call arrival rate λ and mean call holding time $1/\mu$ is offered to a single circuit. When the circuit is occupied the arrival process overflows. When the circuit is free it will become occupied by the next arrival. Hence the overflow process is an IPP with parameters λ, $\gamma = \mu$, $\omega = \lambda$.

6.3.3 Overflow from a group of circuits

The overflow process from a circuit group is not generally an IPP (by our definition) except when the group has only one circuit. The process has negative-exponentially distributed on-times during which there is a Poisson call arrival process with fixed rate. However, the off-times are not exponentially distributed. Each off-period corresponds to an inter-busy period of the circuit group and its length is distributed as a mixture of exponentials.

The overflow traffic process has factorial moments

$$f_k = a^k \frac{\dfrac{a^c}{c!}}{\displaystyle\sum_{j=0}^{c} \binom{k+j-1}{j} \frac{a^{c-j}}{(c-j)!}}$$

where $a = \lambda/\mu$ is the traffic offered to the group, and c is the number of circuits.

The overflow process is a renewal process, i.e., the intervals between overflowing calls are independent and identically distributed.

6.3.4 Combination of overflow streams

In the network depicted in Fig. 6.1, a combination of two overflow streams is offered to the final-choice route AD. In practice there may be many more than two overflow streams, and there may also be a stream of fresh (non-overflow) traffic offered directly to the final-choice route. It is then exceedingly difficult to compute the precise performance of the final-choice route, particularly when the circuit groups are large.

The most widely used approximation method is known as Wilkinson's Equivalent Random Traffic (ERT). The mean and variance of each overflow stream is computed, and these are summed across all the streams to give the mean and variance of the total traffic offered to the final-choice route (this requires that the overflow streams be statistically independent of one another). Using these two moments, the combined traffic can then be modelled as the overflow from a single first-choice circuit group, enabling the final-choice group to be computed easily.

There are three sources of error in this technique

(1) Each overflow stream is characterised by only its first two moments.

(2) The combined traffic stream is not a renewal process, but it is modelled as one.

(3) The individual overflow streams experience differing loss probabilities on the final-choice route, whereas ERT gives only a single average value.

6.3.5 The IPP as a model for overflow traffic

We have already seen that the IPP is the correct model for the overflow from a single circuit. It can also be used as an approximate model for the overflow from a group of circuits. The question then arises as to what values should be used for the IPP parameters. Out of many possibilities, the following two methods should be considered:

(1) The IPP parameters λ, γ, ω are equated with the analogous parameters of the overflow process.

(2) The first three factorial moments of the IPP are equated with those of the overflow process.

These methods may give very different results. For example, if the moment-matching approach is used, then the resulting value of λ may not be close to the call arrival rate at the first-choice group. The moment-matching approach has normally been used in practice for the purpose of dimensioning final-choice groups, but this may not be best for other applications.

The IPP may also be used as a model for a combination of overflow streams, using a moment-matching fit. This has the same shortcomings as the ERT mentioned in Section 6.3.4. (The extension from two to three moments does not, by itself, appear to give any improvement in accuracy.) A more promising approach is to use two or more IPPs to model the individual overflow traffic streams. In the following section we give a method for solving the state equations for the final-choice route, assuming that an arbitrary number of independent IPPs are used to represent overflow traffic streams.

6.4 MATRIX SOLUTION METHOD

6.4.1 State equations

We assume initially that two independent IPPs, in addition to a pure chance traffic stream of intensity a erlangs, are offered to a group of N circuits. The IPPs have parameters $\lambda_1, \gamma_1, \omega_1$, and $\lambda_2, \gamma_2, \omega_2$. All calls are assumed to have unit mean holding time. The state probabilities are $P_{\varepsilon_1\varepsilon_2}(k)$, representing the probability that the first IPP is in state ε_1 $(=0, 1)$, the second IPP is in state ε_2, and k circuits are busy ($k = 0, 1, \ldots, N$).

We shall use the column vector notation

$$P(k) = [P_{00}(k), P_{10}(k), P_{01}(k), P_{11}(k)]'$$

The equilibrium state equations can now be written as

$$H_1\mathbf{P}(k-1) - (H_1 + H_2 + kI)\mathbf{P}(k) + (k+1)\mathbf{P}(k+1) = \mathbf{0}$$

$$(k = 0, 1, \ldots, N-1; \mathbf{P}(-1) = \mathbf{0}) \quad (6.6)$$

where $H_1 = \text{diag}\,(a, \lambda_1 + a, \lambda_2 + a, \lambda_1 + \lambda_2 + a)$

$$H_2 = \begin{bmatrix} \omega_1 + \omega_2 & -\gamma_1 & -\gamma_2 & 0 \\ -\omega_1 & \gamma_1 + \omega_2 & 0 & -\gamma_2 \\ -\omega_2 & 0 & \omega_1 + \gamma_2 & -\gamma_1 \\ 0 & -\omega_2 & -\omega_1 & \gamma_1 + \gamma_2 \end{bmatrix}$$

The normalising condition is given by

$$\sum_{k=0}^{N} \mathbf{P}(k) = \mathbf{V} = [(1-p_1)(1-p_2), p_1(1-p_2), (1-p_1)p_2, p_1p_2]' \quad (6.7)$$

where p_1 = probability that first IPP is in the on-state

$$= \frac{\omega_1}{\omega_1 + \gamma_1}, \text{ etc.}$$

The three traffic streams offered to the circuit group experience distinct loss probabilities, as follows:

Pure chance stream: $P_{00}(N) + P_{10}(N) + P_{01}(N) + P_{11}(N)$

First IPP: $\dfrac{1}{p_1}(P_{10}(N) + P_{11}(N))$

Second IPP: $\dfrac{1}{p_2}(P_{01}(N) + P_{11}(N))$

6.4.2 Recurrence solution

The state eqn. (6.6), and the following solution method, apply for an arbitrary number of IPPs, provided that the coefficient matrices H_1

and H_2 are correctly defined. For a combination of m IPPs these matrices have order 2^m. For a single IPP (together with the pure chance traffic stream of a erlangs) we have

$$\mathbf{P}(k) = [P_0(k), P_1(k)]'$$

$$H_1 = \begin{bmatrix} a & 0 \\ 0 & \lambda + a \end{bmatrix}$$

$$H_2 = \begin{bmatrix} \omega & -\gamma \\ -\omega & \gamma \end{bmatrix}$$

$$\mathbf{V} = [1 - p, p]'$$

Let $\mathbf{P}(k) = Q_k\mathbf{P}(0)$, where Q_k is (for 2 IPPs) a 4×4 matrix. The normalising eqn. (6.7) gives

$$\mathbf{P}(k) = Q_k\left[\sum_{j=0}^{N} Q_j\right]^{-1}\mathbf{V} \tag{6.8}$$

If we substitute Q_k for $\mathbf{P}(k)$ in the state equations (6.6) we obtain the matrix equations

$$H_1Q_{k-1} - (H_1 + H_2 + kI)Q_k + (k+1)Q_{k+1} = 0 \tag{6.9}$$

where

$$Q_0 = I, \quad Q_{-1} = 0$$

Clearly we can obtain the matrices Q_k from the eqn. (6.9) sequentially, starting at $k = 0$. The state probabilities $\mathbf{P}(k)$ can then be obtained from eqn. (6.8). However, this approach is numerically unstable since the Q_k contain large numbers which cancel upon multiplying by the inverse in eqn. (6.8).

Now let

$$X_k = Q_k\left[\sum_{j=0}^{k} Q_j\right]^{-1}$$

Then

$$\mathbf{P}(N) = X_N\mathbf{V}$$

By inverting X_k we obtain the recurrence relation

$$X_k^{-1} = X_{k-1}^{-1}Y_k + I \tag{6.10}$$

where

$$Y_k = Q_{k-1}Q_k^{-1}$$

Right multiplication by Q_k^{-1} in eqn. (6.9) gives

$$Y_{k+1}^{-1} = \frac{1}{k+1}[H_1 + H_2 + kI - H_1 Y_k] \tag{6.11}$$

Hence the two expressions (6.10) and 6.11) can be used to compute Y_k and X_k^{-1} successively, starting from $Y_0 = 0$, $X_0 = I$.

This gives $\mathbf{P}(N) = X_N\mathbf{V}$, which is all that is required to give the loss probabilities. Other state probabilities can be obtained from $\mathbf{P}(N)$ using the relation

$$\mathbf{P}(k-1) = Y_k\mathbf{P}(k)$$

The expression (6.10) is directly analogous to the well-known recurrence relation for the Erlang loss formula

$$\frac{1}{E_N(A)} = \frac{1}{E_{N-1}(A)}\left(\frac{N}{A}\right) + 1$$

6.4.3 Trunk reservation

Alternative routing networks are known to be sensitive to traffic overload. In particular, congestion on hierarchical (final-choice) routes affects most severely the non-overflow (pure chance) traffic that is offered directly to these routes. For this reason it is desirable to restrict the access of overflow traffic to final-choice groups, giving priority to non-overflow traffic.

One method of achieving this is known as trunk reservation. With this method a control parameter R is associated with the final-choice group. Overflowing calls are only allowed to access the group when more than R circuits are free. When there are R or less free circuits, access is allowed only to non-overflow calls. Effective values of R are typically 1, 2 or 3.

Existing dimensioning methods for overflow traffic, such as the Wilkinson ERT, were not capable of handling trunk reservation. The IPP recurrence method described above was developed specifically for this purpose [3]. When trunk reservation is used the state equations for states $k \geqslant N - R$ need to be amended. This can be incorporated very easily in the IPP method by making the appropriate amendments to the matrix H_1 when the recursion reaches $k = N - R$.

6.5 GENERALISATIONS

6.5.1 Renewal theory

The method described above, based on the solution of state equations, provides a detailed and accurate analysis of the performance of a circuit group offered a combination of overflow traffic streams. However, in some networks there may be multiple stages of overflow, for example traffic may overflow from a first-choice to an intermediate route, and then to a final-choice route. In order to analyse such networks we need to be able to determine the characteristics of traffic overflowing from a circuit group that is itself offered a combination of overflow traffic streams.

Manfield and Downs [4] describe a suitable method that is based on a slightly different application of IPPs. This uses a renewal theory approach which relates the moments of the inter-overflow times to the moments of the interarrival times. Individual traffic streams are again modelled by IPPs using a three moment fit.

6.5.2 Phase processes

The IPP may be generalised to a process with a number of phases, having a different Poisson arrival rate in each phase, negative exponentially distributed phase lengths, and an arbitrary matrix of transition probabilities between phases. Neuts has investigated phase processes offered to queueing systems. (See, e.g. [5] which develops an iterative matrix-based solution.) This type of process may be used to represent time-varying arrival and service rates, and is particularly applicable to packet switching networks.

REFERENCES

1. Kuczura, A., 'The interrupted Poisson process as an overflow process', *Bell System Technical J.*, **52**, No. 3, 437–448 (1973)
2. Milne, C. B., 'Transient behaviour of the interrupted Poisson process', *JRSS*, Series B, **44**, Pt. 3, 398–405 (1982)
3. Songhurst, D. J., 'Protection against traffic overload in hierarchical networks employing alternative routing', *Network Planning Symposium*, Paris, October (1980)
4. Manfield, D. R. and Downs, T., 'Decomposition of traffic in loss systems with renewal input', *IEEE Trans. Comm.*, **COM-27**, 44–58 (1979)
5. Neuts, M. F., 'The M/M/1 queue with randomly varying arrival and service rates', *Opsearch*, **15**, No. 4, 139–157 (1978).

7

Generating Functions, Bounds and Approximations in Optical Communications

J. J. O'Reilly

7.1 INTRODUCTION

Statistical modelling of electrical communication systems is well developed and familiar but the problems of optical communications are of a different character and are generally less widely appreciated. In this chapter we show that, with the aid of generating functions, it is possible to obtain a detailed statistical characterisation of the signal and noise processes encountered in direct detection optical receivers.

The optical signal is inherently stochastic and the simplest realistic model for optical detection is that of a filtered Poisson process in which randomly detected photons excite a bandlimiting filter. This model can reasonably be adopted when deriving performance bounds—fundamental limits to optical communication—but must for practical purposes be augmented to allow for further stochastic disturbances of a more complex nature. In particular, multiplication of quantum noise associated with the gain mechanism of an avalanche photodiode, additive coloured Gaussian noise introduced by post-detection signal processing circuits and the influence of filtering on these disturbances must be accommodated in the model. The result is a superposition of a marked (compound) and filtered Poisson process with a coloured Gaussian process—the mixed compound Poisson plus Gaussian regime [1].

The complexity of the signal and noise statistics is such that it is generally not practicable, in anything other than a formal sense, to derive the probability density function. However, as we shall see, the use of generating functions allows us nonetheless to obtain an accurate characterisation. This is complete in that the mapping from characteristic functions to probability density functions is an isomorphism—a probability density function provides neither more nor less information than does the corresponding characteristic function. Indeed, for practical application a generating function characterisation may well be preferable. For example, it has proved practicable to rigorously optimise certain optical communications receivers using bounds couched in terms of generating functions [1, 2] while, in contrast, just the performance evaluation of receivers has proved extremely complex when operating on the probability density function [3]. In view of this we seek here to examine critically the generating function characterisation of optical receivers.

7.2 OPTICAL DIRECT DETECTION MODEL

We are concerned with the direct detection process in which incident photons with energy greater than the bandgap of the semiconductor detector can be absorbed by the creation of hole-electron pairs. The mean creation rate is given by:

$$\lambda(t) = \eta \frac{P(t)}{h\nu} \tag{7.1}$$

where

$\lambda(t)$ = instantaneous mean creation rate
$P(t)$ = incident optical power
h = Planks constant
ν = optical frequency
η = detector quantum efficiency, corresponding to the probability of absorption by this mechanism

We are interested here in the statistics of the output current, which is related to $\lambda(t)$, the generation rate parameter. Concerning outselves with one type of carrier (electrons, say, for the sake of definiteness) the number of carriers generated in a time interval (t_1, t_2) is given by a Poisson distribution:

$$P_n = \frac{m^n e^{-m}}{n!} \tag{7.2}$$

with

$$m = m(t_1, t_2) = \int_{t_1}^{t_2} \lambda(t)\, dt$$

$$= \text{Mean number created in } (t_1, t_2) \tag{7.3}$$

The process is not stationary since $\lambda(t)$ is time varying but eqn. (7.2) is nonetheless recognised as the familiar Poisson distribution. This identifies a nonhomogeneous Poisson process [4].

Let $w(t)$ be the response of a photodiode to the generation of a single hole-electron pair, such that

$$\int_{-\infty}^{\infty} w(t)\,dt = q, \quad \text{the electronic charge} \tag{7.4}$$

Denoting by $\{t_i\}$ the generation times resulting from the received optical signal $P(t)$, the output current can be expressed as

$$X(t) = \sum_i w(t - t_i) \tag{7.5}$$

This is an invariantly filtered nonhomogeneous Poisson process. The characteristic function for the process is readily derived (e.g. [5]) as

$$\phi_X(s\theta) = \exp\left\{\int_0^t \lambda(t)[\exp(j\theta w(z)) - 1]\,dz\right\} \tag{7.6}$$

Further filtering may occur subsequently and is readily accounted for by replacing $w(t)$ by the convolution of $w(t)$ with $h(t)$, the impulse response of the postdetection filter.

We have assumed so far that the filter function is determinate. This is not generally true for optical detection. Random variation of the filter function can arise, for example, from

(1) carriers being generated at different points within the detector giving rise to variation in the current pulse resulting from carrier transport within the device,
(2) carrier multiplication within the device due to an avalanche process can result in each detection event giving rise to a pulse with random amplitude.

These two phenomena can, of course, be experienced in combination. A general model which can allow for random filtering together with additive noise impairments has been described by Hoversten [5], this allows for random filtering by associating a random variable with event times, t_i. Denoting these random variables, which we assume to be independent and identically distributed (iid), as U_i we can express the output process, in the absence of additive impairments, as

$$X(t) = \sum_i w(t - t_i;\, U_i) \tag{7.7}$$

with characteristic function [6]

$$\phi_X(t) = \exp\left[\int_0^t \lambda(\mathrm{t})\mathrm{E}\{\exp(j\theta w(zU)) - 1\}\right] dz \tag{7.8}$$

Fortunately, this rather formidable formulation is not usually necessary for optical communications. Randomness in the filtering arises predominantly from the avalanche effect. Each primary detection event results in an integer number of hole-electron pairs and the avalanche process is of very short duration so that it can be treated as instantaneous. The diode response is thus:

$$X(t) = \sum_i g_i w(t - t_i) \tag{7.9}$$

where each detection event is 'marked' by a random gain g_i corresponding to the number of pairs resulting from primary carrier generation at time t_i. The random variables $\{g_i, t_i\}$ correspond to a marked or compound Poisson process, the $\{t_i\}$ are Poisson distributed in any interval with mean determined by the driving process rate. The $\{g_i\}$ are iid.

For most practical purposes the diode intrinsic response $w(t)$ is negligible compared with subsequent electrical filtering and the filtered diode output can be described by

$$Y(t) = q \sum_i g_i h(t - t_i) \tag{7.10}$$

where $h(t)$ is the filter impulse response. This corresponds to a marked and filtered Poisson process or a filtered compound Poisson process.

The performance of an optical detection system will be governed by the statistics of $Y(t)$ and of any further additive (Gaussian) noise processes introduced in the postdetection circuitry. The probability density function p_y depends on the driving process rate parameter $\lambda(t)$, the filter impulse response $h(t)$ and on the avalanche gain distribution p_g. It is not generally possible to obtain a tractable closed form expression for p_y, but in the next section we show that a generating function description is readily derived.

7.3. STATISTICAL DESCRIPTION FOR THE DETECTED OPTICAL SIGNAL

We wish to derive a moment generating function for the process described by eqn. (7.10). Let the time axis be segmented into elementary intervals and write the contribution to the output current at time instant t due to carriers generated within a sufficiently small Δt centred on t_n as

$$Y_{t_n}(t) = q \sum_i^I g_i h(t - t_i) \tag{7.11}$$

Here I represents the number of detection events within Δt centred on t_n and is Poisson distributed with mean

$$E\{I\} = \lambda(t_n)\,\Delta t$$

We can write a conditional moment generating function (mgf) for Y_{t_n}:

$$\begin{aligned}\{M_{t_n}(s)|I = I_1\} &= E\{e^{sY_{t_n}}\} \\ &= E\left\{\exp\left[\text{sq}\sum_{i=1}^{I_1} g_i h(t - t_n)\right]\right\}\end{aligned} \tag{7.12}$$

where the conditioning is on a given number I_1 of detection events within $\lambda(t)$. Since the g_i are iid we can rewrite eqn. (7.12) as

$$\{M_{t_n}(s)|I = I_1\} = \prod_{i=1}^{I_1} E\{\exp[sqgh(t - t_n v]\} \tag{7.13}$$

Here g without a subscript represents the gain random variable with mgf given by

$$M_g(s) = E\{\exp{(sg)}\} \tag{7.14}$$

Equation (7.13) can thus be written as

$$\{M_{t_n}(s)|I = I_1\} = [M_g(sqh(t - t_n))]^{I_1} \tag{7.15}$$

We now average over I to remove the conditioning with respect to the number of detection events in the time interval Δt. Noting that I is Poisson distributed, we obtain

$$\begin{aligned}M_{t_n}(s) &= \exp\{-\gamma(t_n)\Delta t\}\sum_{i=1}^{\infty}\frac{1}{i!}\{\lambda(t_n)\Delta t\}^i\{M_g(sqh(t - t_n))\}^i \\ &= \exp\{\lambda(t_n)\,\Delta t[M_g(sqh(t - t_n)) - 1]\}\end{aligned} \tag{7.16}$$

Considering now different time intervals, the number of detection events are independent while the avalanche gain associated with a given event is independent of the gain associated with other events. Hence the mgf for the random variable $Y(t)$ corresponding to the sum of the contributions from the various intervals is given by the product of the elementary contributory mgf's:

$$\begin{aligned}M_Y(s) &\simeq \prod_n M_{t_n}(s) \\ &= \prod_n \exp\{\lambda(t_n)\,\Delta t[M_g(sqh(t - t_n)) - 1]\} \\ &= \exp\left(\sum_n \lambda(t_n)\Delta t[M_g(sqh(t - t_n)) - 1]\right)\end{aligned} \tag{7.17}$$

In the limit as $\Delta t \to 0$ the summation becomes an integral giving

$$M_Y(s) = \exp\left\{\int_{-\infty}^{\infty} \lambda(\tau)[M_g(sqh(t-\tau)) - 1]\,d\tau\right\} \tag{7.18}$$

Equation (7.18) provides the statistical description of $Y(t)$ we are seeking: it accommodates inhomogeneity in the driving Poisson process, $\lambda(t)$, compound randomness in the form of an arbitrary avalanche gain mgf, $M_g(s)$, together with deterministic linear filtering, $h(t)$. It is not generally possible to transform eqn. (7.18) to obtain an analytic expression for p_y but this need not concern us unduly since the generating function constitutes as complete a statistical description as does the probability density function. Also, many useful statistics are obtained more readily from a generating function description.

7.4 AVALANCHE GAIN CONSIDERATIONS

Practical application of eqn. (7.18) requires an expression for the avalanche gain mgf. Various models have been proposed and we will examine some of these shortly. As an illustrative preamble we consider first the important non-avalanche gain case. We accommodate this by choosing a gain pdf of the form:

$$p_g = \delta(g-1) \Rightarrow M_g(s) = \exp(s) \tag{7.19}$$

hence

$$M_Y(s) = \exp\left\{\int_{-\infty}^{\infty} \lambda(\tau)[\exp(sqh(t-\tau)) - 1]\,d\tau\right\} \tag{7.20}$$

The avalanche gain process is usually described in terms of an effective value of k, the ratio of the ionization rates for holes and electrons. Physically this is something of an approximation but noise spectral density formulations and secondary carrier number distributions derived on this basis have been found to give good experimental agreement in many situations [7, 8]. Using this gain model Personick has shown that for $k = 0$ the gain distribution is of discrete geometric form [9] with mgf given by

$$M_g(s) = \frac{1}{1 - E\{g\}\{(1 - \exp(-s)\}} \tag{7.21}$$

For most practical purposes the discrete geometric distribution can be replaced here by an exponential gain distribution

$$p_g(g) = \frac{1}{E\{g\}} \exp[-g/E\{g\}]; \quad g > 0 \tag{7.22}$$

with mgf

$$M_g(s) = \frac{1}{1 - sE\{g\}}; \quad s < 1/E\{g\} \tag{7.23}$$

This has been used by Mazo and Salz [10] to estimate bounds on optical receiver performance, $k = 0$ corresponding to a best possible avalanche device.

For the general case of an arbitrary k Personick has obtained an implicit equation for the gain mgf [11]. No closed form solution to this has been reported. Lagrange expansion techniques allow an infinite series solut8on to be obtained [10] but it is not practicable to incorporate this into eqn. (7.18), any further analysis becoming quite intractable. Accordingly, the implicit gain mgf is generally used only as a basis for numerical evaluation.

An alternative to determining M_g numerically is provided by an empirical expression for the output of an avalanche photodiode [12] which has been used in simulation studies [3, 14]. Also, House has transformed this expression [15] and obtained the approximate gain mgf

$$M_g(s) = \frac{a}{(a-1)^2}[1 - \{1 - 2s(a-1)E\{g\}\}^{1/2}] - \frac{1}{(a-1)}sE\{g\} + 1 \tag{7.24}$$

where

$$a = kE\{g\} + \left(2 - \frac{1}{E\{g\}}\right)(1 - k) \tag{7.25}$$

is an excess noise factor for the avalanche process. This last result, albeit approximate and not corresponding to any real gain distribution, has proved valuable for detailed receiver studies [15, 16] since it is readily incorporated into the general expression, eqn. (7.18).

It is interesting to note that none of the avalanche gain mgf's described are particularly well behaved. In all cases the mgf fails to exist for some s. For the exponential gain model the singularity is removable but for both the empirical mgf and that due to Personick the problem is more substantial. Difficulties can arise, for example, when attempting to minimise a Chernoff bound based on either of these models. The difficulty derives from the far-tail behaviour and it seems possible that this is a result of simplifications in the physical model for the avalanche process rather than of fundamental origin. In view of this, some minor modification to the empirical mgf may be feasible without compromising its accuracy. We note also that even in the absence of full knowledge of the gain mgf useful statistics relating to the output process—such as the mean and variance—can be derived from just partial information (see Chapter 1).

7.5 RECEIVER NOISE

The description obtained to this point reflects the inherent stochastic nature of the signal process (eqn. (7.21)) and shows that further impairment may occur due to the random gain mechanism. In addition, Gaussian noise will be encountered but this is readily accommodated in the model. The Gaussian noise is additive and independent of the signal and gain processes. Since addition of independent random variables gives rise to a multiplication of their generating functions the output process

$$Y_0(t) = Y(t) + Y_n(t)$$

has mgf

$$M_{Y_0}(s) = M_Y(s) M_{Y_n}(s) \tag{7.26}$$

where

$$M_{Y_n}(s) = \exp\left\{\frac{\sigma_n^2}{2} s^2\right\}$$

is the mgf for a zero-mean Gaussian process with variance σ_n^2.

Equation (7.26) provides the required stochastic characterisation for signal and noise processes encountered in direct detection optical receivers.

7.6 BINARY OPTICAL SIGNALLING

We have obtained a generating function characterisation for a detected optical signal with deterministic rate parameter $\lambda(t)$ in the presence of additive Gaussian noise. We now consider that $\lambda(t)$ is determined by the data, such that

$$\lambda(t) = \frac{\eta}{h\nu} \sum_i a_i p(t - iT) \tag{7.27}$$

where $\{a_i\}$ is the data sequence and $p(t)$ is the signal element pulse shape. Since $\{a_i\}$ is a random sequence the detected signal is a doubly stochastic marked and filtered Poisson point process. The output signal after filtering by a filter with impulse response $h(t)$ has the form

$$y_s(t) = \sum_k q g_k h(t - t_k) \tag{7.28}$$

where q is the electronic charge and g_k is the gain experienced by a primary carrier generated at time t_k. Equation (7.28) constitutes a specific realisation of a random process $Y_s(t)$ with generating function given by (7.20).

Additive Gaussian noise introduced by the receiver front-end is generally not white but has a power spectral density given by

$$N(\omega) = N_0(1 + \omega^2/\omega_0^2) \tag{7.29}$$

where N_0, ω_0 are constants associated with the receiver. Following filtering, this noise then has variance σ^2 and mgf given by

$$\begin{aligned} M_{y_n}(s) &= \exp\left\{\frac{s^2}{4} N_0 \int_{-\infty}^{\infty} [h^2(t) + h'^2(t)/\omega_0^2]\, dt\right\} \\ &= \exp\left\{\frac{\sigma^2}{2} s^2\right\} \end{aligned} \tag{7.30}$$

The total output is the sum of two independent random processes and this has mgf of the form given in (7.26), corresponding to the product of eqns. (7.20) and (7.30).

To allow for interference between adjacent optical pulses, it is convenient to consider mgf's conditioned on a specific data pattern and further conditioned by $a_0 = 0$ or 1. We then partially remove the conditioning by averaging over all possible sequences while maintaining the symbol conditioning $a_0 = 0, 1$ as detailed in [2]. It is these symbol conditioned mgf's, $M_0(s, h)$, $M_1(s, h)$, which we use in the analysis of binary optical signalling.

7.6.1 Chernoff bound on error probability

The average error probability assuming equal *a priori* element probabilities is given by

$$P_e = \tfrac{1}{2}P_e\{a_0 = 1\} + \tfrac{1}{2}P_e\{a_0 = 0\} \tag{7.31}$$

Bounding the conditional error probabilities by

$$P_e\{a_0 = 1\} \leqslant B_{e_1} = M_1(s_1, h) \exp(-s_1 F), \quad s_1 < 0$$

$$P_e\{a_0 = 0\} \leqslant B_{e_0} = M(s_0, h) \exp(-s_0, F), \quad s_0 > 0$$

where F is the decision threshold such that

$$E\{y \mid a_0 = 0\} < F < E\{y \mid a_1 = 1\}$$

we can obtain a Chernoff bound B_e on P_e given by

$$P_e \leqslant B_e = \tfrac{1}{2}B_{e_1} + \tfrac{1}{2}B_{e_0}$$
$$= \tfrac{1}{2}\{M_1(s_1, h)\exp(-s_1F) + M_0(s_0, h)\exp(-s_0h)\},$$
$$s_1 < 0 < s_0 \quad (7.32)$$

It is generally convenient to adopt a single variable $s = s_0 = -s_1$, such that

$$B_e = \tfrac{1}{2}\{M_1(-s, h)\exp(sF) + M_0(s, h)\exp(-sF)\},$$
$$s > 0 \quad (7.33)$$

This bound may be minimised by choosing appropriately the decision threshold F and filter impulse response $h(t)$.

7.6.2 Modified Chernoff bound

Recently Prabhu [18] has derived a modified Chernoff bound (MCB) applicable to baseband binary electrical communication systems corrupted by additive interference and independent zero-mean noise. This MCB requires only the evaluation or bounding of the moment generating function of the interference. As derived by Prabhu the MCB is not applicable to direct detection optical communication since it does not accommodate the inherently stochastic character of the optical signal process. However, by following a broadly similar approach, the MCB can be extended to render it applicable to optical communications [18, 19], this involves modelling the detected optical signal as outlined above in terms of conditional generating functions. The bound is then derived using principles of analytic continuation and contour integration as follows:

The conditional error probabilities of eqn. (7.31) are given by

$$P_e\{a_0 = 0\} = \int_F^{\infty} p_{Y_0}(y\,|\,0)\,dy$$
$$= \frac{1}{2\pi}\int_F^{\infty} dy \int_{-\infty}^{\infty} \exp(-j\theta y)\phi_{Y_0}(\theta)\,d\theta \quad (7.34)$$

and

$$P_e\{a_0 = 1\} = \int_{-\infty}^{F} p_{Y_1}(y\,|\,1)\,dy$$
$$= \frac{1}{2\pi}\int_{-\infty}^{F} dy \int_{-\infty}^{\infty} \exp(-j\theta y)\phi_{Y_1}(\theta)\,d\theta \quad (7.35)$$

where p_{Y_1} is the conditional probability density function and ϕ_{Y_1} the conditional characteristic function of the signal and noise process at the input to the decision device. Extending the characteristic function into the complex plane by way of analytic continuation allows the conditional probabilities to be obtained by contour integration

$$P_e\{a_0 = i\} = \frac{(-1)^i}{2\pi j} \int_{C_i} \frac{1}{z} \exp(-jFz)\phi_{Y_i}(z)\, dz \qquad (7.36)$$

where C_0, C_1 are contours chosen to be parallel to the real axis in the lower and upper half planes respectively.

Substituting for z in eqn. (7.36) according to

$$z = u - js_0; \quad s_0 > 0, \quad i = 0$$

$$z = u + js_1; \quad |s_1 > 0, \quad i = 1$$

the procedures of eqn. (7.18) can be employed, as detailed in [19], to bound the conditional error probabilities

$$P_e\{a_0 = 0\} \leqslant \frac{1}{\sqrt{(2\pi)}s_0\sigma} \exp(-Fs_0 + \sigma^2 s_0^2/2)M_0(s_0) \quad (7.37a)$$

$$P_e\{a_0 = 1\} \leqslant \frac{1}{\sqrt{(2\pi)}s_1\sigma} \exp(Fs_1 + \sigma^2 s_1^2/2)M_1(-s_1) \quad (7.37b)$$

where here M_1, M_0 represent conditional moment generating functions relating to the signal process and σ^2 is the additive Gaussian noise variance. With the simplification $s = s_0 = s_1$, eqn. (7.37) may be substituted into (7.31) to yield the upper bound:

$$B_{e_m} = \frac{1}{2\sqrt{(2\pi)}s\sigma} \{M_1(-s)\exp(sF) + M_0(s)\exp(-sF)\} \quad (7.38)$$

where σ^2 is the variance of the additive Gaussian noise. For cases of practical interest (7.38) is generally very much tighter than (7.33). This is illustrated in Figs. 7.1 and 7.2 for non-avalanche gain and avalanche gain receivers respectively. The tightness of the bound is sensitive to the ratio of signal dependent to additive noise and thus varies with the mean avalanche gain; this dependence is illustrated in Fig. 7.3.

7.6.3 Gram–Charlier series

The characteristic function description for the signal and noise processes may also be employed to obtain approximations for P_e. Approximations based on conditional Gaussian statistics with means

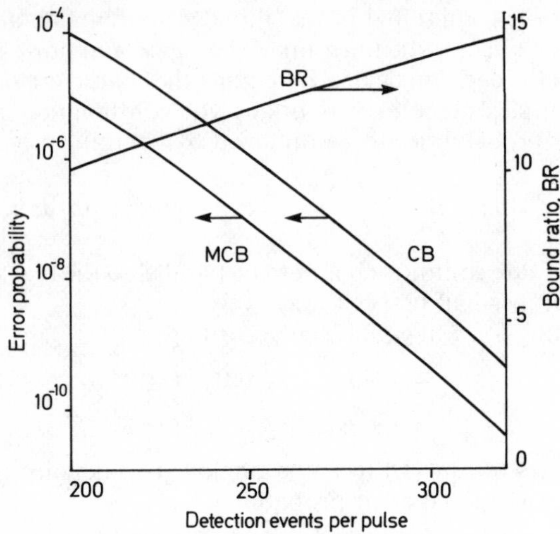

Fig. 7.1 Comparison of Chernoff and modified bounds for non-avalanche receiver

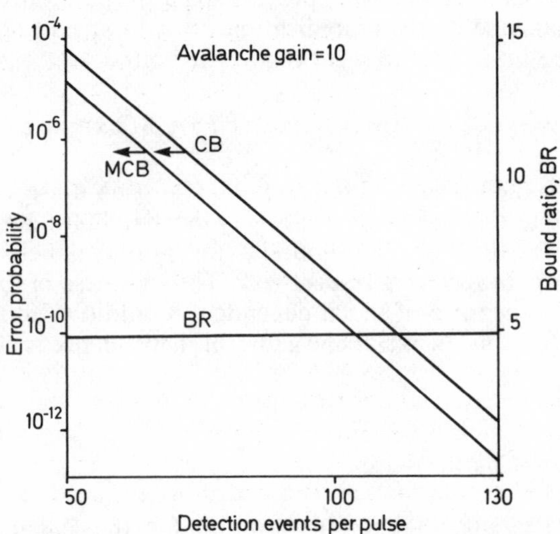

Fig. 7.2 Comparison of Chernoff and modified bounds for a receiver with avalanche gain

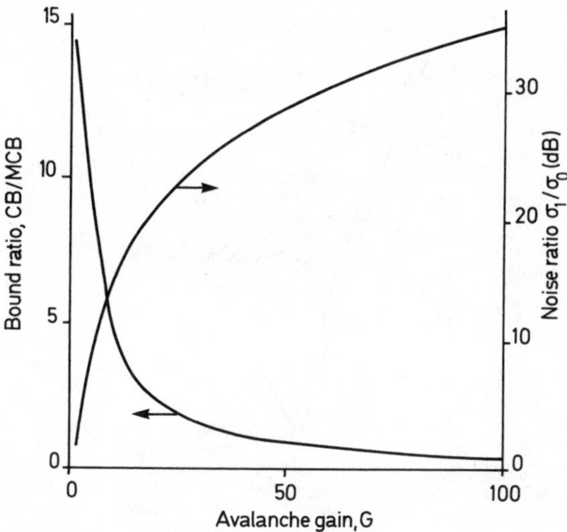

Fig. 7.3 Influence of avalanche gain modified bound tightness and noise ratio

and variances determined from M_0, M_1 are useful for design but for precise performance evaluation more accurate methods are required. The Gram–Charlier series provides a useful approach to this problem.

The method, reviewed briefly in Chapter 4, can be particularly appropriate in the present context since the cumulants of the signal plus noise process, required in the computation of the Gram–Charlier coefficients, are readily obtained from the log of the moment generating function, the log mgf being a cumulant generating function.

The power of the technique is illustrated in Fig. 7.4, taken from [15], where the Chernoff bound, Gaussian approximation and Gram–Charlier series are compared for a pure Poisson channel. Broadly speaking, the Gram–Charlier series has been found useful for performance evaluation in optical communication systems dominated by additive Gaussian noise or for a pure Poisson plus Gaussian channel. This occurs since adding a Gaussian process to some other underlying process, such as a Poisson, increases the variance but does not affect the higher-order cumulants.† The influence of higher-order cumulants is thus reduced and convergence enhanced. The compound randomness introduced by avalanche devices, however, is less tractable.

† Cumulants are additive for sums of independent random variables and cumulants of order 3 and higher are zero for a Gaussian random variable.

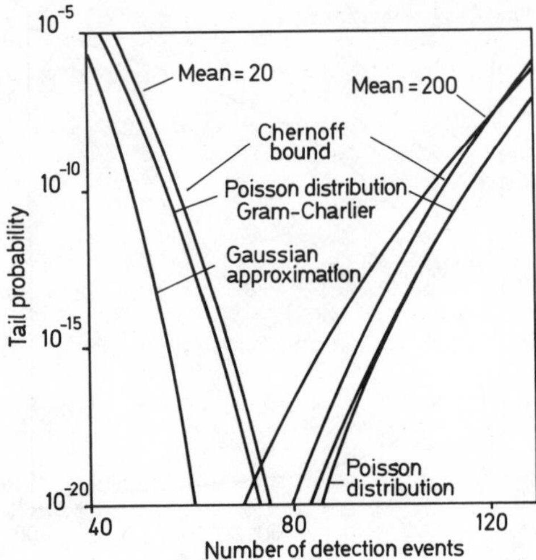

Fig. 7.4 Comparison of bounds and approximation for Poisson distributions

Accurate performance assessment in the presence of avalanche gain can be achieved with the aid of Monte Carlo methods and the acceleration provided by importance sampling has proved useful in this context [3, 15].

7.7 CONCLUSION

The aim of this chapter has been to show how, with the aid of generating functions, a stochastic representation for signal and noise processes encountered in optical receivers can be obtained. The statistics have been shown to be those of a non-homogeneous marked and filtered Poisson process, further corrupted by additive Gaussian noise. The application to binary optical signalling has been outlined and the relative merits of various bounds and approximations discussed.

REFERENCES

1. Rocha, J. R. F. and O'Reilly, J. J., 'Optical communication—the mixed compound Poisson plus Gaussian regime', *Digest of IEEE International Symposium on Information Theory*, Les Arcs, France, 175, June (1982)
2. Rocha, J. R. F. and O'Reilly, J. J., 'The Chernoff bound as a basis for optical fibre receiver design', *Digest of Sixteenth International Conference on Information Sciences and Systems*, Princeton, U.S.A., 159–163, March (1982)

3. Balaban, P., 'Statistical evaluation of the error rate of the fibre guide repeater using importance sampling', *Bell Syst. Tech. J.*, **55**, No. 6, 745–766 (1976)
4. Davenport, W. B., *Probability and Random Processes*, McGraw-Hill, 464–466 (1970)
5. Hoversten, E. V., 'Optical communication theory', in Arecchi, F. T. and Schultz, E. O. (eds.), *Laser Handbook*, North Holland, 1805–1862 (1972)
6. Snyder, D. L., *Random Point Processes*, Wiley, 164–185 (1975)
7. McIntyre, R. J., 'The distribution of gains in uniformly multiplying avalanche photodiodes: Theory', *IEEE Trans. Electron. Dev.*, **ED-19**, No. 6, 703–713 (1972)
8. Conradi, J., 'ibid: Experimental', 713–718
9. Personick, S. D., 'New results on avalanche multiplication statistics with applications to optical detection', *Bell Syst. Tech. J.*, **50**, No. 1, 167–189 (1971)
10. Mazo, J. E. and Salz, J., 'On optical communication via direct detection of light impulses', *Bell Syst. Tech. J.*, **55**, No. 3, 347–369 (1976)
11. Personick, S. D., 'Statistics of a general class of avalanche detectors with applications fo optical communication', *Bell Syst. Tech. J.*, **50**, No. 10, 3075–3095 (1971)
12. Webb, P. P., McIntyre, R. J., and Conradi, J., 'Properties of avalanche photodiodes', *RCA Review*, **35**, 235–276 (1974)
13. Balaban, P., Fleischer, P. E. and Zucker, H., 'The probability distribution of gains in avalanche photodiodes', *IEEE Trans. Electron. Dev.*, **ED-00**, 1189–1190 (1976)
14. Personick, S. D. *et al.*, 'A detailed comparison of four approaches to the calculation of the sensitivity of optical receivers', *IEEE Trans. Comm.*, **COM-00**, 541–548 (1977)
15. House, K. E., Ph.D. Thesis, University of Essex (1979)
16. Rocha, J. R. F. and O'Reilly, J. J., 'Optimum filters for binary optical telecommunication', *Digest of IEE Colloquium: Filters for Telecommunications* (1982)
17. Prabhu, V. K., 'Modified Chernoff bounds for PAM systems with noise and interference', *IEEE Trans. Inform. Theory*, **IT-22**, 95–100 (1982)
18. Rocha, J. R. F. and O'Reilly, J. J., 'A modified Chernoff bound for binary optical communication', *Electron. Lett.*, **18**, No. 16, 708–710 (1982)
19. Rocha, J. R. F. and O'Reilly, J. J., 'Error probability bounds for optical communication', Telecommunication Systems Group Report No. 176, Optical Communications Laboratory, University of Essex (1982)
20. Cramer, H., *Mathematical Methods of Statistics*, Princeton University Press, 17.6, 17.7 (1946)

8

Modelling Burst Errors in Digital Transmission

G. G. Pullum

8.1 INTRODUCTION

Interference in various electrical forms has created and always will create problems which must be solved if satisfactory communication links are to be established. When analogue signals are being transmitted, the occurrence in the system of an isolated noise spike or pulse often passes unnoticed and very rarely gives cause for concern. When information is transmitted digitally, the presence of even the smallest amounts of interference can sometimes be disastrous. Great care has therefore to be taken in the design of networks and equipment to minimise the effects of corrupted 'bits' and to this end a knowledge of the statistical distribution of such bits is invaluable. One of the most popular distributions used to describe the occurrence of errors in digital communication links is the Poisson distribution. However, it is known that errors, when they occur, sometimes do so in 'bursts' rather than as single isolated events, so that the straightforward Poisson distribution is then no longer consistent with observations. It might be argued that the occurrence of the bursts themselves are described reasonably well by a Poisson distribution with parameter m_1, say, and that the number of errors within a burst by another Poisson distribution with parameter m_2.† The question therefore arises as to whether it is feasible to combine both of these attributes in a single distribution. The answer of course is in the affirmative. In fact, just such a distribution was described by Neyman [1] in 1939 and is known as Neyman's Type A Contagious Distribution (referred to from now on in this paper as the NTA distribution). Studies in STL in 1979 utilising the NTA distribution in formulating the error performance objectives for integrated services

† The error process is then a special case of the compound Poisson process described in Chapter 1, Section 1.3.

digital networks (ISDN) resulted in the submission on 4th January 1980 of proposals to CCITT Study Group XVIII (Digital Networks). In 1981, Jones and Pullum [2] presented a paper based upon the NTA distribution at the second international conference on telecommunication transmission.

8.2. THE NTA DISTRIBUTION

8.2.1 Underlying mathematical concepts

The NTA distribution can be regarded as a compound distribution [3, 4], it being made up of two separate discrete distributions as follows:

(1) a Poisson distribution with parameter m_1 describing the occurrence of bursts or clusters of errors
(2) a Poisson distribution with parameter m_2 describing the number of errors within a cluster.

In the context of errors in digital transmission, we might consider a sample of 'n' bits so that m_1 becomes the expected number of clusters per n bits and m_2 the expected number of errors per cluster. Thus the expected number of errors per n bits is simply $m_1 m_2$.

The probability (p_r) of observing exactly 'r' errors per n bits as given by the NTA distribution is

$$p_r = e^{-m_1} \frac{m_2^r}{r!} \sum_{t=0}^{\infty} \frac{z^t}{t!} t^r \tag{8.1}$$

where $z = m_1 e^{-m_2}$ and $m_1, m_2 > 0$. In particular

$$p_0 = \exp\{-m_1(1 - e^{-m_2})\} \tag{8.2.0}$$

$$p_1 = m_1 m_2 e^{-m_2} p_0 \tag{8.2.1}$$

$$p_2 = \frac{m_1 m_2}{2} e^{-m_2}(p_1 + m_2 p_0) \tag{8.2.2}$$

$$\vdots \qquad\qquad \vdots$$

$$p_{(n+1)} = \frac{m_1 m_2 e^{-m_2}}{n+1} \sum_{t=0}^{n} \frac{m_2 t}{t!} p_{(n-t)}$$

The above relationships are very useful for computational purposes and are due to Beall [5].

8.2.2 Moments and other properties

As mentioned above, the expected value of an NTA variate x is given by

$$E(x) = m_1 m_2 \tag{8.3}$$

In addition

$$\mu_2 = \text{var}\,(x) = m_1 m_2(1 + m_2) \tag{8.4}$$

and

$$\mu_3 = m_1 m_2(1 + 3m_2 + m_2^2) \tag{8.5}$$

There exists an extremely useful recurrence relationship among the cumulants [6] which, allowing for a misprint in the reference, is

$$K_{r+1} = m_2\left(K_r + \frac{\partial K_r}{\partial m_2}\right) \tag{8.6}$$

from which, in addition to expressions (8.3) to (8.5) above, one can obtain (remembering that $E(x) = K_1$; $\mu_2 = K_2$ and $\mu_3 = K_3$ but $\mu_4 \neq K_4, \ldots$)

$$\begin{aligned}
K_4 &= m_1 m_2(1 + 7m_2 + 6m_2^2 + m_2^3) \\
K_5 &= m_1 m_2(1 + 15m_2 + 25m_2^2 + 10m_2^3 + m_2^4) \\
K_6 &= m_1 m_2(1 + 31m_2 + 90m_2^2 + 65m_2^3 + 15m_2^4 + m_2^5) \\
&\vdots \qquad\qquad\qquad \vdots
\end{aligned}$$

It is interesting to note that the numbers in parentheses are in fact Stirling numbers of the second kind and can thus be obtained from tables [7].

It is possible for the distribution to have three or more modes: the modal values being approximately multiples of m_2.

As the parameter m_2 decreases such that $m_1 m_2$ remains finite, the (NTA) distribution tends to a Poisson distribution with expected value $m_1 m_2$.

8.3 FITTING DATA

8.3.1 General considerations

Estimators of the parameters m_1 and m_2 (denoted by $\hat{m}_1$ and $\hat{m}_2$ respectively) frequently used in practice [8] are

$$\hat{m}_1 = \bar{x}/\hat{m}_2 \tag{8.7a}$$

and

$$\hat{m}_2 = (s^2 - \bar{x})/\bar{x} \tag{8.7b}$$

where $\bar{x}$ is the mean of a set of n independent random variables $x_1, x_2, \ldots, x_n$ and

$$s^2 = \frac{\sum^{n} (x_i - \bar{x})^2}{n - 1}$$

An alternative method of estimation uses the sample mean and observed proportion of zeroes. The equations for the estimators are then

$$\bar{x} = \hat{m}_1 \hat{m}_2 \tag{8.8a}$$

and

$$p_0 = \exp\{-\hat{m}_1(1 - e^{-\hat{m}_2})\} \tag{8.8b}$$

For further details the reader is referred to Johnson and Kotz [9].

8.3.2 Numerical example

The data in the first two columns of Table 8.1 are taken from Contribution No. 208 to CCITT Study Group XVIII dated August 1978. From Table 8.1 we find that the expected value of x, $\{E(x)\}$, is 1.1296×10^{-2} and the variance is 3.6014×10^{-2}; hence using eqns. (8.7a) and (8.7b) we obtain

$$\hat{m}_2 = 2.1882 \quad \text{and} \quad \hat{m}_1 = 5.1625 \times 10^{-3}$$

Table 8.1 2048 KBIT/S SYSTEM IN TRUNK JUNCTION NETWORK: PATH LENGTH 52 KM, OBSERVATION TIME 6 DAYS (518400 SEC)

Errors/s (x)	*No. of seconds (observed)* ($f(x)$)	*Expected no. of seconds based on* *Poisson*	*NTA*
0	516150	512577	516029
1, . . . , 2†	1496	5822	1369
3	432	0	522
4	187	0	286
5	65	0	126
6	43	0	46
7	15	0	15
8	5	0	4
>9	7	0	3
	Total: 518400	Total: 518399	Total: 518400

Total no. of errors: 5856

† Treated as '2' since total number of errors is quoted as 5856.

On substituting these values for m_1 and m_2 in expressions (8.2), corresponding values for $p_0, p_1, p_2, \ldots$ can be obtained, which when multiplied by 518400 yields the entries in the last column of the table. The entries in the penultimate column are, of course, values of e^{-K}, Ke^{-K}, $(K^2/2!)e^{-K}, \ldots$ (where $K = 1.1296 \times 10^{-2}$) all multiplied by 518400.

On comparing the entries in the third and fourth columns of Tables 8.1 with corresponding entries in the second, it is obvious that the NTA distribution is a better fit to the observations than is the Poisson. This is hardly surprising since in the limit as $m_2 \rightarrow 0$, the NTA distribution tends to the Poisson anyway. What is more important is to find answers to the following questions:

(1) to what extent is the NTA distribution an acceptable model for the observed occurrence of errors?
(2) given that the NTA is used to model the occurrence of errors, what is the best way of estimating m_1 and m_2?

With regard to (1), the question is best answered by taking many more measurements of the kind described in Table 8.1 using different communication paths, bit rates, and so on. For those sets of results where the predicted errors appear close to the observed errors, goodness-of-fit tests might usefully be undertaken. With regard to (2), it should in theory make no difference to the measured values of $\hat{m}_1$ and $\hat{m}_2$ whether the interval chosen (the second in the case of Table 8.1) was 1 s, 10 s, 1 min or any other value. However, in practice, the interval chosen (quadrat size) will almost certainly affect the results obtained. Reasons for this are put forward in a paper by Pielou [10] to which the reader is referred.

8.4. NETWORK PERFORMANCE IN RELATION TO THE NTA DISTRIBUTION

The International Telegraph and Telephone Consultative Committee (CCITT) have put forward recommendations regarding criteria that should be met if the performance of a network is to be judged as acceptable. An example of such criteria as applied to the error performance of an international digital connection forming part of an integrated services digital network, is recommendation G 821. In this recommendation the performance objectives refer to a 64 kbit/s connection used for voice traffic or as a bearer channel for data type services. The performance objective is described in Table 8.2.

In the table the abbreviation BER stands for bit error ratio and is the ratio of the number of bits incorrectly received to the total number of bits sent.

Table 8.2

BER in 1 minute	*Percentage of available minutes*	*BER in 1 second*	*Percentage of available seconds*
Worse than 1.10^{-6}	Less than 10%	>0	Less than 8%
Better than 1.10^{-6}	More than 90%	0	More than 92%

If we consider first of all an interval of one second, then at a transmission rate of 64 kbit/s, a total of $n = 64000$ bits will have been transmitted. Suppose p denotes the long-term BER. We have the expected number of errors in a sample of size n given simply by np. Thus

$$m_1 m_2 = np$$

Hence, if n is fixed (e.g., $n = 64 \times 10^3$) and p is fixed (e.g., $p = 3 \times 10^{-6}$) then, for any one value of m_2, there exists a corresponding value of m_1. Hence, by making use of expression (8.2.0), values of p_0 may be plotted against m_2 for various values of p, as illustrated in Fig. 8.1. From this figure we can see that, if the distribution of errors is described reasonably well by an NTA distribution, then (for $p = 3 \times 10^{-6}$), 92% of available seconds will have a BER in one second = 0 (i.e., 92% of seconds will be error free) provided $m_2 \geqslant 2$.

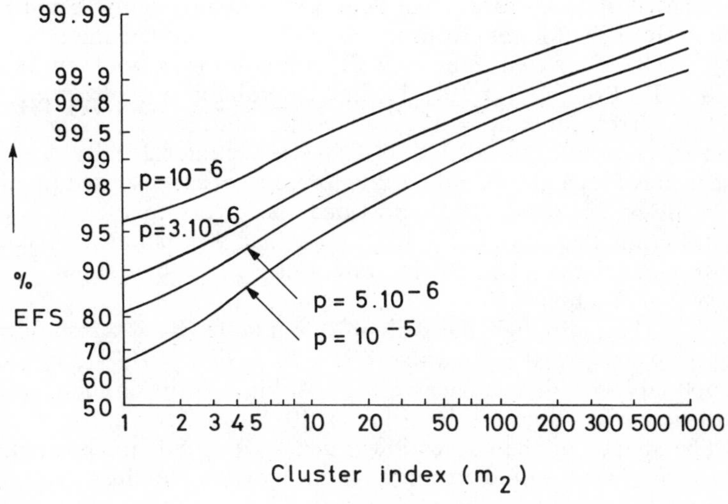

Fig. 8.1

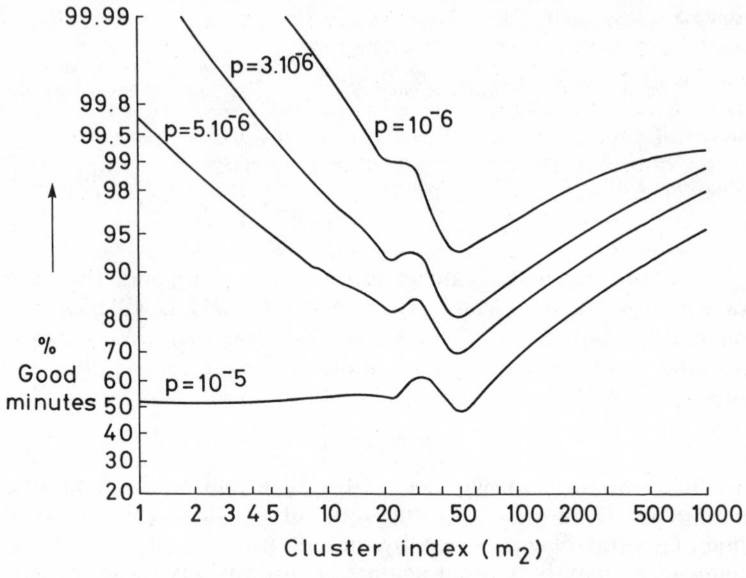

Fig. 8.2

If now we consider an interval of 1 minute, $n = 60 \times 64 \times 10^3 = 3.84 \times 10^6$ bits and so for $p = 3.10^{-6}$, $m_1 m_2 = np = 11.52$. Thus the expected number of errors per minute will be 11.52. If, in fact, three or fewer errors are measured in an interval of 1 minute we say the BER for that minute is better than 3 in 3.84×10^6 (roughly 1 in 10^6). Similarly, if 38 or fewer errors are observed in a 1 minute interval, we say the BER for that minute (BER threshold) is better than 38 in 3.84×10^6 (roughly 1 in 10^5). To find the probability of observing 38 or fewer errors for $m_1 m_2 = 11.52$ we select a value for m_2 (say, $m_2 = 2$) and sum terms (8.2.0), (8.2.1)... (8.2.38) (see Section 8.2.1). By varying the value of m_2 and proceeding as before, curves as illustrated by Fig. 8.2 can be obtained. Thus, provided the errors follow an NTA distribution, then for $p = 3 \times 10^{-6}$, more than 90% minutes will have a BER better than 1 in 10^5 for values of m_2 less than about 35 or greater than about 110.

Note that, although the BER in one minute (BER threshold) is specified as 1×10^{-6} in Table 8.2, a BER of 1×10^{-5} has been proposed as an alternative and it is possible that the ultimate value may lie in the range 1×10^{-5} to 1×10^{-6}.

The shapes of the curves illustrated by Fig. 8.1 are not really surprising. One would expect that the greater the clustering, the greater too would be the probability of an error-free second. The curves illustrated by Fig. 8.2, on the other hand, may not at all be what one would expect. For the curves having $p < 1 \times 10^{-5}$ the general trend is that as m_2 increases the probability of observing 38 or fewer errors (i.e., the percentage of 'good' minutes) falls at first then the curves, after passing over a 'hump', eventually rise. With the exception of the 'hump' one might explain the trend as follows. Suppose we considered a particular set of minutes such that for some small value of m_2 each minute had a number of errors in it which was just below the threshold of 38. Then all such minutes under consideration would be 'good' (i.e., would have a BER better than 1 in 10^5). Now suppose we could in some way increase the clustering in that same set of minutes in such a way as to leave the total number of errors unchanged. Then some minutes which previously were 'good' would become 'bad' as errors from adjacent good minutes were 'swept' into them. Thus the probability of 'good' minutes would fall. This trend would continue until eventually all the errors would be clustered into one 'bad' minute and just before this occurred the probability of finding 'good' minutes can be expected to increase as m_2 increases. This, of course, is only a possible pictorial explanation for the general shape of the curves; no attempt has been made to explain the 'hump'.

8.5 COMMUNICATION LINKS IN TANDEM

Consider two communication links (x and y) in tandem and suppose the errors in one link occur independently of those in the other. In addition suppose that the errors observed at the receiving end of the tandem connection are simply the sums of the errors from the individual links. Then for link x the cumulants are (see Section 8.2.2)

$$
\begin{aligned}
K_{1x} &= m_{1x} m_{2x} \\
K_{2x} &= m_{1x} m_{2x}(1 + m_{2x}) \\
K_{3x} &= m_{1x} m_{2x}(1 + 3m_{2x} + m_{2x}^2) \\
&\vdots
\end{aligned}
$$

Similarly for link y we have the above relationships with suffix x replaced by suffix y. The cumulants of the final distribution of errors (from both links combined) will be

$$
\begin{aligned}
K_{1f} &= K_{1x} + K_{1y} = m_{1x} m_{2x} + m_{1y} m_{2y} \\
K_{2f} &= K_{2x} + K_{2y} = m_{1x} m_{2x}(1 + m_{2x}) + m_{1y} m_{2y}(1 + m_{2y}) \\
K_{3f} &= K_{3x} + K_{3y} = m_{1x} m_{2x}(1 + 3m_{2x} + m_{2x}^2) + m_{1y} m_{2y}(1 + 3m_{2y} + m_2^{2y}) \\
&\vdots
\end{aligned}
$$

From the above it can be seen that if $m_{2x} = m_{2y} = m_2$, say, then the cumulants $K_{1f}, K_{2f}, K_{3f}, \ldots$ *will follow the same pattern as those for the individual links* x and *y*. Thus the distribution of the errors from the two links combined will also have an NTA distribution with parameters $(m_{1x} + m_{1y})$ and m_2. The question now arises as to how much m_2 may be allowed to vary from link to link such that the overall distribution of errors can still be described reasonably well by an NTA distribution. At this point we need to qualify what we mean by 'reasonably well'. However, we now seem to be racing too far ahead. We need first and foremost more raw data to see whether the NTA distribution is acceptable for single links and then to see how best to measure m_1 and m_2. For this we must wait until the results of many more measurements are available.

8.6 CONCLUSIONS

It is believed that, although the NTA distribution has been used to model events far removed from those for which the distribution was originally developed, Jones and Pullum [2] are the first to describe an attempt to use the distribution as a model for errors occurring in digital transmission links. So far only a small amount of data from actual in-service digital networks has been obtained. Many more measurements need to be made and analysed in order merely to begin to answer questions like (1) and (2) in Section 8.3.2 or the question relating to variations in m_2 from link to link posed in Section 8.5. It is hoped that sufficient interest in using the NTA distribution to model errors in digital transmission systems has been aroused that useful data will soon be forthcoming. Meanwhile it may be remarked that there are doubtless many other models that might serve equally well to model errors. The important point about the NTA distribution is that it is especially suited to the measurement of events which occur in clusters and although there appears to be little data available in the form as illustrated in Section 8.3.2, there is widespread agreement as to the clustering nature of the errors themselves.

REFERENCES

1. Neyman, J., 'On a new class of contagious distributions, applicable in entomology and bacteriology', *Ann. Math. Statist.*, **10**, No. 35 (1939)
2. Jones, W. T. and Pullum, G. G., 'Error Performance Objectives for Integrated Services Digital Networks Exhibiting Error Clustering', *IEE Conf. Publication No. 193—2nd International Conference on Telecommunication Transmission*, London (1981)

3. Kendall, M. G. and Stuart, A., *The Advanced Theory of Statistics*, Vol. 1, p. 144, Ex. 5.7 and 5.8, Griffin (1969)
4. Johnson, N. L. and Kotz, S., *Discrete Distributions*, Wiley, Chapter 8 (1969)
5 .Beall, G., 'The fit and significance of contagious distributions when applied to observations of larvae insects', *Ecology*, **21**, 460–474 (1940)
6. Johnson, N. L. and Kotz, S., *Discrete Distributions*, Wiley, Chapter 9, p. 218 (1969)
7. Abromwitz and Stegun, *Handbook of Mathematical Functions*, Dover, p. 835 (1970)
8. Johnson, N. L. and Kotz, S., *Discrete Distributions*, Wiley, Chapter 9, p. 223 (1969)
9. Johnson, N. L. and Kotz, S., *Discrete Distributions*', Wiley, Chapter 9, p. 224 (1969)
10. Pielou, E. C., 'The effect on quadrat size on the estimation of the parameters of Neyman's and Thomas' distribution', *J. Ecology*, **45**, 31–47 (1957)

9

Non-Linear Operations on Stochastic Processes

K. W. Cattermole

9.1 INTRODUCTION

The theory of linear systems, both determinate and stochastic, is highly developed and fairly well known. Non-linear operations are usually considered as 'difficult', not just because the mathematical analysis may become more complicated but rather because there seems to be no one procedure of general applicability. There are *ad hoc* methods for many specific problems, together with nominally general methods which may or may not work out easily in a particular case. For some problems of practical importance, analytical methods do not seem to yield tractable solutions, and so resort is had to numerical methods and to simulations. The balance of advantage between the several approaches depends on the application: methods common in one field may be unused in another.

It would be the height of presumption to suppose that within this present chapter we can even survey the topic thoroughly, let alone impose order upon it; so we must delimit our treatment. Its main features are as follows.

(1) The stochastic processes considered are those which can usefully model signals, or noise, or signals-plus-noise, in a communication channel. The typical non-linearities are those which arise in practical channels: quadratic and cubic distortions, quadratic and other rectifying characteristics as used for detection or timing extraction, overload and limiting characteristics, quantization and clipping. The emphasis is on instantaneous non-linearities.

(2) Even within this limited field, it is necessary that several different methods be given. The aim is, however, to concentrate mainly on methods which are in principle fairly general, and to

exhibit such coherence as the field possesses. The unifying concept is that our several methods depend upon specific representations of the non-linearity which, though superficially diverse, are all linear transformations of each other.

(3) There is a deliberate emphasis on the theory of diagonal processes and their associated orthogonal expansions. This class includes the Gaussian processes, but is a good deal wider, and is known to include other processes of practical interest. It is perhaps the widest class for which a fairly tractable general theory is known at present. Although quite well established in the specialised research literature, it does not seem to be widely known, perhaps in part for want of a clear tutorial exposition.

9.2 SOME DEFINITIONS AND BASIC THEORY

The purpose of this section is to state, rather than develop, some concepts for later use. The reader is assumed to have a general knowledge of probability and random process theory to about the level of Davenport's introductory text [1]. The concepts of this section should all be fairly familiar to anyone who has studied statistical communication theory.

9.2.1 Univariate probability distributions

We denote a *random variable* by a capital letter, and a specific value in its sample space by the corresponding lower-case letter, for example X, x. Our random variables will all be continuous, with sample space either the real line $-\infty < x < +\infty$ or an interval thereof. *Probability distributions* will be defined by *densities*, for example

$$P(x < X < x + dx) \equiv g(x)\,dx \tag{9.1a}$$

or the corresponding *cumulative distributions*

$$P(X < x) \equiv \mathbf{G}(x) = \int_{-\infty}^{x} g(\xi)\,d\xi \tag{9.1b}$$

A *statistic* of the random variable X is the expectation of some function of X,

$$E\{f(X)\} = \int f(x)g(x)\,dx \tag{9.2}$$

The common statistics are the *mean* and *variance*

$$m \equiv E\{X\} \tag{9.3}$$

$$\sigma^2 \equiv V\{X\} \equiv E\{(X - m)^2\} \tag{9.4}$$

We shall also make extensive use of the *characteristic function*

$$E\{e^{j\theta X}\} = \int_{-\infty}^{\infty} g(x)e^{j\theta x}\, dx \equiv G(\theta) \tag{9.5}$$

This is a statistic of the variable X, as shown by the left-hand side. Equally, it is a linear transformation of the density; it will be recognised that $g(x)$, $G(\theta)$ are related by the Fourier transform (save for a scaling factor of 2π which is conventionally inserted in signal theory but omitted in probability theory). Expansion in Taylor series shows that

$$G(\theta) = 1 + j\theta m - \tfrac{1}{2}\theta^2(m^2 + \sigma^2) + \mathrm{O}(\theta^3) \tag{9.6}$$

and that

$$\log\{G(\theta)\} = j\theta m - \tfrac{1}{2}\theta^2\sigma^2 + \mathrm{O}(\theta^3) \tag{9.7}$$

The foregoing relationships are easily verified in the *Gaussian distribution* for which

$$g(x) = \frac{1}{(2\pi)^{1/2}\sigma} \exp\left\{\frac{-(x - m)^2}{2\sigma^2}\right\} \tag{9.8a}$$

$$G(\theta) = \exp\{j\theta m - \tfrac{1}{2}\theta^2\sigma^2\} \tag{9.8b}$$

9.2.2 Bivariate probability distributions

Several random variables will have a *joint distribution* which can be considered as a multi-dimensional distribution with the appropriate multi-dimensional moments, characteristic function, etc. We shall make extensive use of *bivariate distributions.* Two random variables X_1, X_2 will have a two-dimensional density

$$P(x_1 < X_1 < x_1 + dx_1, x_2 < X_2 < x_2 + dx_2 = g(x_1, x_2)\, dx_1\, dx_2 \tag{9.9}$$

Unconditional statistics are found by integrating over the two-dimensional sample space.

$$E\{f(X_1, X_1)\} = \iint f(x_1, x_2)g(x_1, x_2)\, dx_1\, dx_2 \tag{9.10}$$

Conditional statistics are found by integrating over a one-dimensional slice, for example

$$E\{f(X_1)\,|\,X_2 = c\} = \frac{\int f(x_1)g(x_1, c)\,dx_1}{\int g(x_1, c)\,dx_1} \tag{9.11}$$

The scaling factor in the denominator of eqn. (9.11) is a value of the *marginal density*

$$g_2(x_2) = \int g(x_1, x_2)\,dx_1 \tag{9.12}$$

which is the density of X_2 averaged over all X_1: it may be considered as a projection of the two-dimensional distribution into one dimension. The algebraic variables are separable, with (for example) $g(x_1, x_2) = g_1(x_1)g_2(x_2)$. if and only if the random variables are independent. The statistics include the mean and variance m_i and σ_i^2 of individual variables X_i, and also the expectations of cross-products such as $E\{X_1X_2\}$. A particularly important joint statistic is the *covariance*

$$\mathrm{Cov}\,(X_1, X_2) = E\{(X_1 - m_1)(X_2 - m_2)\} \tag{9.13a}$$

which is divided by a normalising factor to give the *correlation coefficient*

$$\rho = \frac{E\{(X_1 - m_1)(X_2 - m_2)\}}{\sigma_1\sigma_2} \tag{9.13b}$$

The latter is zero for independent variables, and is bounded by $|\rho| < 1$ (the extreme values being attained only if there is a one-to-one relationship between values of X_1 and X_2, i.e. perfect correlation). We shall make extensive use of the *bivariate characteristic function*

$$E\{e^{j(\theta_1X_1 + \theta_2X_2)}\} = \int_{-\infty}^{\infty}\int_{-\infty}^{\infty} g(x_1, x_2)e^{j(\theta_1x_1 + \theta_2x_2)}\,dx_1\,dx_2$$
$$\equiv G(\theta_1, \theta_2) \tag{9.14}$$

which is a Fourier transform in two dimensions. Expansion in Taylor series shows that

$$G(\theta_1, \theta_2) = 1 + j\theta_1m_1 + j\theta_2m_2 - \tfrac{1}{2}\theta_1^2(m_1^2 + \sigma_1^2)$$
$$- \tfrac{1}{2}\theta_2^2(m_2^2 + \sigma_2^2) - \theta_1\theta_2(m_1m_2 + \rho\sigma_1\sigma_2) + \mathrm{O}(\theta^3) \tag{9.15}$$

and that

$$\log\{G(\theta_1, \theta_2)\} = j\theta_1m_1 + j\theta_2m_2 - \tfrac{1}{2}\sigma_1^2\sigma_1^2$$
$$- \tfrac{1}{2}\theta_2^2\sigma_2^2 - \theta_1\theta_2\rho\sigma_1\sigma_2 + \mathrm{O}(\theta^3) \tag{9.16}$$

As an example we quote the *bivariate Gaussian* distribution with zero mean

$$g(x_1, x_2) = \frac{1}{2\pi\sigma_1\sigma_2(1-\rho^2)^{1/2}} \times \exp\left\{\frac{-1}{2(1-\rho^2)}\left(\frac{x_1^2}{\sigma_1^2} - \frac{2\rho x_1 x_2}{\sigma_1\sigma_2} + \frac{x_2^2}{\sigma_2^2}\right)\right\} \tag{9.17a}$$

$$G(\theta_1, \theta_2) = \exp\{-\tfrac{1}{2}(\sigma_1^2\theta_1^2 + 2\rho\sigma_1\sigma_2\theta_1\theta_2 + \sigma_2^2\theta_2^2)\} \tag{9.17b}$$

The two marginal distributions, and all conditional distributions, are one-dimensional Gaussian as can easily be verified.

9.2.3 Random processes

A *stochastic process* or *random process* is, in our present context, a random phenomenon which takes place over an extended period of time. More formally, we define an index set T, such that for every $t \in T$ there is a random variable $X(t)$ which is a sample of the random process. The process is described statistically by the joint distribution of the $X(t)$ for all $t \in T$. For any given t, X takes values in a sample space which in this context is known as the *state space*. Given a finite set $(t_1, t_2, \ldots)$ the random vector $[X(t_1), X(t_2), \ldots]$ can be described and analysed by means of multivariate statistics. Alternatively, we can focus attention on one particular sequence of values over an interval on the index set: a specific noise waveform in a time interval, for example. This is known as a *realisation* of a random process, which we can consider as being drawn randomly from a set of all possible realisations. Statistics can in principle be formed either by averaging over all possible realisations (the ensemble average) or by averaging over an indefinitly long period of time (the temporal average). If statistics do not change with time, the process is *stationary*. If also the temporal and ensemble averages yield the same statistics, the process is *ergodic*. Many stochastic processes of practical interest are stationary and ergodic; an even larger proportion of textbook examples have these convenient properties. One class of non-stationary processes arises rather commonly in communication systems: the *cyclo-stationary processes*, whose statistics vary periodically. Familiar examples are (1) a periodically sampled stationary random process; (2) a sinusoidal carrier modulated by a stationary random process; (3) either of the foregoing modified by passage through a linear time-invariant filter. The classical treatment of cyclo-stationarity appears in Franks' textbook of signal theory [2]. We shall not develop cyclo-stationary theory explicitly, but some of our general methods (especially Section 9.8) are applicable with only minor changes either to stationary or to cyclo-stationary processes.

Strictly speaking, a process is stationary if and only if all its statistics are time-invariant: but how often do we investigate, either in theory or practice, the joint statistics of a large number of samples? Usually only statistics up to second order are taken into account: that is to say, the mean and variance of samples $X(t_i)$, together with the covariance of two samples $X(t_1)$, $X(t_2)$. If these statistics are time-invariant, the process is said to be *covariance-stationary*. (There is no implication that higher-order statistics are non-stationary: just that we know nothing about them. The tacit assumption is usually, that they would turn out to be stationary if only we were able to investigate them.) The usual emphasis on covaraince-stationary theory can be justified by several positive factors, in addition to the sheer difficulty of a treatment involving higher-order statistics. (1) It gives useful results in practice. (2) It leads to the spectral properties of random processes, which are readily measured or modified by means of linear filters and are the basis of much analysis and design. (3) Stationary Gaussian processes, an obviously important class, are fully described by their second-order statistics. (4) Stationary diagonal processes, a larger class and very important for reasons to be presented later, are described by their sample distributions plus the covariance between samples.

If a process $X(t)$ is covariance-stationary, then (1) statistics of single samples are invariant with time; (2) joint statistics of two samples $X(t_1)$, $X(t_2)$ depend only on the time difference $t_2 - t_1$. We may therefore define an *autocovariance function*

$$R_X(\tau) = \text{Cov}\,\{X(t), X(t+\tau)\} \tag{9.18a}$$

It is also usual to define an *autocorrelation function*

$$\mathbf{R}_X(\tau) = E\{X(t)X(t+\tau)\} \tag{9.18b}$$

$$= R_X(\tau) + [E(X)]^2 \tag{9.18c}$$

(The difference in connotation of 'covariance' and 'correlation' is not the same here as in expressions (9.13): however, these usages are so well entrenched that it would only confuse matters further if we departed from them for the sake of consistency.)

Let us consider a covariance-stationary Gaussian process, which is described by the joint statistics of $X_1 \equiv X(t)$ and $X_2 \equiv X(t+\tau)$. For simplicity we will take a zero-mean process, such as electrical noise. The joint distribution is the bivariate Gaussian, expressions (9.17). Putting $\sigma^2 = V(X_1) = V(X_2)$ and $\rho = R_X(\tau)/\sigma^2$, we have

$$g(x_1, x_2) = \frac{1}{2\pi\sigma^2(1-\rho^2)^{1/2}} \exp\left\{\frac{-(x_1^2 - 2\rho x_1 x_2 + x_2^2)}{2\sigma^2(1-\rho^2)}\right\} \tag{9.19a}$$

$$G(\theta_1, \theta_2) = \exp\{-\tfrac{1}{2}\sigma^2(\theta_1^2 + 2\rho\theta_1\theta_2 + \theta_2^2)\} \tag{9.19b}$$

This joint distribution is fully described by its marginal distributions (both Gaussian with variance σ^2) and the single additional parameter ρ. Thus Gaussian noise is fully specified by one parameter (its power) and one function of time difference (the autocovaraince). As we shall see in Section 9.8, this is not necessarily true of non-Gaussian processes.

9.2.4 Spectral theory

An absolutely integrable signal $g(t)$ and its spectrum $G(f)$ are related by the usual Fourier transform expressions

$$G(f) = \int_{-\infty}^{\infty} g(t)e^{-j2\pi ft}\,dt \tag{9.20a}$$

$$g(t) = \int_{-\infty}^{\infty} G(f)e^{j2\pi ft}\,df \tag{9.20b}$$

which we shall symbolise by $g(t) \rightarrow G(f)$. Products in either domain transform into convolutions in the other

$$g(t)h(t) \rightarrow G(f) * H(f) \tag{9.21a}$$

$$g(t) * h(t) \rightarrow G(f)H(f) \tag{9.21b}$$

The autocorrelation of a finite-energy signal is

$$\mathbf{R}_g(\tau) = \int_{-\infty}^{\infty} g(x)g(x+\tau)\,dx \tag{9.22a}$$

$$= g(t) * g(-t) \tag{9.22b}$$

where the second version follows on comparison with the usual convolution integral. The Fourier transform of the finite autocorrelation function is, using (9.21b) and (9.22b)

$$\mathbf{R}_g(\tau) \rightarrow G(f)G(-f) = |G(f)|^2 \tag{9.23}$$

To extend spectral theory to the realisations of a stationary random process, which of course are not integrable, it is necessary to perform a limiting operation on the time average, so that the autocorrelation function of an infinitely extended waveform $h(t)$ is

$$\mathbf{R}_h(\tau) = \underset{T\to\infty}{\text{Limit}}\, T^{-1} \int_{-\frac{1}{2}T}^{\frac{1}{2}T} h(x)h(x+\tau)\,dx \tag{9.24}$$

A corresponding operation in the frequency domain gives the *power spectrum*

$$\mathbf{P}_h(f) = \underset{T\to\infty}{\mathit{Limit}}\, T^{-1}|H(f) * T \operatorname{sinc} fT|^2 \tag{9.25}$$

and an extension of the reasoning in eqn. (9.23) gives

$$\mathbf{R}_h(\tau) \rightarrow \mathbf{P}_h(f) \tag{9.26a}$$

For an ergodic random process it is clear that the time-average eqn. (9.24) over any realisation gives the expectation (9.18b), so the reasoning remains valid. The relationship

$$\mathbf{R}_X(\tau) \rightarrow \mathbf{P}_X(f) \tag{9.26b}$$

for a stationary random process $X(t)$ is the *Wiener–Khintchine theorem.* It follows that the autocorrelation function and the power spectrum are equivalent statistics. In particular, a covariance-stationary Gaussian process is fully defined by its power spectrum.

Passage through a linear filter is treated very readily in spectral terms. Let a covariance-stationary process $X(t)$ be applied to a linear filter of impulse response $g(t) \rightarrow G(f)$. Then the output is a covariance-stationary process $Y(t)$ such that

$$\mathbf{R}_Y(\tau) = \mathbf{R}_X(\tau) * \mathbf{R}_g(\tau) \tag{9.27a}$$

where $\mathbf{R}_g(\tau)$ is the finite autocorrelation (9.22): and on Fourier transformation

$$\mathbf{P}_Y(f) = \mathbf{P}_X(f)|G(f)|^2 \tag{9.27b}$$

If the input process is uncorrelated (for example an ideal Poisson point process) then the output correlation and spectral shaping are due entirely to the filter. If the input process is Gaussian, then so is the output (since it is formed by a linear superposition, and the Gaussian distribution is stable). The power spectrum is then a complete statistical description of both input and output. This is not in general true if the operation is non-linear, or if either process is non-Gaussian. The power spectrum may nevertheless be a useful statistic, since the spectral distribution of distortion, interference etc. is usually significant in a practical communication system.

9.3 REPRESENTATION OF NON-LINEARITIES

9.3.1 Types of non-linearity considered

We shall deal only with instantaneous non-linearities, where the output $Y(t)$ depends only on the simultaneous input $X(t)$ and not on the past history. Such a transfer characteristic is specified by a function $y = \phi(x)$ which is single-valued; multiple values arise only in connection with hysteresis, which is excluded from the present treatment.

The types of non-linearity commonly encountered include (see Fig. 9.1)

(1) smooth small departures from linearity, giving rise to harmonic distortion and intermodulation of low order (Fig. 9.1a)
(2) clipping or limiting characteristics either hard or soft (Figs. 9.1b, c) and may be symmetrical (as shown) or asymmetrical
(3) rectifiers, either half-wave (Figs. 9.1d, g) or full-wave (Figs. 9.1e, f); sharp (Figs. 9.1d, e), quadratic (Fig. 9.1f) or exponential (Fig. 9.1g)
(4) amplitude compression and expansion characteristics, logarithmic and exponential (Fig. 9.1g)
(5) discontinuities, either single (Fig. 9.1h) or multiple as in quantisation characteristics (Figs. 9.1i, j)
(6) combinations, such as distortion or quantisation together with limiting.

The specification of the non-linearity may take many forms, including

(1) explicitly defined analytic functions with tractable properties, having derivatives of all orders (e.g. square law, exponential, etc.)
(2) explicitly defined functions of other forms (e.g. including magnitude or slope discontinuities)
(3) analytic but implicit definitions, the output y being the solution of a non-linear equation (e.g. current through a series combination of a linear resistor and an exponential semiconductor device)
(4) empirical relationship given only as a set of numerical measurements (often the case with electronic devices operated in a near-limiting regime for which no precise theory is known, e.g. high-power travelling-wave amplifiers).

However the characteristic may be specified, we have normally a choice of several different representations, as we shall see in the next section.

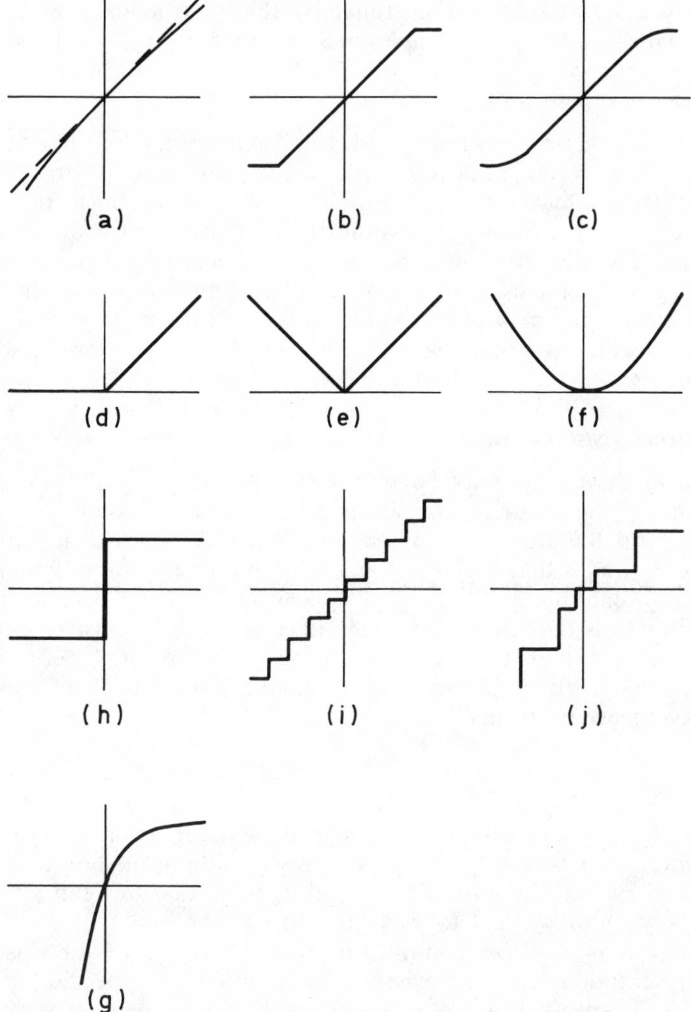

Fig. 9.1 Types of non-linearity

9.3.2 Types of representation

Direct representation

If the characteristic is specified by an explicit function, then we may be able to use this directly. For example, the amplitude distribution of the output may be a simple transformation of the input distribution. This is an obvious approach whenever it is analytically tractable; unfortunately, this is true only in a few special cases. We give examples in Section 9.4.

Power series

Analytic function can be expanded in Taylor series, and in suitable cases the series may converge sufficiently rapidly to justify truncation after a few terms. Empirical smooth non-linearities may be approximated by a polynomial. Certain special devices (e.g. square law detectors) may be precisely modelled by a polynomial. We give examples in Section 9.5. The limitations are that (1) practical series may converge too slowly; (2) the method is totally inapplicable to non-linearities with amplitude or slope discontinuities.

Characteristic functions

The function $y(x)$ may be represented by an integral transform, such as the Fourier transform (characteristic function, in the statistical literature) or Laplace transform. A powerful feature of this method is that it gives an analytical representation of functions with discontinuities. Examples appear in Section 9.6. The main limitation is that closed-form solutions may still be elusive, so that recourse is had to series expansions; we are not necessarily confined to simple power series, as above, but again there may be convergence problems.

Orthogonal functions

The foregoing methods are defined without reference to the random variables in question. A representation of the non-linearity is constructed; the implied assumption being that this can operate on any input process to yield an output whose statistics we can then estimate. Now it turns out that a very powerful class of representations can be devised, each adapted to a specific type of random process. We take a specific probability density as a weight function, construct a set of functions which are orthogonal with the given weighting, and expand the non-linear characteristic in a series each term of which is a member of the orthogonal set. If the input process has the given density as its amplitude distribution, then certain statistics of the output process can be calculated quite easily. This procedure applies to a large class of non-linearities both continuous and discontinuous, analytical and empirical. For many practical random processes, the orthogonal functions required are well known: and there is a general theory from which others could be derived. The theory is developed, and some examples are given, in Section 9.7. This procedure is especially powerful for the class known as diagonal processes; these have bivariate distributions which admit a one-dimensional expansion, so that second-order statistics may be calculated in a relatively simple way. The theory of diagonal processes is developed, with examples, in Section 9.8.

9.3.3 A note on the literature

The literature of this subject is large but scattered. The only specialised textbook known to the present author is Deutsch [3], which despite some limitations is a very useful source. The only other book devoted to the subject is Kuznetsov *et al.* [4]. This is a collection of articles by the editors and others, originally appearing in Russian journals; it is neither well organised or easy to follow, but appears to contain a good deal of relevant material. Some, but by no means all, textbooks of statistical communication theory contain sections on non-linear operations; notably Middleton [5] (which despite its 'introductory' title is about the most substantial and comprehensive book), Harman [6], Blachman [7] and Beckmann [8]; each of these contains interesting material but is far from comprehensive. A classical treatment, highly recommended both for this and for other statistical material, is the lengthy paper by Rice [9] which is the primary source for the characteristic-function method. The orthogonal-function approach, despite some earlier adumbrations, effectively derives from the work of Barrett and Lampard [10, 11] which in the communication field has been applied and extended by Blachman [12] and others. In the statistical literature, a useful general treatment is that of Lancaster [13, 14]. There is an extensive technical literature on specific problems, and also some mathematical literature on orthogonal functions, from which selected references will be given in appropriate places.

9.4 DIRECT TRANSFORMATION

9.4.1 Univariate statistics

We consider first a single random variable X with cumulative distribution $\mathbf{G}_X(x)$, and a single-valued function $\phi(\;)$. Then $Y=\phi(X)$ is a random variable. To find its distribution $\mathbf{G}_Y(y)$ we have to identify the subset B_y of sample space such that if $x \in B_y$, $\phi(x) \leqslant y$: then

$$\mathbf{G}_Y(y) = E\{I_{B_y}(X)\} \tag{9.28}$$

where I_{By} is the appropriate indicator function. If $\phi(\ )$ is monotonic, then B_y is a single semi-infinite interval: also $\phi(\ )$ has an inverse $\phi^{-1}(\ )$. For a non-decreasing function

$$\mathbf{G}_Y(y) = P(Y \leqslant y) = P(X \leqslant \phi^{-1}(y)) = \mathbf{G}_X[\phi^{-1}(y)] \tag{9.29}$$

Differentiating to find the density

$$g_Y(y) = g_X[\phi^{-1}(y)] \frac{d}{dy} \phi^{-1}(y) \tag{9.30a}$$

which is perhaps most memorable in the form

$$g_Y(y)\, dy = g_X(x)\, dx \tag{9.30b}$$

If $\phi(\ )$ is not monotonic, then the sample space of X must be partitioned into intervals within each of which the function is monotonic: the distribution of Y is then the sum of several terms each derived by the above method.

If we know the transformed distribution, then we have the choice of two methods for calculating statistics of the form (9.2), namely

$$E\{f(Y)\} = \int f(y) g_Y(y)\, dy \tag{9.31a}$$

$$= \int f[\phi(x)] g_X(x)\, dx \tag{9.31b}$$

Normally we will choose whichever is the easier to evaluate: in the examples which follow we use eqn. (9.31b), largely because the input distribution has well-known properties.

Simple analytic functions $\phi(x)$ often yield relationships between well-known distributions.

Examples

(1) Consider the exponential function $y = e^x$, with inverse $x = \log y$ ($y > 0$). If X has the Gaussian (normal) density (9.8a) then Y has the lognormal density

$$g_Y(y) = \frac{1}{(2\pi)^{1/2}\sigma y} \exp\left\{\frac{-(\log y - m)^2}{2\sigma^2}\right\} \tag{9.32}$$

Some statistics follow readily from the properties of the Gaussian distribution: for example, from (9.8b) the moments are

$$E\{Y^r\} = E\{e^{rX}\} = \exp\{rm + \tfrac{1}{2}r^2\sigma^2\} \qquad (9.33)$$

(2) Consider the quadratic function $y = x^2$. This is not monotonic, and formally we must make the transformation separately for positive and negative x: by symmetry, it can be carried out for positive x only and the resulting y-density doubled. Some statistics follow readily, for example the moments

$$E(Y^r) = E(X^{2r}) \qquad (9.34)$$

Let X be Gaussian with zero mean and unit variance: then

$$g_Y(y) = \frac{e^{-(1/2)y}}{(2\pi y)^{1/2}} = \frac{(\frac{1}{2})^{1/2} y^{-1/2} e^{-(1/2)y}}{\Gamma(\frac{1}{2})} \qquad (9.35)$$

which is a gamma distribution of order $\frac{1}{2}$. The first two moments are

$$\begin{aligned} E(Y) &= E(X^2) = 1 \\ E(Y^2) &= E(X^4) = 3 \end{aligned} \qquad (9.36)$$

If X has a Rayleigh distribution, then Y is negative-exponential; in this case the form (9.31a) would be the easier to evaluate, in general.

9.4.2 Bivariate statistics

Formally, we can make the appropriate transformation in the bivariate density and then obtain expressions for two-dimensional densities and statistics. Again, there are two equivalent expressions for any statistic, namely the two-dimensional equivalents of expressions (9.31a and 9.31b). For instance, we use the second form to express the autocorrelation of the output

$$\mathbf{R}_Y(\tau) = E\{Y_1 Y_2\} = E\{\phi(X_1)\phi(X_2)\} \qquad (9.37a)$$

$$= \iint \phi(x_1)\phi(x_2) g_X(x_1, x_2)\, dx_1\, dx_2 \qquad (9.37b)$$

(both integrals being between $\pm\infty$). This can sometimes yield a simple relationship.

9.4.3 Symmetrical clipping

Consider as an example the case of band-limited Gaussian noise passed through an ideal symmetrical clipper with a transfer characteristic like Fig. 9.1h: the output is ± 1 according as the input is $\lessgtr 0$. A single sample of the output is an equiprobable binary variable. The autocorrelation of the output can be found from an integral of the form (9.37b). Without loss of generality we can take $g_X(x_1, x_2)$ as (9.19a) with unit variance. Then $\mathbf{R}_X(\tau) = \rho$ and

$$\mathbf{R}_Y(\tau) = \iint \frac{\operatorname{sgn}(x_1 x_2)}{2\pi(1-\rho^2)^{1/2}} \times \exp\left\{\frac{-(x_1^2 - 2\rho x_1 x_2 + x_2^2)}{2(1-\rho^2)}\right\} dx_1\, dx_2 \quad (9.38)$$

Now $\phi(x_1)\phi(x_2) = \operatorname{sgn}(x_1 x_2)$ is $+1$ in the first and third quadrants, and -1 in the second and fourth quadrants. On conversion to polar coordinates, the Gaussian can be integrated over a quadrant. Harman [6] gives the result as

$$\mathbf{R}_Y(\tau) = \frac{1}{\pi} \sin^{-1} \mathbf{R}_X(\tau) \quad (9.39)$$

The effect is that autocorrelation is reduced for almost all τ; hence on Fourier transformation we would expect the power spectrum to be broadened by the clipping operation.

9.5 POWER SERIES

9.5.1 Origin of the series

The methods of this section require us to express the non-linearity in the form

$$y \equiv \phi(x) = \textstyle\sum_i a_i x_i \quad (i = 0,1,2\ldots) \quad (9.40)$$

and to be of practical use the series must either terminate or converge fairly rapidly. We can envisage four sources of such expressions.

(1) A non-linear device may be precisely characterised by a polynomial, for example a square-law detector.

(2) A non-linearity may be characterised by an explicit analytical function which can be expanded in Taylor series. For example, the current/voltage relationship of many semi-conductor devices is well represented over the working range by the (theoretically supported) expression

$$\phi(x) = e^{cx} - 1 \quad (9.41)$$

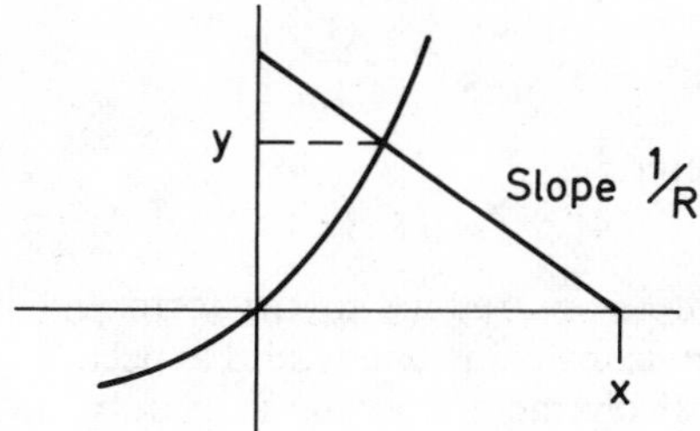

Fig. 9.2 Exponential device with non-zero impedance

assuming suitably normalised scales for x and y. If the static working point is x_0, then the behaviour for small signals (excursions $<c^{-1}$) is approximated by the first few terms of the series

$$\phi(x) = (e^{cx_0} - 1) + e^{cx_0} \sum_{i=1}^{\infty} \frac{c^i(x - x_0)^i}{i!} \tag{9.42a}$$

To evaluate the distortion of small signals, we can normalise relative to a_1, obtaining

$$\frac{a_i}{a_1} = \frac{c^{i-1}}{i!} \quad (i = 2, 3, \ldots) \tag{9.42b}$$

(3) A non-linearity may be defined implicitly by an equation whose solution can be expanded in power series. For example, suppose that a semiconductor device with the above exponential characteristic is driven from a source of internal resistance R (Fig. 9.2). The output y is related to x by the equation

$$x = yR + \frac{1}{c} \log (1 + y) \tag{9.43}$$

If the static working point is x_0, y_0 then we can postulate a series expansion for small signals

$$y - y_0 = \sum_{i=1}^{\infty} a_i (x - x_0)^i \tag{9.44}$$

Substitute into eqn. (9.43) and equate coefficients on either side. The resulting equations can be solved for coefficients $a_1, a_2, \ldots$ in turn. The first two normalised distortion coefficients are

$$\frac{a_2}{a_1} = \frac{c}{2(1 + z)^2} \tag{9.45a}$$

$$\frac{a_3}{a_1} = \frac{c^2(1 - 2z)}{6(1 + z)^4} \tag{9.45b}$$

where the parameter z is

$$z = cR(1 + y_0) \tag{9.45c}$$

Clearly these expressions reduce to the form (9.42b) if $R = 0$, but diminish with increasing R, as one would expect.

(4) An empirical characteristic showing only small and smooth departures from linearity may be approximated by a polynomial. Small quadratic and cubic terms may be estimated by measuring harmonic or intermodulation distortions, using for example the well-known relationships

$$\cos^2 \alpha t = \tfrac{1}{2}(\cos 2\alpha t + 1) \tag{9.46a}$$

$$\cos^3 \alpha t = \tfrac{1}{4}(\cos 3\alpha t + 3 \cos \alpha t) \tag{9.46b}$$

These are consistent with the spectral theory developed below, from which the vehaviour with other excitations, either determinate or stochastic, may be calculated.

9.5.2 Spectral theory

First we consider the response of a power-law non-linearity to a determinate signal $g(t) \rightarrow G(f)$. The output of a quadratic characteristic is

$$[g(t)]^2 \rightarrow G(f) * G(f) \tag{9.47}$$

using the convolution theorem (9.21). Continuing the product, the output of an nth power law has a spectrum in the form of an n-fold convolution. Summing such terms, we find that the output of the non-linearity (9.40) is

$$\sum_n a_n[g(t)]^n \rightarrow \sum_n a_n[G(f)]^{*n} \tag{9.48}$$

the notation on the right-hand side signifying a convolution of n similar factors. Equations (9.46) are easily verified in the frequency domain by this method (Fig. 9.3a).

A similar argument can be developed for the power spectrum of a stochastic signal, with one significant difference which will emerge. Consider the integral representation of the convolution (eqn. 9.47), which is

$$[G(f)]^{*2} = \int G(f - x)G(x)\, dx \tag{9.49}$$

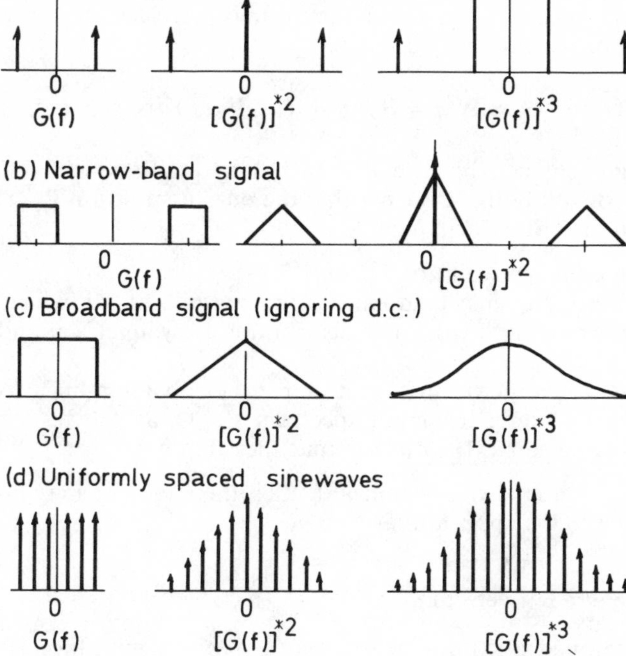
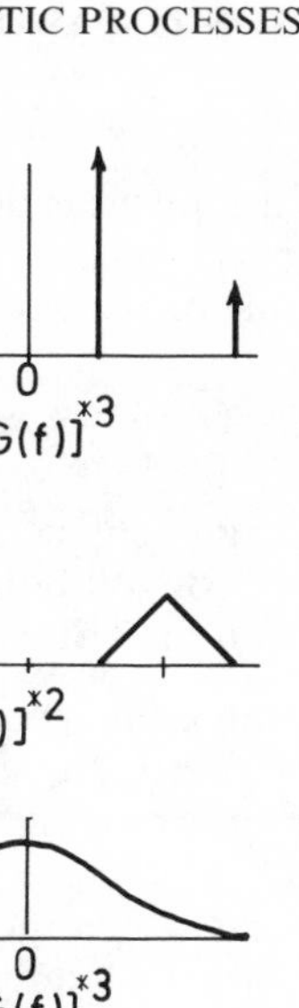

Fig. 9.3 Convolution of frequency spectra

The output at frequency f is the superposition of sum frequencies generated by all input frequency pairs $x, f - x$: the generation of sum frequencies as a result of multiplication in the time domain is a well-known result of the equation

$$e^{j2\pi xt}\, e^{j2\pi(f-x)t} = e^{j2\pi ft} \tag{9.50}$$

Now a random signal will similarly generate sum frequencies in respect of all frequency pairs $x, f - x$: but the components will (with one exception) be incoherent and random in phase. Consequently, we add their powers rather than their magnitudes: and this leads to a convolution of power spectra. The exception is that zero-frequency components ($f = 0$ at the output) add coherently, which results in a spectral impulse at the origin. This in electrical terms is the dc component, or in statistical terms the expectation of the output. Thus the power spectrum of the output of a square-law device, given an input power spectrum $\mathbf{P}_X(f)$, is

$$\mathbf{P}_Y(f) = [\mathbf{P}_X(f)]^{*2} + E\{X^2\}\delta(f) \tag{9.51}$$

This argument is developed at length in Rice's classic paper [9]. By Parseval's theorem, the mean square may be calculated in either the time or the frequency domain.

$$E\{X^2\} = \mathbf{R}_X(0) = \int_{-\infty}^{\infty} \mathbf{P}_X(f)\,df \tag{9.52}$$

The argument can be extended to cubic, quadratic etc. laws by further convolutions. It is clear that the output spectrum $\mathbf{P}_Y(f)$ due to a non-linearity x^n will include

(1) an n-fold convolution $[\mathbf{P}_X(f)]^{*n}$
(2) a zero-frequency component of magnitude $E\{X^n\}$, which may be zero notably if the input distribution is symmetrical and n is odd
(3) possibly other terms; for example the cubic non-linearity has a term including the input spectrum $\mathbf{P}_X(f)$, as can be seen on convolving eqn. (51) with the input spectrum.

The first item is always significant: the others may or may not be, according to the application.

9.5.3 Square-law detectors

The pure quadratic non-linearity has a special significance, since it is used in two devices of widespread use in communication systems: (1) the square-law detector as used in analogue carrier systems; (2) the square-law timing extractor as used in digital communications.

The foregoing theory shows that, if a narrow-band random signal centred on a frequency α be applied to a square-law device, there will be outputs in bands centred on zero-frequency and on 2α. Figure 9.3b shows the form of power spectrum, in the absence of a carrier wave: it is interesting to consider the effect of adding a carrier to the random signal (which is of course the usual practical case). The subject is discussed quite thoroughly by Rice [9], Blachman [7] and Harman [6]: also, we revert to it in Section 9.6.4.

The timing-recovery case is somewhat different in that the input is not a stationary random process, but rather is cyclo-stationary. Again, however, the significant feature is the recovery of energy in the band around the second harmonic of the peak in spectral density: because of the cyclo-stationarity, there may be a line at the second harmonic frequency. Timing recovery is discussed in Chapter 10.

9.5.4 Distortion in carrier systems

In frequency-division-multiplex carrier systems, the combined spectrum of a fully-loaded set of channels is often modelled as a broadband random signal. The transmission path, such as a coaxial cable with a chain of repeaters, is designed to be very linear: departures from linearity are small and smooth (provided the system is not overloaded) and so the transfer characteristic can be a series (9.40) with large a_1 (the linear term), small but non-trivial a_2 and a_3 (second and third-order intermodulation) and still smaller, probably negligible, terms of higher order.

The power spectrum is effectively broadband, as Fig. 3(c), with the reservation that there is usually a low-frequency cut-off ignored in the illustration; so the second and third-order intermodulation spectra are approximately as shown. Some of the distortion power is outside the useful spectrum and can be ignored in practice.

It is also possible to model a multi-carrier system as a set of sinusoids (Fig. 9.3d). Bennett [15] uses a combinatorial approach which is essentially a discrete-component analogue of the continuous-spectrum theory; his method does have the advantage of distinguishing components which may accumulate in different ways along a repeater chain. For this topic, see also [16].

In frequency-modulated microwave radio systems carrying fdm telephony, intermodulation may be caused by delay distortion. If the group delay as a function of frequency deviation be represented by a power series, then the foregoing theory applies with little change: the main difference is that: (1) frequency spectra such as Fig. 9.3c are all tilted upwards at high frequencies (6 dB/octave) in the course of fm demodulation; (2) the signal power spectra may be deliverately altered by pre-emphasis.

9.6 THE CHARACTERISTIC-FUNCTION METHOD

9.6.1 The characteristic-function representation

There are many important practical problems to which power-series are inapplicable, and so a representation is needed which exists, and is practically tractable, for a larger class of non-linearities. The characteristic-function method introduced by Rice [9] is in principle quite general: is equivalent to the power-series method when the series exists: and is applicable in many cases where series do not exist or do not readily converge.

The non-linearity $\phi(x)$ is represented by the integral transform

$$\phi(x) = \int_c \Phi(u)e^{jux}\,du \tag{9.53}$$

The path of integration c may be along the real line between $\pm\infty$, in which case $\phi(x)$ and $\Phi(u)$ are effectively Fourier transforms; or alternatively there may be indentation or displacement of the path in the complex plane to avoid singularities, in which case the relationship is usually that of the Laplace transform. Many examples are given by Rice [9], Middleton [5] and Deutsch [3].

We have seen that for a covariance-stationary process the most important statistic is the autocorrelation function (9.37). Substitution of the representation (9.53) in eqn. (9.37b) gives

$$\mathbf{R}_Y(\tau) = \int_{-\infty}^{\infty}\int_{-\infty}^{\infty}\left[\int_c \Phi(u_1)e^{ju_1x_1}\,du_1 \times \int_c \Phi(u_2)e^{ju_2x_2}\,du_2\right] g_X(x_1, x_2)\,dx_1\,dx_2 \tag{9.54a}$$

Rearrangement of the order of integration gives

$$\mathbf{R}_Y(\tau) = \int_c\int_c \Phi(u_1)\Phi(u_2) \times\left[\int_{-\infty}^{\infty}\int_{-\infty}^{\infty} g_X(x_1, x_2)^{ej(u_1x_1+u_2x_2)}\,dx_1\,dx_2\right] du_1\,du_2 \tag{9.54b}$$

The expression in brackets is recognisable as a joint characteristic function of the form (9.14), so

$$\mathbf{R}_Y(\tau) = \int_c\int_c \Phi(u_1)\Phi(u_2)G_X(u_1, u_2)\,du_1\,du_2 \tag{9.54c}$$

This is similar in structure to eqn (9.37b) except that each function in the integrand of either is replaced in the other by its transform. (Note that the functional dependence on τ is contained implicitly in g_X and G_X.) This alternative version may lead to a closed-form solution in some cases, or it may lend itself to expansion in series.

9.6.2 Power-series in ρ as solutions

For illustrative purposes we will take an example whose solution we know by other means*: the ideal symmetrical clipper, Fig. 9.1h.

* Deutsch remarks, wryly but truthfully, in the preface to his book [3]: 'The number of solved problems does not tend to increase with the number of available techniques . . .'.

The transformation (9.53) in this case is

$$\phi(x) = \operatorname{sgn}(x), \quad \Phi(u) = \frac{1}{j\pi u} \tag{9.55}$$

with integration along the real line (i.e. the usual Fourier transform). For Gaussian noise we can without loss of generality take the characteristic function as (9.19b) with unit variance. The output autocorrelation is then, from eqn. (9.54c)

$$\mathbf{R}_Y(\tau) = \int_{-\infty}^{\infty}\int_{-\infty}^{\infty} \frac{-1}{\pi^2 u_1 u_2} \exp\{-\tfrac{1}{2}(u_1^2 + 2\rho u_1 u_2 + u_2^2\}\, du_1\, du_2 \tag{9.56a}$$

This is soluble if the cross-product term $\exp\{-\rho u_1 u_2\}$ be expanded in power series: the autocorrelation becomes

$$\frac{-1}{\pi^2} \sum_{n=0}^{\infty} \left[\int_{-\infty}^{\infty} u^{n-1} e^{-(1/2)u^2}\, du\right]^2 \tag{9.56b}$$

The integral vanishes for even n: for $n = 1$ it takes the value $(2\pi)^{1/2}$ and for $n = 2m + 1$ it is $(2m - 1)!!\,(2\pi)^{1/2}$. This leads to the solution

$$\mathbf{R}_Y(\tau) = \frac{2}{\pi}\left[\rho + \frac{\rho^3}{3!} + \frac{\rho^5 3^2}{5!} + \frac{\rho^7 15^2}{7!} + \cdots\right] \tag{9.56c}$$

$$= \frac{2}{\pi}\sin^{-1}\rho = \frac{2}{\pi}\sin^{-1}\mathbf{R}_X(\tau) \tag{9.56d}$$

consistent with our previous solution (9.39). This example, though simple, illustrates three points of wider significance. Firstly, we can deal with discontinuities, which though not capable of being represented by power series or other algebraic expressions often have simple algebraic Fourier or Laplace transforms. Secondly, it is often possible and useful to expand (9.54) as a series whose terms contain powers of the input autocorrelation. Thirdly, for Gaussian inputs the resulting integrals are often tractable, so that the series coefficients can be evaluated: the series may then be recognisable as a hypergeometric function. Examples are given by Middleton [5] and Deutsch [3].

The terms of a series such as (9.56c) have an interesting physical interpretation. From the Wiener–Khintchine theorem (9.26b) and the convolution theorem (9.21), it follows immediately that the nth power of an autocorrelation function transforms into the nth self-convolution of a power spectrum. So if

$$\mathbf{R}_Y(\tau) = \sum_n K_n[\mathbf{R}_X(\tau)]^n \tag{9.57a}$$

then

$$\mathbf{P}_Y(f) = \sum_n K_n[\mathbf{P}_X(f)]^{*n} \tag{9.57b}$$

As with the power-series method of Section 9.5, we have arrived at a series of self-convolutions which may, especially with narrow band processes, yield clearly distinguishable components in the output spectrum.

9.6.3 Price's theorem

The characteristic-function theory lends itself to another technique for the calculation of the output autocorrelation, in the important special case where the input is Gaussian. We shall develop here a restricted form of a more general result due to Price [17]; for a fuller discussion see either Deutsch [3] or the original paper.

Let us write out eqn. (9.54c) for the special case of a unit-variance Gaussian input,

$$\mathbf{R}_Y = \iint \Phi(u_1)\Phi(u_2) \exp\{-\tfrac{1}{2}(u_1^2 + 2\rho u_1 u_2 + u_2^2)\, du_1\, du_2 \tag{9.58}$$

Differentiating with respect to ρ,

$$\frac{d\mathbf{R}_Y}{d\rho} = \iint \Phi(u_1)\Phi(u_2)(-u_1 u_2) \times \exp\{-\tfrac{1}{2}(u^2 + 2\rho u_1 u_2 + u_2^2)\}\, du_1\, du_2 \tag{9.59}$$

But from the transform relationship (9.53) it is clear that $ju\Phi(u)$ is the transform of $\phi'(x)$. Consequently, eqn. (9.59) is the autocorrelation of a Gaussian process which is subject to a non-linearity $\phi'(x)$, and we can write this as

$$\frac{d\mathbf{R}_Y}{d\rho} = E\{\phi'(X_1)\phi'(X_2)\} \tag{9.60a}$$

Differentiating n times gives, by continuation of the same reasoning

$$\frac{d^n\mathbf{R}_Y}{d\rho^n} = E\{\phi^{(n)}(X_1)\phi^{(n)}(X_2)\} \tag{9.60b}$$

which is the useful result. Effectively, by using characteristic functions we have been able to invoke the differentiation property of Fourier transforms.

The value of this approach is that such non-linearities as discontinuities, or power-law segments with discontinuous slope or amplitude, have simple derivatives: differentiate enough times, and you obtain impulsive singularities. Consider the ideal symmetrical clipper, Fig. 9.1h, with transform (9.55). Its derivative is $2\delta(x)$. Consequently

$$\frac{d\mathbf{R}_Y}{d\rho} = 4 \iint \frac{\delta(x_1)\delta(x_2)}{2\pi(1-\rho^2)^{1/2}} \times \exp\left\{\frac{-(x_1^2 - 2\rho x_1 x_2 + x_2^2)}{2(1-\rho^2)}\right\} dx_1\, dx_2 \qquad (9.61a)$$

$$= \frac{2}{\pi(1-\rho^2)^{1/2}} \qquad (9.61b)$$

Integration (with the obvious boundary condition $\mathbf{R}_Y = 0$ when $\rho = 0$) gives

$$\mathbf{R}_Y = \frac{2}{\pi} \sin^{-1} \rho \qquad (9.62)$$

consistent with our previous solutions (9.39) and (9.56).

Price's theorem applies only to Gaussian inputs, and indeed the original proof covers also the converse, that if the stated property holds the input must be Gaussian. However, McGraw and Wagner [18] have shown that a similar but more complicated relationship can be derived for a larger class of bivariate distributions having the property of elliptical symmetry. These distributions are related to the diagonal processes which we shall describe in Section 9.8, and we revert to the topic there.

9.6.4 Sinusoidal signal plus Gaussian noise

A classical application of the characteristic-function technique, from Rice's original work [9] onwards, has been to the superposition of a signal (such as a sinusoidal carrier with some form of amplitude, phase or frequency modulation) and noise. Now if the signal and noise are independent (as is usual in an additive-noise channel) the characteristic function of their sum is just the product of their individual characteristic functions. The characteristic function of Gaussian noise is simple enough (eqn. 9.17b) but that for the sine wave introduces analytical difficulties even in the simplest case of constant amplitude and uniformly random phase. If the sinusoid is $A \cos \beta t$, then the characteristic function is

$$E\{\exp\,[j\theta_1 A \cos \beta t + j\theta_2 A \cos \beta(t+\tau)]\} \qquad (9.63)$$

We use the well-known series

$$\exp\{j\theta A\cos\beta t\} = J_0(\theta A) + 2\sum_{n=1}^{\infty} j^n J_n(\theta A)\cos n\beta t \qquad (9.64)$$

Time-averaging the product of two such series gives the characteristic function of the sine wave as

$$J_0(\theta_1 A)J_0(\theta_2 A) + 2\sum_{n=1}^{\infty}(-1)^n J_n(\theta_1 A)J_n(\theta_2 A)\cos n\beta\tau \qquad (9.65)$$

The product of this with

$$\exp\{-\tfrac{1}{2}(\theta_1^2 + 2\rho\theta_1\theta_2 + \theta_2^2)\} = \exp\{-\tfrac{1}{2}(\theta_1^2 + \theta_2^2)\}\sum_{k=0}^{\infty}\frac{(-\rho\theta_1\theta_2)^k}{k!} \qquad (9.66)$$

is to be taken as G_X in expression (9.54c). This gives a doubly-infinite series for the output autocorrelation, of the form

$$\mathbf{R}_Y(\tau) = \sum_{n=0}^{\infty}\sum_{k=0}^{\infty} c_{mk}[\rho(\tau)]^k \cos n\beta\tau \qquad (9.67a)$$

with coefficients

$$c_{0k} = \frac{(-1)^k}{k!}\left\{\int_c \Phi(u)J_0(uA)u^k e^{-(1/2)u^2}\,du\right\}^2 \qquad (9.67b)$$

$$C_{nk} = \frac{2(-1)^{n+k}}{k!}\left\{\int_c \Phi(u)J_n(uA)u^k e^{-(1/2)u^2}\,du\right\}^2 \qquad (9.67c)$$

These latter expressions look rather complicated, but can be evaluated analytically in many cases of interest: for further information see references [5], [6], [7] and [9]. Our present concern is with the general form of (9.67a), which has a clear physical interpretation. One index of the double series defines powers of the input noise autocorrelation, hence self-convolutions of the input noise spectrum; these are now familiar from previous examples, and arise whenever a stationary random process is applied to a non-linearity. The other index defines harmonics of the input sine wave: thus the set of noise bands are translated to the vicinity of each harmonic, by a mixing process in the non-linear device (Fig. 9.4). Modulation of the sine wave complicates the details, without much change of principle. There is a substantial literature on this topic in the context of satellite communications, where a transponder with significant non-linearity is driven by a superposition of several modulated carriers and band-pass noise:

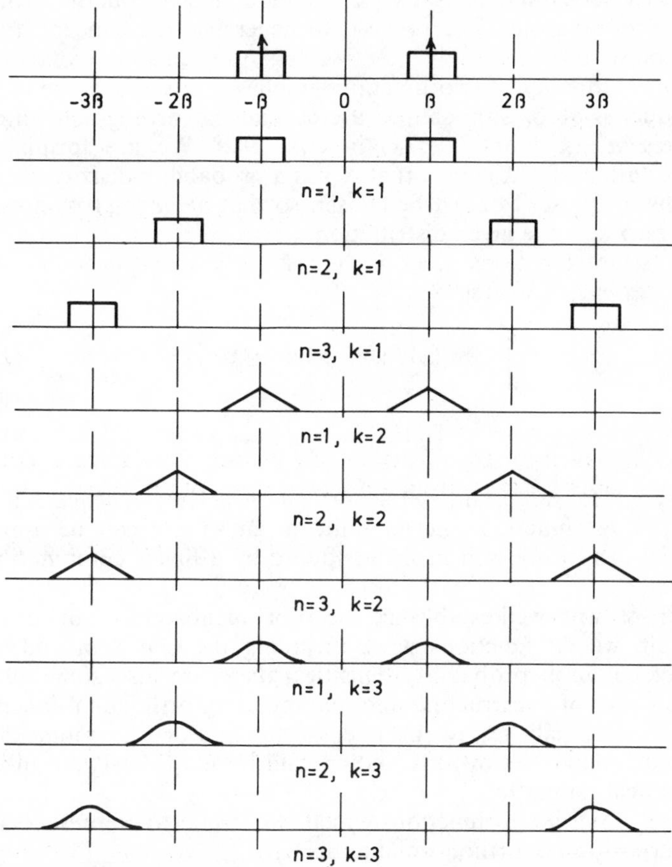

Fig. 9.4 Distortion of sinewave plus noise (for notation see Equation 9.67)

see for instance Shimbo [19], Fuenzalida *et al.* [20] and Chapter 13. The non-linearities may sometimes be modelled by analytic approximations; it is also common for the coefficients in series such as (9.67) to be evaluated numerically.

9.7 ORTHOGONAL-FUNCTION EXPANSIONS

9.7.1 Orthogonality properties

We are accustomed to the idea that signals and other variables can be represented as a series of mutually orthogonal terms. The

Fourier series and integral, for example, depend on the orthogonality of sinusoids. We proceed to generalise the concept, to suit our present topic.

Orthogonality of two functions implies that the average of their product is zero. Suppose we are seeking an orthogonal series to represent the function $f(x)$ in eqn. (9.2). We are forming the expectation of $f(X)$ given that X has a probability density $g(x)$: the terms of the series must be chosen so that average cross-products are zero with the given distribution.

A set of functions $q_i(x)$, $i = 0, 1, 2, \ldots$ is *orthogonal* under the *weight function* $w(x)$ if

$$\int_{-\infty}^{\infty} q_i(x)q_j(x)w(x)\,dx = 0, \quad i \neq j \tag{9.68a}$$

$$\int_{-\infty}^{\infty} [q_i(x)]^2 w(x)\,dx = k_i \neq 0 \tag{9.68b}$$

If $k_i = 1$ for all i, the set is *orthonormal*. It is *complete* if any continuous function $f(x)$ of finite mean square can be approximated arbitrarily well in mean square by a linear combination of the $q_i(x)$.

There are well-established sets of orthogonal functions for certain weight functions $w(x)$, including the Gaussian and some other common probability densities; there are also some general techniques of constructing and manipulating orthogonal functions, which we shall review in later sections. For the moment, we suppose that a set $q_i(x)$ is known, and develop its significance for statistical estimation.

Let $f(x)$ be a function which we wish to represent as a superposition of orthonormal terms

$$f(x) = \sum_i c_i q_i(x) \tag{9.69}$$

Then the coefficients c_i are given by

$$c_i = \int_{-\infty}^{\infty} q_i(x)f(x)w(x)\,dx \tag{9.70}$$

The difficult parts of the proof relate to completeness and convergence, which we leave to the specialised treatises. Assuming a series representation, it is clear that the coefficients are as stated, since the integral (9.70) is

$$\sum_j c_j \int_{-\infty}^{\infty} q_i(x)q_j(x)w(x)\,dx$$

and by the orthogonality only the term $i = j$ is non-zero. The series (9.69) approximates $f(x)$ in mean square, so that for any given truncation of the series $i < n$ no other choice of coefficients can reduce the mean squared error.

Let us suppose that two functions $f_1(x)$, $f_2(x)$ are represented by series of the form (9.69) with coefficients b_i, c_i respectively. Then

$$E\{f_1(X)f_2(X)\} = \sum_i \sum_j b_i c_j \int_{-\infty}^{\infty} q_i(x)q_j(x)w(x)\,dx \quad (9.71a)$$

$$= \sum_i b_i c_i \quad (9.71b)$$

since by orthogonality only the terms $i = j$ are non-zero.

As a special case of (9.71), consider the output from a non-linearity

$$\phi(x) = \sum_i c_i q_i(x) \quad (9.72)$$

where the q_i are orthonormal with respect to the input distribution. The mean squared output is

$$E\{Y^2\} = E\{[\phi(X)]^2\} = \sum_i c_i^2 \quad (9.73)$$

and we may hope for rapid convergence so that the first few c_i carry most of the significance.

Familiar orthogonal functions include transcendental functions such as sinusoids, Bessel functions etc. but in the present context our orthogonal functions are polynomial. They are conventionally ordered by the highest degree, so that

$$q_i(x) = \sum_{j=0}^{i} q_{ij}x^j \quad (9.74)$$

the highest-degree coefficient q_{ii} always being non-zero though some other coefficients may vanish. We shall sometimes find it convenient to represent a set of orthogonal polynomials by its generating function

$$Q(z, x) = \sum_{n=0}^{\infty} q_n(x)z^n \quad (9.75)$$

where the q_n are not necessarily orthonormal ($k_i = 1$ in eqn. 9.68b). It should be remarked that many types of orthogonal polynomial have well-entrenched definitions and notations with non-unity k_i, which implies that some scaling of expressions such as (9.71b) and (9.73) is necessary.

9.7.2 The Hermite polynomials

This is not the place for an exhaustive treatise on orthogonal polynomials. We shall exhibit the properties of the Hermite polynomials in some detail, as an example: and then indicate briefly what other relevant sets are well known, and how further sets may be constructed. More information may be found in many mathematical texts and handbooks; we draw attention particularly to Szego [21], which is perhaps the most comprehensive standard text, and Lavrentev and Schabat [22], which contains a relatively brief but very powerful treatment including some results which are not easily found elsewhere.

The Hermite polynomials $H_n(x)$ are orthogonal with respect to the Gaussian distribution, which gives them a central place in probability theory. It is convenient to start with the generating function

$$Q(z, x) \equiv \sum_{n=0}^{\infty} \frac{H_n(x)}{n!} z^n = e^{-(1/2)z^2 + zx} \tag{9.76}$$

Some fairly simple manipulation shows that

$$\frac{1}{(2\pi)^{1/2}} \int_{-\infty}^{\infty} Q(z, x)Q(u, x)e^{-(1/2)x^2}\, dx = e^{uz} \tag{9.77}$$

We then expand both sides as power series in u and z, and equate coefficients, with the result

$$\frac{1}{(2\pi)^{1/2}} \int_{-\infty}^{\infty} H_n(x)H_m(x)e^{-(1/2)x^2}\, dx = n!, \quad m = n \tag{9.78a}$$

$$= 0, \quad m \neq n \tag{9.78b}$$

which proves orthogonality and also yields the normalising constants k_i of eqn. (9.68b).

It is useful to have the Fourier transform of $H_n(x)$ weighted by the Gaussian density (it is not strictly a characteristic function because the weighted $H_n(x)$ are not valid probability densities). Transforming the weighted generating function

$$(2\pi)^{-1/2}Q(z, x)e^{-(1/2)x^2} \rightarrow e^{j\theta z - (1/2)\theta^2} \tag{9.79a}$$

Expanding as power series in z and equating coefficients

$$(2\pi)^{-1/2}H_n(x)e^{-(1/2)x^2} \rightarrow (j\theta)^n e^{-(1/2)\theta^2} \tag{9.79b}$$

It follows from the differentiation theorem for the Fourier

transform that

$$H_n(x)e^{-(1/2)x^2} = (-1)^n\left(\frac{d}{dx}\right)^n e^{-(1/2)x^2} \tag{9.80}$$

This is often taken as a defining property, and is consistent with the method of construction described in Section 9.7.4.

Polynomials may be calculated from eqns. (9.76) and (9.80) or from the recursion formulae

$$H_{n+1}(x) = xH_n(x) - nH_{n-1}(x) \tag{9.81}$$

The first few are

$$\begin{aligned} H_0(x) &= 1 \qquad & H_1(x) &= x \\ H_2(x) &= x^2 - 1 \qquad & H_3(x) &= x^3 - 3x \end{aligned} \tag{9.82}$$

We will now look at the use of Hermite polynomials to estimate the statistics of the output of non-linear devices with a Gaussian input. Consider an approximation to the non-linear characteristic

$$\phi(x) \approx \hat{\phi}_N(x) \equiv \sum_{n=0}^{N} c_n H_n(x) \tag{9.83}$$

Typically, as in the following examples, this series converges in the significant range, though non-uniformly. Let us further denote an approximation to the output by

$$\hat{Y}_N = \hat{\phi}_N(X) = \sum_{n=0}^{N} c_n H_n(X) \tag{9.84}$$

The mean-squared output may be estimated by the mean square of this approximation, that is

$$E\{Y^2\} \approx E\{\hat{Y}^2\} = \sum_{n=0}^{N} n!\, c_n^2 \tag{9.85}$$

(the factor $n!$ being the normalising constant k_n in this case). Typically, the series (9.85) converges rather more rapidly than (9.83).

Examples

(1) Consider the exponential non-linearity (9.41) with Gaussian input of variance σ^2. It can be shown that in this case

$$\hat{\phi}_N(x) = (e^{(1/2)c^2r^2} - 1) + e^{(1/2)c^2\sigma^2} \sum_{n=1}^{N} \frac{(c\sigma)^n}{n!} H_n\left(\frac{x}{\sigma}\right) \tag{9.86}$$

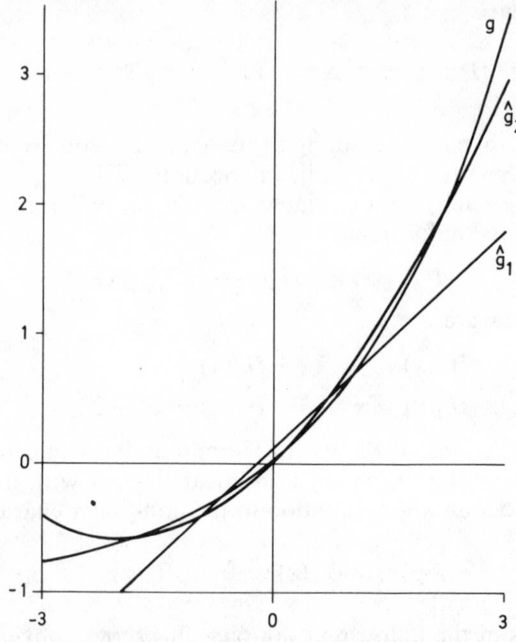

Fig. 9.5 Approximation by Hermite polynomials

A numerical example with $c = \frac{1}{2}$, $\sigma = 1$ is shown in Fig. 9.5; the approximation $\hat{\phi}_4(x)$ is on this scale indistinguishable from the true $\phi(x)$ over the range $\pm 3\sigma$ shown (which is not true for the Taylor series of corresponding degree). The estimator of the mean square is

$$E\{\hat{Y}_N^2\} = (1 - 2e^{(1/2)c^2\sigma^2}) + e^{c^2\sigma^2} \sum_{n=0}^{N} \frac{(c\sigma)^{2n}}{n!} \tag{9.87a}$$

which converges to the true mean-squared value

$$E\{Y^2\} = 1 - 2e^{(1/2)c^2\sigma^2} + e^{2c^2\sigma^2} \tag{9.87b}$$

In the numerical example ($c\sigma = \frac{1}{2}$) the true value is 0.3824 (to four decimal places): successive estimates for $N = 1, 2, 3, 4$ are 0.3387, 0.3789, 0.3822 and 0.3824 respectively.

(2) Consider the symmetrical clipper of Fig. 9.1h and eqn. (9.55), with a Gaussian input which without loss of generality we can take to have unit variance. It can be shown that in this case

$$\hat{\phi}_{2N+1}(x) = \left(\frac{2}{\pi}\right)^{1/2} \sum_{m=0}^{N} \frac{(-1)^m H_{2m+1}(x)}{m!\, 2^m (2m+1)} \tag{9.88}$$

So we have a sequence of polynomial approximations to a discontinuity, which could not have been obtained by Taylor series. Convergence is, however, both non-uniform and very slow. The estimator of the mean-squared output is

$$E\{\hat{Y}^2_{2n+1}\} = \frac{2}{\pi}\sum_{m=0}^{N}\frac{(2m)!}{(m!)^2 2^{2m}(2m+1)} \tag{9.89a}$$

This converges to the correct value

$$E\{Y^2\} = \frac{2}{\pi}\sum_{m=0}^{\infty}\frac{(2m)!}{(m!)^2 2^{2m}(2m+1)} \tag{9.89b}$$

$$= \frac{2}{\pi}\sin^{-1}(1) = 1 \tag{9.89c}$$

but only very slowly. The numerical values for $2N+1=1, 13, 25, 37, 49$ are (to four decimal places) 0.6366, 0.8635, 0.9001, 0.9174 and 0.9280 respectively.

As these examples suggest, the technique is applicable to a great variety of non-linearities (including both analytically and empirically specified characteristics) but convergence may be slow if the characteristics include discontinuities. Studies of quantising distortion in particular have given rise to very long series, but have also led to techniques for dealing with them: see Chapter 11 and Lever and Cattermole [23].

9.7.3 A brief survey of orthogonal polynomials

The principal information which it seems appropriate to include here is a list of probability distributions for which orthogonal polynomials have been defined, with some selected references to further information. This is given in Table 9.1. It will be clear that number 2 is a special case of number 3: numbers 4, 5 and 6 are special cases of number 7: numbers 8 and 9 are related: number 10 is a limiting case of number 11.

9.7.4 Procedures for deriving orthogonal polynomials

Once it is assumed that a set of orthogonal functions does consist of polynomials of increasing degree, it will be clear that a direct, though laborious, procedure is to begin with $q_0(x) = 1$ and then

Table 9.1. ORTHOGONAL POLYNOMIALS

	Probability distribution	*Weight function*	*Conventional designation of polynomials*	*References*
1	Gaussian	$e^{-(1/2)x^2}$	Hermite	21, 22, 24, 25
2	Negative-exponential	$e^{-x}, 0 < x$	Laguerre	21, 22, 24, 25
3	Gamma	$x^k e^{-x}, 0 < x$	Generalised Laguerre	21, 22, 24, 25
4	Rectangular	$1, -1 < x < +1$	Legendre	21, 22, 24, 25
5	Arcsine	$(1 - x^2)^{-1/2}, -1 < x < +1$	Chebyshev	21, 22, 24, 25
6		$(1 - x^2)^{k-1/2}, -1 < x < +1$	Gegenbauer	21, 24, 25
7		$(1 - x)^{\lambda}(1 + x)^{\mu}, -1 < x < +1$	Jacobi	21, 22, 24, 25
8		$x^{-k \log x}, 0 < x$	Stieltjes–Wigert	21
9	Lognormal	$\frac{1}{(2\pi)^{1/2}\sigma x} \exp\left\{\frac{-(\log x - m)^2}{2\sigma^2}\right\}$		26
10	Poisson	$\sum_{n=0}^{\infty} \frac{\lambda^n e^{-\lambda}}{n!} \delta(x - n)$	Poisson–Charlier	21
11	Binomial	$\sum_{n=0}^{N} \binom{N}{n} p^n (1 - p)^{N-n} \delta(x - n)$	Krawtchouk	21

develop the $q_i(x)$ in sequence, adjusting coefficients at each step so that $q_n(x)$ is orthogonal to all $q_i(x)$ for $i < n$.

A variety of methods are described in the literature, but our main purpose here is to draw attention to a method which appears to be powerful but not widely known. The only proof which the present author has seen is in Lavrentev and Schabat [22]. Beckmann [8] gives a statement without proof, but unfortunately his version has suffered a typographical error and is incorrect. The following statement has been checked carefully and should be correct!

Consider a class of weight functions $w(x)$ which satisfy the following conditions

(1) $$\frac{w'(x)}{w(x)} = \frac{\alpha_0 + \alpha_1 x}{\beta_0 + \beta_1 x + \beta_2 x^2} \equiv \frac{\alpha(x)}{\beta(x)}$$

(2) $w(x)\beta(x) = 0$ at both limits of the range of x.

Then a set of polynomials $P_n(x)$ defined as below is orthogonal with respect to $w(x)$; namely

$$P_n(x) = \frac{1}{w(x)}\left(\frac{d}{dx}\right)^n \{w(x)[\beta(x)]^n\} \qquad (9.90)$$

It is easy to show that eqn. (9.80) for the Hermite polynomials is consistent with (9.90). There are also several other cases in which (9.90) agrees with the Rodrigues formula for a set of orthogonal polynomials as given, for example, in references [21] and [24].

9.8 DIAGONAL PROCESSES

9.8.1 Bivariate distributions of diagonal type

We have seen that a significant application of bivariate probability theory is to describe the joint properties of two samples $X_1 = X(t)$ and $X_2 = X(t + \tau)$ from a covariance-stationary random process. The marginal distributions of X_1 and X_2 are identical, but as τ changes the covariance alters. To model such a process, we require a bivariate distribution with given (identical) marginals, and covariance or correlation as a parameter.

Unfortunately, this information is not, in general, enough to specify a bivariate distribution. The Gaussian is an exception: to obtain other bivariate distributions one must impose some further property.

Another problem is that, with a general bivariate distribution, the calculation of expectations $E\{\phi(X_1)\phi(X_2)\}$ may be quite difficult even for apparently simple functions $\phi(x)$. We would like to use some approximating technique, such as the expansion in terms of orthogonal functions which was developed for univariate problems in Section 9.7. However, the prospect of summing multiply-infinite series for everything is not too enticing, and we would like the approximations to be fairly tractable.

It turns out that there is a class of bivariate distributions, including but not limited to the Gaussian, which enable two-dimensional statistics to be developed as a one-dimensional series. We shall expound some general properties of the class, develop the appropriate Gaussian theory in some detail, and give a few illustrative examples.

The first step is to represent a bivariate density $g(x_1, x_2)$ as a series of orthogonal terms. It is perhaps plausible that there may be a representation of the form

$$g(x_1, x_2) = g_1(x_1)g_2(x_2) \sum_{m=0}^{\infty} \sum_{n=0}^{\infty} a_{mn} q_m(x_1) q_n(x_2) \tag{9.91}$$

where all cross-products of the form $q_m(x_1)q_n(x_2)$ are admitted. A bivariate distribution is of *diagonal type* if cross-products for $m \neq n$ are not required: such a distribution admits the series representation

$$g(x_1, x_2) = w(x_1)w(x_2) \sum_{m=0}^{\infty} a_m q_m(x_1) q_m(x_2) \tag{9.92}$$

where the $q_m(x)$ are orthogonal with a weight function $w(x)$ equal to both marginal densities. The concept can be extended to cover bivariate distributions with different marginals, but we shall confine our attention here to the simpler case.

To apply the diagonal series to the calculation of statistics, we postulate series of the form (9.69) for functions $f_1(x_1)$, $f_2(x_2)$ with coefficients b_i, c_i respectively. Then if X_1, X_2 have the joint distribution (9.92) and the $q_i(\ )$ are orthonormal, a joint statistic takes the form

$$E\{f_1(X_1)f_2(X_2)\}$$

$$= \sum_{m=0}^{\infty} a_m \sum_i b_i \int q_i(x_1) q_m(x_1) w(x_1)\, dx_1$$

$$\times \sum_j c_j \int q_j(x_2) q_m(x_2) w(x_2)\, dx_2 \tag{9.93a}$$

$$= \sum_{m=0}^{\infty} a_m b_m c_m \tag{9.93b}$$

since by orthogonality only the terms $i = j = m$ are non-zero.

This theory offers the prospect of calculating the second-order statistics of covaraince-stationary processes by means of one-dimensional expansions. Before surveying diagonal processes in general, we treat some specific examples.

9.8.2 Non-linear operations on Gaussian processes

The bivariate Gaussian distribution is of diagonal type. Its characteristic function (9.17b) with unit variance can be expanded in the series

$$G(\theta_1, \theta_2) = e^{-(1/2)(\theta_1^2+\theta_2^2)} \sum_{m=0}^{\infty} \frac{(-\rho\theta_1\theta_2)^m}{m!} \tag{9.94}$$

Each term of this expansion is separable in θ_1 and θ_2, so can be Fourier-transformed into a separable expression in x_1 and x_2. The transformation (9.79) applied to each dimension of each term gives directly

$$g(x_1, x_2) = \frac{1}{2\pi} e^{-(1/2)(x_1^2+x_2^2)} \sum_{m=0}^{\infty} \frac{\rho^m}{m!} H_m(x_1)H_m(x_2) \tag{9.95}$$

where the $H_m(\ )$ are Hermite polynomials.

Let a Gaussian process of variance σ^2 be applied to a non-linearity (9.83). Then the output autocorrelation can be represented as

$$\mathbf{R}_Y(\tau) = E\{Y_1 Y_2\} = E\{\phi(X_1)\phi(X_2)\} \tag{9.96a}$$

$$= \sum_{m=0}^{\infty} m!\,\rho^m c_m^2 \tag{9.96b}$$

(the factor $m!$ being the normalising constant for the Hermite polynomials). Note that for unit input correlation ρ, this reduces to expression (9.85) for $E\{Y^2\}$.

Examples

(1) Consider the exponential non-linearity (9.41) for which we have already used the expansion (9.86). For this case

$$\mathbf{R}_Y(\tau) = 1 - 2e^{(1/2)c^2\sigma^2} + e^{c^2\sigma^2(1+\rho)} \tag{9.97}$$

which reduces to (9.87b) for unit input correlation. It is also interesting to find the correlation coefficient between output samples. This is

$$\frac{\text{Cov}\,(Y_1, Y_2)}{V(Y)} = \frac{e^{\rho c^2\sigma^2} - 1}{e^{c^2\sigma^2} - 1} \leqslant \rho \tag{9.98}$$

It is approximately equal to the input correlation coefficient ρ for small $c\sigma$ (i.e. for small excursions such that the transfer characteristic is almost linear) but diminishes as the excursions increase and the non-linearity becomes more significant.

(2) Consider the symmetrical clipper of Fig. 9.1h and eqn. (9.55), for which we have already used the expansion (9.88). For this case

$$\mathbf{R}_Y(\tau) = \frac{2}{\pi} \sum_{m=0}^{\infty} \frac{(2m)!\, \rho^{2m+1}}{(m!)^2 2^{2m}(2^m + 1)} \tag{9.99a}$$

$$= \frac{2}{\pi} \sin^{-1} \rho \tag{9.99b}$$

Note that this expression reduces to eqn. (9.89) for unit input correlation. More importantly, this solution is identical with the solution (9.56) obtained by the characteristic-function method, not only in the final result but in the series summed to attain it. This arose because both our characteristic-function solution, and our diagonal bivariate expansion, made use of an expansion of the joint characteristic function into separable terms. There is in fact a general relationship between the two methods, though it does not always emerge so directly as in this example.

9.8.3 A brief survey of diagonal processes

Table 9.2 lists the diagonal processes known to the author. This subject has not yet got into the textbooks (save for a few references to Mehler's expansion of the bivariate Gaussian distribution, which is effectively our eqn. (9.95) though usually derived in a less perspicuous manner) and so fairly complete references to the periodical literature are given in this section.

The subject of diagonal processes effectively begins with the work of Barrett and Lampard [11], who developed some general theory and gave as examples the Gaussian, Rayleigh and arc-sine processes (numbers 1, 2 and 5 in Table 9.2). They showed *inter alia*, that of the coefficients a_m in eqn. (9.92), $a_0 = 1$ and $a_1 = \rho$, the input correlation coefficient. They also discussed cross-correlation of two processes, which we have not considered here. Blachman [12] showed that the output was a sum of uncorrelated terms having a

Table 9.2. DIAGONAL PROCESSES AND DISTRIBUTIONS

	Name of process or distribution	*Weight function (univariate)*	*Type of polynomial*	*References*
1	Gaussian	$e^{-(1/2)x^2}$	Hermite	11, 12, 27, 29, 31, 18, 33
2	Negative-exponential, Rayleigh	$e^{-x}, 0 < x$	Laguerre	11
3	Gamma, chi-squared, generalised Rayleigh	$x^k e^{-x}, 0 < x$	Generalised Laguerre	28, 29, 32
4	—	$1, -1 < x < 1$	Legendre	13
5	Arcsine	$(1 - x^2)^{-1/2}, -1 < x < 1$	Chebyshev	11, 18
6	—	$(1 - x^2)^{k-1/2}, -1 < x < 1$	Gegenbauer	30, 18
7	—	$(1 - x)^{\lambda}(1 + x)^{\mu}, -1 < x < 1$	Jacobi	29

clear relationship to certain series expansions derived by Rice's characteristic-function method. Lancaster [13] developed similar ideas about the structure of bivariate distributions from a statistical standpoint, and exhibited the relationship with Fisher's 'canonical variables'. Brown [34] gave a criterion for a bivariate distribution to be of diagonal type: the conditional moments $E\{X_1^k \mid X_2 = x_2\}$ must be polynomials in x_2 of degree not exceeding k (and similarly with the variables interchanged). Miller *et al.* [28] introduced the generalised Rayleigh process (number 3 in Table 9.2). Wong and Thomas [29] introduced the bivariate distribution based on Jacobi polynomials (number 7): McFadden [30] that based on Gegenbauer polynomials (number 6). McGraw and Wagner [18] developed the concept of elliptically symmetrical distributions, which although in some ways reminiscent of diagonal distributions (and also useful in non-linear problems) are not the same thing. They showed that only the Gaussian, arc-sine and Gegenbauer-polynomial types were common to both classes. Lever and Cattermole [23], investigating the effect of quantization on Gaussian processes, found that the diagonal-process analysis gave rise to a very large number of terms, but that this series had a higher-order structure based on clusters of intermodulation components.

All our previous theory and citations relate to instantaneous non-linearities. An interesting recent development is a corresponding theory for dealing with non-linear devices having an amplitude-dependent phase shift. Blachman [35] and Chie [36] use diagonal expansions in four variables, namely the amplitude and phase of two complex samples of a bandpass random process. There had, of course, been previous work on joint amplitude and phase problems, notably by Shimbo [19], using characteristic-function methods.

9.8.4 Superposition of random processes

Much, though not all, of our analysis has related to an input process described by a single distribution, such as Gaussian noise or signal. Practical interest often relates to a multiplicity of inputs: signal plus noise, or several random signals, or several signals plus noise.

The superposition of random processes can also be dealt with by an extension of the methods described here. Our treatment of sine-wave plus noise in Section 9.6.4 is a guide to the general form of the phenomena: terms of various orders due to each input, together with cross-products due to interaction between inputs. Blachman [37] discusses this in general terms under the truly explanatory title 'The signal × signal, noise × noise and signal × noise output of a non-linearity'. Much recent work, especially that carried out in the context of communication satellites, is mainly concerned with the multiple-input case [19], [20], [35], [36], [38], [39] and Chapter 13. But the problem is really quite an old one, going back to the early studies of intermodulation in carrier systems [15]. It remains a rather complex topic, in which it is hard to see the wood for the trees unless one has the underlying concepts clearly in mind.

9.9 CONCLUSION

The aim of this chapter has been to give a clear tutorial exposition of the principal means of analysing the effect of non-linear operations on random processes. The unifying theme has been the relationship between the various ways of describing the non-linearity: by explicit formulae or curves, by power series, by characteristic functions and by series of orthogonal polynomials. The methods are less *ad hoc*, and more closely related, than is sometimes suggested. We have seen similar concepts emerge from many approaches: self-convolutions of power spectra, frequency translation of components, series of uncorrelated components. It is hoped that, with this mental picture, the reader will be better equipped to find his way through the obfuscating complexity of detailed practical studies.

REFERENCES

1. Davenport, W. B., *Probability and Random Processes*, McGraw-Hill (1970)
2. Franks, L. E., *Signal Theory*, Prentice-Hall (1969)
3. Deutsch, R., *Non-Linear Transformations of Random Processes*, Prentice-Hall (1962)
4. Kuznetsov, P. I., Stratonovich, R. C. and Tikhonov, V. I., *Non-Linear Transformations of Stochastic Processes* (English translation), Pergamon Press (1965)
5. Middleton, D., *An Introduction to Statistical Communication Theory*, McGraw-Hill (1960)
6. Harman, W. W., *Principles of the Statistical Theory of Communications*, McGraw-Hill (1963)
7. Blachman, N. M., *Noise and its Effect on Communication*, McGraw-Hill (1966)
8. Beckmann, P., *Probability in Communication Engineering*, Harcourt, Brace and World (1967)
9. Rice, S. O., 'Mathematical analysis of random noise', *Bell System Tech. J.*, **23**, 282–332 (1944) and **24**, 46–156 (1945). Reprinted in Wax, N. (ed.), *Selected Papers on Noise and Stochastic Processes*, Dover (1954)

10. Lampard, D. G., 'Some theoretical and experimental investigations of random electrical fluctuations', Ph.D. dissertation, University of Cambridge (1954)
11. Barrett, J. F. and Lampard, D. G., 'An expansion for some second-order probability distributions and its applications to noise problems', *Trans. IRE*, **IT-1**, 10–15 (1955)
12. Blachman, N. M., 'The uncorrelated output components of a non-linearity', *Trans. IEEE*, **IT-14**, 250–255 (1968)
13. Lancaster, H. O., 'The structure of bivariate distributions', *Ann. Math. Stats.*, **29**, 719–736 (1958)
14. Lancaster, H. O., 'Correlations and canonical forms of bivariate distributions', *Ann. Math. Stats.*, **34**, 532–538 (1963)
15. Bennett, W. R., 'Cross-modulation requirements on multi-channel amplifiers below overload', *Bell System Tech. J.*, **19**, 587–610 (1940)
16. Bell Telephone Laboratories, *Transmission Systems for Communication*, Bell Telephone Laboratories (1964)
17. Price, R., 'A useful theorem for non-linear devices having Gaussian inputs', *Trans. IRE*, **IT-4**, 69 (1958)
18. McGraw, D. K. and Wagner, J. F., 'Elliptically symmetric distributions', *Trans. IEEE*, **IT-14**, 110–120 (1968)
19. Shimbo, O., 'Effects of intermodulation, AM–PM conversion and additive noise in multicarrier TWT systems', *Proc. IEEE*, **59**, 230–238 (1971)
20. Fuenzalida, J. C., Shimbo, O. and Cook, W. C., 'Time-domain analysis of intermodulation effects caused by non-linear amplifiers', *Comsat Tech. Review*, **3**, 89–127 (1973)
21. Szego, G., 'Orthogonal polynomials', *American Math. Soc. Colloquium Publication*, Vol. XXIII, 3rd ed. (1967)
22. Lawrentjew, M. A. and Schabat, B. W., 'Methoden der komplexen Funktionen Theorie', Deutsches Verlag der Wissenschaften, Berlin (1967). (This is a German translation of a book originally published in Russian: Metody Teorii Funktsii Komplexnogo Peremennogo, Moskva (1951). At the time of writing there is no English translation.)
23. Lever, K. V. and Cattermole, K. W., 'Quantising noise spectra', *Proc. IEEE*, **121**, 545–954 (1974)
24. Abramowitz, M. and Stegun, I., *Handbook of Mathematical Functions*, National Bureau of Standards (1964)
25. Gradshteyn, I. S. and Ryzhik, I. M., *Table of Integrals, Series and Products*, 4th ed., Academic Press (1965)
26. Klose, D. R. and Kurz, L., 'A new representation theory and detection procedure for a class of non-Gaussian channels', *Trans. IEEE*, **COM-17**, 225–234 (1969)
27. Thomson, W. E., 'The response of a non-linear system to random noise', *Proc. IEEE*, **102C**, 46–48 (1955)
28. Miller, K. S., Bernstein, R. I. and Blumenson, L. E., 'Generalised Rayleigh processes', *Quarterly Appl. Maths.*, **16**, 137–145 (1958)
29. Wong, E. and Thomas, J. B., 'On polynomial expansions of second-order distributions', *S.I.A.M. J.*, **10**, 507–516 (1962)
30. McFadden, J. A., 'A diagonal expansion in Gegenbauer polynomials for a class of second-order probability densities', *S.I.A.M. J. Appl. Maths.*, **14**, 1433–1436 (1966)
31. Sarmonov, O. V. and Bratoeva, Z. N., 'Probability properties of bilinear expansions of Hermite polynomials', *Theory Probab. Applic.*, **12**, 470–481 (1967)
32. Watson, G. N., 'Notes on generating functions of polynomials: (1) Laguerre polynomials', *J. London Math. Soc.*, **8**, 189–192 (1933)

33. Watson, G. N., 'Notes on generating functions of polynomials: (2) Hermite polynomials', *J. London Math. Soc.*, **8**, 194–199 (1933)
34. Brown, J. L., 'A criterion for the diagonal expansion of a second-order probability density in orthogonal polynomials', *Trans. IRE*, **IT-4**, 172 (1958)
35. Blachman, N. M., 'Output signals and noise from a non-linearity with amplitude-dependent phase shift', *Trans. IEEE*, **IT-25**, 77–79 (1979)
36. Chie, C. M., 'A modified Barrett–Lampard expansion and its application to bandpass non-linearities with both AM–AM and AM–PM conversion', *Trans. IEEE*, **COM-28**, 1859–1866 (1980)
37. Blachman, N. M., 'The signal × signal, noise × noise and signal × noise output of a non-linearity', *Trans. IEEE*, **IT-14**, 21–27 (1968)
38. Lyons, R. G., 'A stochastic analysis of signal sharing in a bandpass non-linearity', *Trans. IEEE*, **COM-22**, 1778–1788 (1974)
39. Lyons, R. G., 'Signal and interference output of a bandpass non-linearity', *Trans. IEEE*, **COM-27**, 888–891 (1979)

10

Timing Extraction for Baseband Digital Transmission

J. J. O'Reilly

10.1 INTRODUCTION

For digital communication it is necessary to synchronise the receiver clock with the incoming data and this is commonly achieved by operating on the digital signal in such a way as to extract a suitable timing wave. A widely used technique appropriate to baseband pulse amplitude modulated (PAM) signalling is shown in Fig. 10.1. The PAM signal is filtered and applied to a non-linear circuit, the output of which is then bandpass filtered or applied to a phase-locked loop (PLL). The output of the bandpass filter or PLL provides the requisite timing wave at the signalling rate.

Various authors have discussed this scheme, e.g. [1–5]. The possibility of recovering a timing wave devoid of data pattern-dependent jitter has been considered [2] and for the special case of a square-law non-linearity, conditions for the extraction of a jitter-free timing signal have been published [3, 5]. However, the complexity of these analyses is such that the process of timing extraction is still, perhaps, not as well understood as its ubiquity in digital communications deserves. In this chapter a simplified analysis is presented which establishes conditions for zero pattern-dependent jitter. The result presented here represents a modest generalisation of that obtained [3] via quite a different route, in agreement with [5]. The significant feature of the present analysis, however, is its simplicity—the simplification being in large measure achieved by concentrating on a single realisation of the digital signal. It is considered that this approach facilitates an appreciation of the underlying physical mechanisms involved in the timing recovery process. Ultimately, however, the inherent stochastic nature of the problem must be accommodated. Some care is needed here since the digital signal is not truly *stationary* but *cyclostationary* [1, 6].

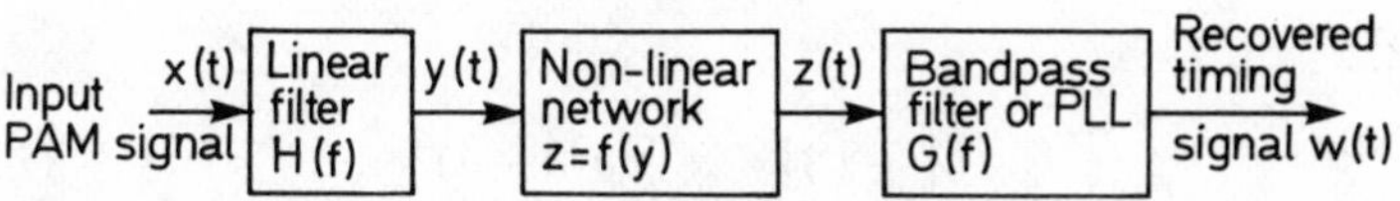

Fig. 10.1 Timing extraction filter

10.2 AN APPARENT PARADOX

Consider a PAM signal given by

$$x(t) = \sum_r a_r p(t - r) \tag{10.1}$$

where $\{a_r\}$ is the message (assumed to derive from a zero mean stationary process with independent elements), $p(t)$ is the signal pulse shape and without loss of generality, unit signalling interval is assumed. It is readily shown, e.g. [7], that the power spectrum for such a signal has a null at the signalling frequency. For example, if we consider full-width rectangular signalling† such that

$$p(t) = \text{rect}\ (t) \triangleq \begin{cases} 1, & |t| < \frac{1}{2} \\ 0, & \text{elsewhere} \end{cases} \tag{10.2a}$$

$$\Rightarrow P(f) = \text{sinc}\ (f) \triangleq \frac{\sin \pi f}{\pi f} \tag{10.2b}$$

the power spectrum $S_x(f)$ has the form

$$S_x(f) = \text{sinc}^2\ (f) \tag{10.3}$$

and

$$S_x(f)|_{f=1} = \text{sinc}^2\ (1) = 0$$

Since $S_x(f)$ is zero at the signalling rate, $x(t)$ does not directly contain a timing component. Further, and perhaps more seriously, $S_x(f)$ is a *continuous* spectrum; it contains no discrete spectral line components. Nevertheless, observation of the time signal $x(t)$—with data transitions occurring at integral time intervals—assures us of the presence of timing information. Considering the ensemble of all possible realisations of $x(t)$ we note that, while

$$E[X(t)] = E[a]\sum_r p(t - r) \tag{10.4}$$

is identically zero, the square of the PAM signal has a periodic mean

† Time domain signals are denoted by lower-case letters and Fourier spectra by upper-case letters, $x(t) \Leftrightarrow X(f)$ representing a Fourier transform pair.

value

$$E[X^2(t)] = \overline{a^2}\sum_r p^2(t-r) \tag{10.5}$$

the magnitude of the periodic component depending on the pulse shape,†† $p(t)$. Similarly, it is found experimentally that if $x(t)$ is appropriately filtered and the resultant is applied to a square-law device the output contains a spectral line at the signalling rate.

The presence of timing information in $x(t)$, despite the absence of a line component in $S_x(f)$, and the emergence of a line component on non-linear processing, may at first sight seem paradoxical. We shall resolve this apparent paradox shortly but it is instructive at this stage to note the qualitative explanation given by Bylanski and Ingram [5]. These authors point out that the continuous spectrum is not that of a 'truly random' signal but 'has hidden phase relationships between its different components'. For a proper understanding of the timing recovery process we must discover the nature of these 'hidden phase relationships'. We note here, however, that it is a general property of cyclostationary processes that there may be correlation between components in different frequency bands [8].

10.3 SPECTRAL SYMMETRY IN DIGITAL SIGNALS

The PAM signal of (10.1) may be rewritten as

$$\begin{aligned} x(t) &= \sum_r a_r p(t-r) \\ &= p(t) * \sum_r a_r \delta(t-r) \\ &= p(t) * s(t) \end{aligned} \tag{10.6}$$

where $*$ on the line denotes convolution. Note that $s(t)$ has the form of a regularly sampled signal containing all information relating to the message $\{a_r\}$, which is, in turn, simply one specific realisation of a discrete stochastic process. Since $s(t)$ contains the message information we can temporarily neglect $p(t)$ by letting

$$p(t) = \delta(t) \Rightarrow p(t) * s(t) = s(t)$$

†† Equations (10.4) and (10.5) are consequences of the fact that the PAM signal of (10.1) constitutes a zero mean (wide-sense) cyclostationary process. A cyclostationary (or periodic non-stationary) process has statistical moments which are periodic in time, rather than constant as for a stationary process.

corresponding to impulse signalling. A practical signal pulse shape can readily be reintroduced at a later stage by way of the convolution theorem.

In virtue of the regularly sampled signal format we can associate with $s(t)$ a real time signal $a(t)$ strictly band limited to $|f| < \frac{1}{2}$, where we are in essence invoking a converse of the sampling theorem [9]. Further, subject to the convergence of the Fourier integrals,† we can identify a Fourier spectrum $A(f)$ such that $a(t) \Leftrightarrow A(f)$ and $A(f) = 0$ for $|f| \geqslant \frac{1}{2}$. Introducing the notation

$$\text{rep}\,\{f(t)\} \triangleq \sum_r f(t-r)$$

we can write

$$\begin{aligned} s(t) &= \sum_r a_r s(t-r) \\ &= a(t)\,\text{rep}\,\{\delta(t)\} \qquad (10.7a) \\ \Rightarrow S(f) &= \text{rep}\,\{A(f)\} \qquad (10.7b) \end{aligned}$$

A representative message spectrum $A(f)$ and corresponding replicated spectrum $S(f)$ is shown in Fig. 10.2. For convenience of representation, real, finite energy functions are depicted, although no such restriction is implied for the analysis. In general, $A(f)$ and $S(f)$ will be complex functions of frequency, but since $\{a_r\} \Rightarrow a(t)$ is real, $A(f)$ is Hermitian with $A(f) = A^*(-f)$. Here, * as a superscript denotes complex conjugation. Hence $S(f)$ is both Hermitian and periodic, $S(f) = S^*(-f)$ and $S(f) = S(f-1)$, whence $S(f)$ exhibits Hermitian symmetry about $f = \pm\frac{1}{2}$. (There are many other symmetries to $S(f)$ but these need not concern us here.)

10.4 RECOVERY OF A JITTER-FREE TIMING SIGNAL

Let us now examine the effect on $S(f)$ of a linear filtering operation, $H(f)$. Let $H(f)$ be chosen such that its positive frequency part

$$H_+(f) \triangleq H(f) \cdot U(f)$$

† It should be noted that unless $a(t)$ is finite the convergence of the Fourier integral is not generally guaranteed. The use of generalised functions (distributions) [10] admits periodic and almost periodic functions [10], however, and by delving deeper into functional analysis we can define a *generalised* Fourier transform [11] which extends the class of admissible functions yet further, to the extent required for the present discussion [12].

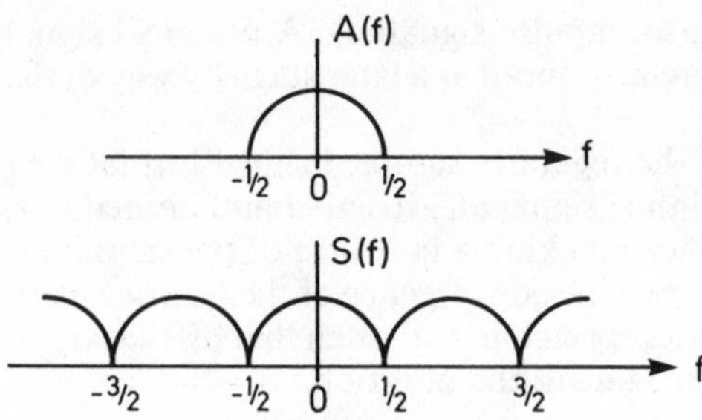

Fig. 10.2 Illustrative message spectrum A(f) and corresponding replicated spectrum S(f)

with $U(f)$ the unit step function, exhibits Hermitian symmetry about $f = +\frac{1}{2}$, i.e. $H_+(f+\frac{1}{2}) = H_+^*(\frac{1}{2}-f)$. It follows that the negative frequency part $H_-(f)$ exhibits similar symmetry about $f = -\frac{1}{2}$. For convenience such functions† will henceforth be described as *bandpass Hermitian* about $f = \pm\frac{1}{2}$.

The output from the filter, $y(t) \Leftrightarrow Y(f)$ is

$$y(t) = s(t) * h(t)$$

$$\Rightarrow Y(f) = S(f) \cdot H(f) \tag{10.8}$$

so that $Y(f)$ is also bandpass Hermitian about $f = \pm\frac{1}{2}$. This is illustrated in Fig. 10.3, where again real functions are assumed for representational convenience. In virtue of these symmetries we observe that $y(t)$ has precisely the form of a double-sideband suppressed carrier amplitude-modulated signal, the (suppressed) carrier frequency being one half of the signalling rate. Hence we have

$$y(t) = r(t) \cos \pi t \tag{10.9}$$

where $r(t) = 2\mathscr{F}^{-1}\{H_+(f+\frac{1}{2})S(f+\frac{1}{2})\}$ is a real function of time with zero mean value, $\bar{r}(t) = 0$.

It is appropriate now to reconsider the influence of the signal pulse shape, $p(t) \Leftrightarrow P(f)$. At the input to the filter we have, from (10.6)

$$p(t) * s(t) \Leftrightarrow P(f)S(f)$$

so that the output is

$$y(t) = p(t) * s(t) * h(t) \Leftrightarrow Y(f) = P(f) \cdot S(f) \cdot H(f)$$

† Those familiar with the complex envelope concept will recognise these constraints as ensuring that the corresponding complex envelope is a *real* signal. We will not, however, make explicit use of the complex envelope formalism.

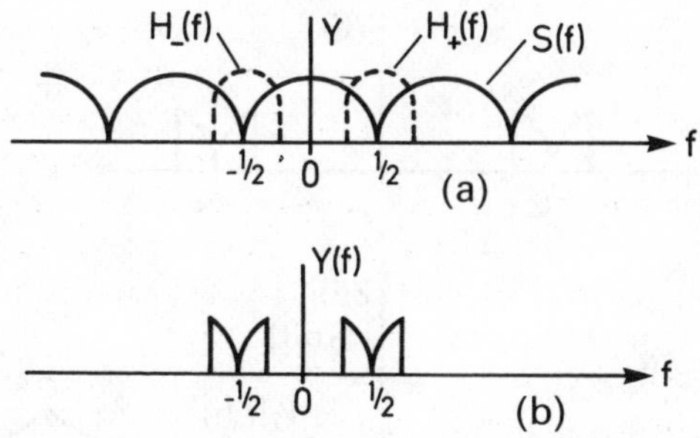

Fig. 10.3 Filtering of S(f) to obtain Y(f) with the desired symmetries. (a) Replicated spectrum and filter response, (b) bandpass spectrum Y(f)

Since $S(f)$ is Hermitian symmetric about $f = +\frac{1}{2}, f = -\frac{1}{2}$, $Y(f)$ is bandpass Hermitian about $f = \pm\frac{1}{2}$ if and only if $P(f)H(f)$ exhibits such symmetry. This then is the joint constraint on $p(t)$, $h(t) \Leftrightarrow P(f)$, $H(f)$ required to preserve the double-sideband signal format of (10.9).

Following Fig. 10.1, $y(t)$ is now applied to the non-linear network to yield $z(t)$:

$$\begin{aligned} z(t) &= f\{y(t)\} \\ &= y^2(t) \quad \text{for a square law non-linearity} \end{aligned} \tag{10.10}$$

and application of the convolution theorem then suggests

$$Z(f) = \int_{-\infty}^{\infty} Y(u)Y(f-u)\,du = Y^{*2}(f) \tag{10.11}$$

where ^{*n} as a superscript denotes n-fold self-convolution of a function. However, we wish to consider $y(t)$ existing throughout time, corresponding to an infinite energy, finite average power signal. It was to admit this possibility that we invoked the generalised Fourier transform in (10.7). It is possible to proceed with the analysis in the frequency domain, but some effort is required to establish the convergence of the integral [12] in eqn. (10.11). Fortunately a time-domain approach enables us to circumvent these difficulties.

Considering $z(t)$ in the time domain, we have

$$z(t) = r^2(t)\cos^2(\pi t) = \tfrac{1}{2}r^2(t)(1 + \cos 2\pi t) \tag{10.12}$$

where $r^2(t)$ is non-negative with non-zero mean $\overline{r^2}(t)$. Hence we can write

$$r^2(t) = r_0(t) + \overline{r^2}(t) \tag{10.13}$$

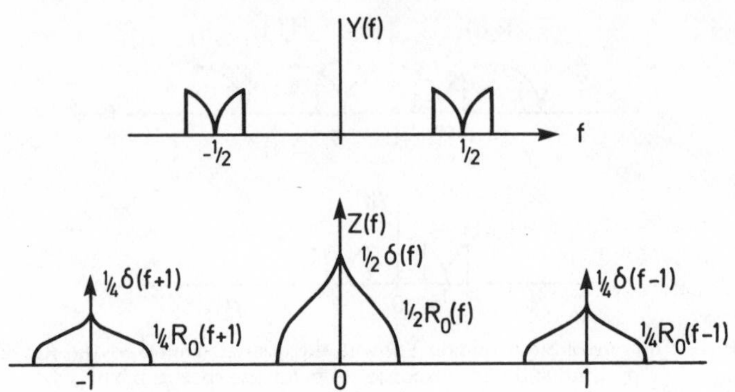

Fig. 10.4 The spectrum Z(f) = Y*(f), including discrete spectral lines at f = 0, ±1

with $r_0(t)$ a zero mean signal corresponding to $r^2(t)$ with the dc term, $\overline{r^2}(t)$ eliminated. We now have

$$z(t) = \tfrac{1}{2}(r_0(t) + \overline{r^2}(t))(1 + \cos 2\pi t)$$

$$\Rightarrow Z(f) = \tfrac{1}{2}R_0(f) + \tfrac{1}{4}R_0(f+1) + \tfrac{1}{4}R_0(f-1)$$

$$+ \tfrac{1}{2}\overline{r^2}(t)\,\delta(f) + \tfrac{1}{4}\overline{r^2}(t)\,\delta(f+1) + \tfrac{1}{4}\overline{r^2}(t)\,\delta(f-1) \quad (10.14)$$

The spectrum $Z(f)$ and its relation to $Y(f)$ is shown schematically in Fig. 10.4. Note in particular the line components, corresponding to the last three terms in eqn. (10.14). The spectral line pair at $f = \pm 1$ corresponds to the wanted timing component, $\tfrac{1}{2}\overline{r^2}(t)$ cos $2\pi t$, which may be extracted from $z(t)$ by bandpass filtering. However, any finite bandwidth filter will admit some portion of the other terms in (10.14), and if we wish to minimize this contribution a very narrowband filter must be employed. On the other hand, the use of a narrowband filter will limit the signalling speed tolerance of the timing recovery circuit and may lead to large timing phase errors in the presence of parameter variations with, for example, temperature or age. Hence a relatively modest bandwidth filter may be preferred.

Closer examination of eqn. (10.14) is rewarding in this respect. If there is no spectral overlap between $R_0(f)$ and $R_0(f \pm 1)$ it is in principle possible to eliminate the dc and baseband components

$$\tfrac{1}{2}\overline{r^2}(t)\delta(f) + \tfrac{1}{2}R_0(f)$$

whilst preserving unimpaired the components at and centered on $f = \pm 1$. Denoting the output of such a filter by $w(t) \Leftrightarrow W(f)$ we have

$$W(f) = \tfrac{1}{4}\{R_0(f \pm 1) + \overline{r^2}(t)\,\delta(f \pm 1)\}$$

$$\Rightarrow w(t) = \tfrac{1}{2}r_0(t)\cos 2\pi t + \tfrac{1}{2}\overline{r^2}(t)\cos 2\pi t$$

$$= \tfrac{1}{2}r^2(t)\cos 2\pi t \qquad (10.15)$$

where $r^2(t) = r_0(t) + \overline{r^2}(t)$ is non-negative. Hence $w(t)$ is an envelope modulated wave with regularly spaced zero crossings. The *envelope* of *$w(t)$ is dependent upon the data* $\{a_r\}$ but the phase is data independent. Hence a down (or up) crossing detector operating on $w(t)$ will provide the requisite re-timing signal.

Returning now to eqn. (10.15), it should be noted that in requiring no spectral overlap between $R_0(f)$, $R_0(f \pm 1)$ and that the output filter reject $R_0(f)$ but pass $R_0(f \pm 1)$ unmodified, we are being unnecessarily restrictive. Less stringent, yet sufficient, conditions are as follows:

(1) The output bandpass filter $G(f)$ must eliminate the dc and baseband components, $\Rightarrow R_0(f) \cdot G(f) = 0$ for all f.
(2) $G(f)$ must be bandpass Hermitian about $f = \pm 1$.
(3) $G(\pm 1) \geqslant G(f)$ for all f.

The first two conditions ensure that the output $w(t)$ has the form of a double-sideband signal but do not guarantee a non-negative real envelope. This is obtained by imposing the third condition which is practically unrestrictive albeit greater than strictly necessary.

The output $w(t) \Leftrightarrow W(f)$ is of envelope modulated form, with data-independent zero crossings, as shown in Fig. 10.5. Reducing the bandwidth of the output filter $G(f)$ reduces the data dependent envelope fluctuations. This will ease the demands on the zero crossing detector but will also, as noted previously, restrict the tolerance of the system to signalling speed variations and mistuning. It is possible to balance the filtering requirement between $H(f)$ and $G(f)$, a good compromise being to require $P(f)H(f)$ to have a (finite support) bandwidth $B_H < \tfrac{1}{2}$ to guarantee rejection of the baseband components of $z(t)$. Under these circumstances $w(t)$ has a real non-negative envelope, the zero crossings are regularly spaced and independent of the data sequence $\{a_r\}$, and the timing circuit is not unduly sensitive to speed variation or tuning errors.

10.5 STATISTICAL PROPERTIES OF THE PAM SIGNAL AND TIMING WAVE

In the foregoing we have considered a single realisation for the PAM signal and have exploited certain spectral symmetries to establish

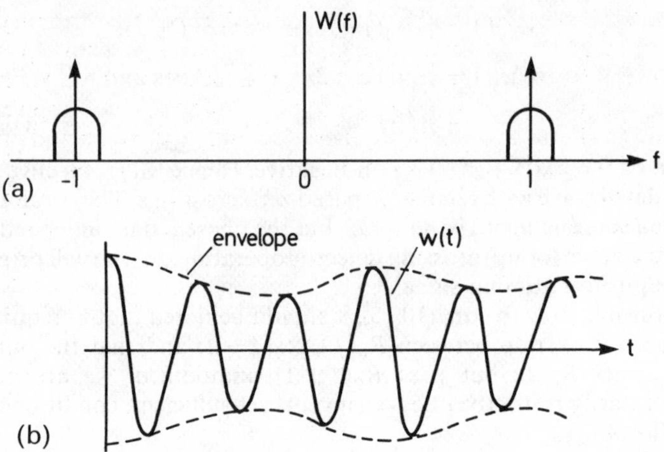

Fig. 10.5 Output timing signals w(f) and spectrum W(f). (a) Representative output spectrum, (b) representative output time-domain signal with regularly spaced zero crossings

criteria for the recovery of a timing wave with data-independent zero-crossings. This concentration on a single realisation may seem tantamount to treating the problem deterministically. It should be noted, however, that the message $\{a_r\}$ is selected arbitrarily from the set of admissible messages. It is appropriate, therefore, that we now examine more carefully the stochastic aspects of the problem.

10.5.1 Cyclostationarity

While the statistical characteristics of a stationary process are invariant under arbitrary time shifts, those of a cyclostationary process† are only invariant under shifts by an integer multiple of a period, T. Consequently the statistical characteristics—probability densities, moments, functions, autocorrelation functions, etc.—depend on absolute time as well as on the time difference. However, the absolute time dependence is periodic. In particular, a random process is said to be wide-sense cyclostationary if and only if its mean and autocorrelation function exhibit periodicity [6]

$$M_x(t) \triangleq E[x(t)] = M_x(t+T) \quad \text{for all } t$$
$$R_{xx}(t,\tau) \triangleq E[x(t), x^*(\tau)] = R_{xx}(t+T, \tau+T) \quad \text{for all } t, \tau \tag{10.16}$$

† Such processes are also termed 'periodic non-stationary' [13].

We noted earlier that the PAM signal of (10.1) is cyclostationary, yet asserted that it is possible to compute a power spectrum $S_x(f)$. In view of (10.16) this latter statement deserves an explanation. With $x(t)$ non-stationary the usual spectral density concept is not applicable; we cannot simply invoke the Wiener–Khintchine theorem and obtain $S_x(f)$ as the Fourier transform of $R_{xx}(t, \tau)$ due to the absolute time dependence. However, a stationary process can be derived from $x(t)$ by phase randomising [6], that is, by averaging the statistical parameters over one cycle

$$\tilde{M}_x = \frac{1}{T}\int_{-T/2}^{T/2} M_x(t)\,dt \tag{10.17a}$$

$$\tilde{R}_{xx}(\tau) = \frac{1}{T}\int_{-T/2}^{T/2} R_{xx}(t, \tau)\,dt \tag{10.17b}$$

Fourier transformation of eqn. (10.17b) then yields a power spectrum $\tilde{S}_x(f)$. Strictly this is the power spectrum of the wide sense stationary process derived from $x(t)$, but it is often referred to simply as the power spectrum $S_x(f)$ of $x(t)$. If $x(t)$ is linearly filtered, as in Fig. 10.1, the resultant $y(t)$ is still cyclostationary, although the degree of cyclostationarity may be enhanced or reduced by the filtering operation. For timing extraction purposes we wish to enhance the degree of cyclostationarity and this is achieved by bandpass filtering at one-half the signalling rate. To derive a power spectrum $S_y(f)$ for the filtered process it is necessary, once again, to employ phase randomising.

We wish now to examine the effect on a cyclostationary process of a non-linearity $f(\cdot)$. It is sometimes possible to compute the spectral density at the output of a non-linear device in terms of multiple convolutions of the input spectral density (Chapter 9 and [15]). This is not the case here, however. Cyclostationary processes in contrast with stationary processes generally exhibit correlation between components in different frequency bands. By phase randomising in order to obtain a power spectrum we generate a stationary process and lose trace of these correlations, which are in essence the 'hidden phase relationships' quoted earlier from [4]. In earlier sections of this chapter these difficulties were circumvented by considering a single realisation $x(t)$ and employing a generalised Fourier integral to generate a spectrum. The spectrum $X(f)$ obtained by this means is quite different in character from $S_x(f)$. It is a complex function of frequency and so explicitly contains the requisite phase information. Note, however, that $X(f)$ is itself a single realisation of a random process—a random function of the frequency variable [4]—and so does not represent a characterisation of the *process*.

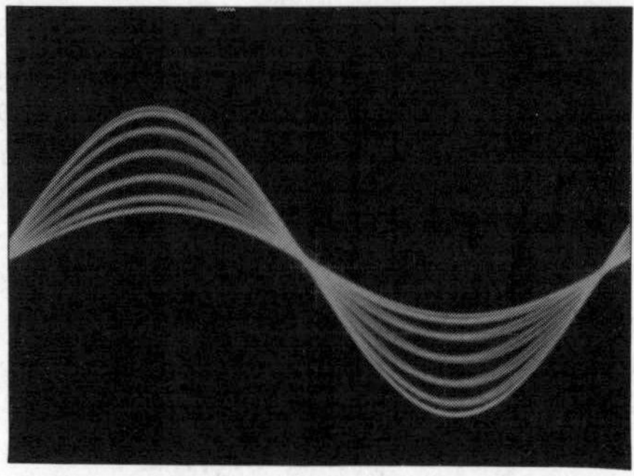

Fig. 10.6 Recovered timing signal

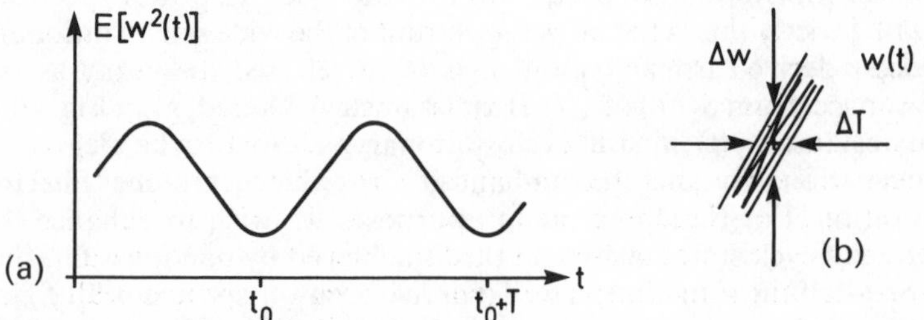

Fig. 10.7 Variation of the mean square value of the timing wave and its relation to timing jitter (after [3]). (a) Variation of $E[w^2(t)]$, t_0 is the mean zero crossing of w(t); (b) relation between time and amplitude variations near t_0

Time-invariant operations preserve cyclostationarity [8], although, again, the *degree* may be altered. In the present context this means that the output of the squarer, $z(t)$, and the recovered timing wave, $w(t)$, represent cyclostationary processes. In contrast to the inputs $x(t)$, $y(t)$ the output processes $z(t)$, $w(t)$ exhibit a periodic mean value. This is apparent from the oscilloscope display of Fig. 10.6 in which the nearly sinusoidal timing wave—with slowly varying amplitude and, in general, small phase perturbations—is displayed by synchronising the horizontal trace with the data 'clock' signal. The mean square value of the timing wave is periodic and non-negative, as illustrated in Fig. 10.7a. The minimum value of $E[z^2(t)]$ has been used as a measure of residual timing jitter [3, 5], a justification for this metric is provided by Fig. 10.7b. It can be seen that $E[w^2(t)]$ goes to zero when the zero crossings of $w(t)$ are jitter free.

10.5.2 The ergodic hypothesis

In translating from the distribution of zero crossings of a single realisation to the distribution of zero crossings in the vicinity of some time t_0 for the ensemble of possible realisations, we are essentially invoking some form of ergodicity. So far as any particular system is concerned it is perhaps the behaviour of a single realisation that is of interest, but, given that the data $\{a_r\}$ cannot be known *a priori*, it is necessary to consider the ensemble of all possible messages. For an ergodic process, time and ensemble averages are equal and it suffices to evaluate whichever is the more convenient. If we wish to apply the idea of ergodicity here, however, we must proceed with caution—ergodic processes are a sub-class of stationary processes and we are not dealing with stationary processes.

The essential property of an ergodic process is that if we know a single realisation over the whole time axis $t \in (-\infty, \infty)$, then by making timeshifts we can obtain an infinite statistical ensemble of realisations. We can thus determine the statistical characteristics of an ergodic process from a single realisation. With this perspective it is possible to conceive a form of ergodicity applicable to cyclostationary processes if we restrict the time shifts to being integer multiples of the period and can thereby generate the ensemble from a single realisation.† Ensemble statistics evaluated at a specific time t_0 would then equate to statistics computed from samples of a single realisation taken on the set: $\{t \mid t = t_0 + rT, r \text{ integer}\}$. With this assumption we can compute the jitter measure $E[w^2(t)]$ by ensemble averaging or by time-sample averaging. The former has been found convenient in the past [3, 5] but in extending the present treatment to jitter evaluation the latter is more appropriate. We will not give the details here but simply note that the problem is equivalent to that of computing intersymbol interference (isi), except that we operate on the timing wave $w(t)$ rather than on the information bearing signal $x(t)$. Calculation of isi variance in terms of a signal spectrum is well documented in the literature [e.g. 17, 18].

10.6 CONCLUSIONS

In this chapter conditions for the extraction from a digital signal of a timing wave free of data-dependent jitter have been established. The result is not new, but the method of derivation is not known to have been published previously and is thought to have certain pedagogical

† The term 'cycloergodic' would seem appropriate for processes possessing this property.

advantages. The importance of the cyclostationary nature of the stochastic processes involved in digital transmission has been noted and it is suggested that it is this very cyclostationarity which allows the recovery of a timing wave. Finally, an indication of how the analysis may be extended to allow for the computation of timing-jitter when the jitter-free conditions are not satisfied (as must always be the case in practice) has been outlined. It remains to be seen whether this approach yields any computational advantage compared with the conventional method, rather it is likely that the most significant feature of the present method of analysis is the insight it provides into the timing recovery process.

REFERENCES

1. Bennett, W. R., 'Statistics of regenerative digital transmission', *Bell Syst. Tech. J.*, **37**, 1501–1542 (1958)
2. Takasaki, Y., 'Timing extraction in baseband pulse transmission', *IEEE Trans. Commun.*, **COM-20**, 877–884 (1972)
3. Franks, L. E. and Babrouski, J. R., 'Statistical properties of timing jitter in a PAM timing recovery scheme', *IEEE Trans. Commun.*, **COM-22**, 913–920 (1974)
4. Bylanski, P. and Ingram, D. G. W., *Digital Transmission Systems*, Peter Peregrinus Ltd., pp. 194–204 (1965)
5. Muñoz Rodriguez, D., 'Digital Gommunication—Relationship Between Signal Design and System Performance', Thesis for the degree of Ph.D., University of Essex (1979)
6. Franks, L. E., *Signal Theory*, Prentice-Hall (1969)
7. Bennett, W. R. and Davey, J. R., *Data Transmission*, McGraw-Hill, pp. 316–321 (1965)
8. Gardner, F. M. and Franks, L. E., 'Characterisation of cyclostationary random signal processes', *IEEE Trans. Inf. Th.*, **IT-21**, 4–14 (1977)
9. Shannon, C. E., 'Communication in the presence of noise', *Proc. IRE*, **37**, 10–21 (1949)
10. Gel'fand, I. M. and Shilov, G. E., *Generalized Functions*, *Vol. 1*, Academic Press (1964)
11. Papoulis, A., *The Fourier Integral and its Applications*, McGraw-Hill, pp. 259–264 (1962)
12. Milne, R. D., *Applied Functional Analysis*, Pitman (1980)
13. Stratonovich, R. L., *Topics in the Theory of Random Noise*, Vol. 1, Gordon and Breach, pp. 139–141 (1963)
14. Ibid., pp. 27–31
15. Cattermole, K. W., 'Stochastic problems of multiple communication', in *Communication Systems and Random Process Theory*, Skwirzynski, J. K. (ed.), Sijthoff and Noordhoff (1978)
16. Rice, S. O., 'Mathematical analysis of random noise', *Bell System Tech. J.*, **23**, 282–332 (1944) and **24**, 146–156 (1945)
17. Lucky, R. W., Salz, J. and Weldon, E. J., *Principles of Data Communication*, McGraw-Hill, pp. 75–79 (1968)
18. O'Mahony, M. J., 'Signal design for optical fibre communication', in *Optimisation Methods in Electronics and Communications*, Cattermole, K. W. and O'Reilly, J. J. (eds.), Vol. 1 of Mathematical Topics in Telecommunications Series, Pentech Press (1983)

11

Quantising Noise Spectra

K. V. Lever

11.1 INTRODUCTION

The operation of quantisation is familiar to telecommunication engineers as the source of distortion which, in the case of a well-engineered system, will limit the performance of a pulse code modulation (pcm) system [1]. The effect is equally important in other digital systems not employing a communication channel to link remote terminals—such as digital filters for analogue-to-analogue signal processing. Systems of both types find application in a large number of military and civilian telecommunication networks and indeed, the use of digital techniques is now so widespread that applications outside the area of telecommunications are also numerous. The universality of the need to interface analogue and digital subsystems, which is where the deterioration due to quantisation occurs, makes it important that we (1) obtain a clear theoretical and practical grasp of the nature of the quantisation mechanism (and its interaction with sampling) and (2) translate this understanding into simple design procedures that can be conveniently and routinely applied by engineers designing systems requiring analogue/digital interfaces. In 1970, when the work to be described in this chapter was undertaken as part of his postgraduate industrial training [2] at the GEC Hirst Research Centre, it struck the author as curious that, while the state of the art was adequate in the respect to item (2), it was not so in respect to item (1). That is, accurate design formulae had been derived and satisfactorily applied for a long time, a situation that was consolidated by the publication in 1969 of Cattermole's exposition [1] of the principles of pcm. And yet it was clear from a literature survey of publications up to 1970 that there were still some gaps, inconsistencies and apparent contradictions in our basic theoretical knowledge of the quantisation mechanism. It was the purpose of this case study to expose these problem areas, fill the gaps and reconcile the contradictions. To a certain extent these objectives were achieved, and the results of the study were published in 1974 [3].

Specifically there seemed to be two streams of thought on how to analyse the behaviour of a quantiser. Bennett's early paper [4] presented a numerical method, based on Rice's characteristic function technique [5] and Chapter 9 of this volume, for calculating the spectrum of quantising noise (and other derived parameters such as total inband noise power) for an uncompanded pcm system with a Gaussian input signal.† Bennett's analysis, though somewhat roundabout, led to a simple, easily computed expression for the quantising noise spectrum, the celebrated B-series, which gives close numerical agreement with experimental results, yet does not correspond in any obvious way to the intermodulation mechanism causing the distortion. Indeed, one of Bennett's critics, Velichkin, commented [6] that in Bennett's analysis '... the level quantisation and the time quantisation were analysed jointly, and this led to such complex expressions for the correlation function that the spectral density could only be determined approximately'. This is true, but Velichkin's alternative method is also susceptible to criticism. His approach (which appears to be very similar to that adopted independently and simultaneously by Shimbo, Ohira and Nitadori [7]) formulates the analysis in terms of a recognisable decomposition of the nonlinearity into a wanted signal component plus unwanted intermodulation products, but unfortunately he made no attempt to overcome the problem presented by the immense burden of numerical computation that is encountered in cases of practical interest. Robertson subsequently took up this challenge by adopting a brute-force number-crunching approach, and, in an excellent paper [8], announced (amongst other things) that the intermodulation series required up to 10,000 terms for accurate computation in the case of a 5-bit 32-level system. This would seem to imply that, for higher resolution systems having 7 or 8 bits, of practical interest to telephone engineers, the computation would be prohibitively expensive. One is forced to the conclusion that, despite the theoretical soundness, there is something seriously wrong with the Velichkin–Robertson approach: after all, Bennett's method, though unclear, was adequate. The point is emphasised by another publication by Richards [9], using a third approach (regression analysis) to derive satisfactorily compact formulae for computing quantising noise-to-signal ratio

† The Gaussian assumption is relevant: fdm assemblies in fdm pcm systems are close to Gaussian, and such systems are routinely tested by means of test sets incorporating a broad band white noise process simulating the multiplex. Single-channel speech systems are tested by means of narrow band (pseudo) Gaussian wave forms, because this provides a reproducible measurement as well as imitating to a limited extent the main formant of speech. The Gaussian test wave form is not meant to be an accurate simulation of speech, which is known to be far from Gaussian.

(NSR), an adequate measure of subjective quality in the case of a single-channel speech system, sampling close to the Nyquist rate.

What was needed was obviously an attack upon the theoretical modelling of the quantisation operation via a more general theory of nonlinear distortion. A number of attempts [10–13], had been made to formulate such a general theory, but Blachman's landmark papers [14, 15] crystallised this line of thought. It was clear that by applying Blachman's method to the quantisation nonlinearity it would be possible to resolve the theoretical difficulties previously encountered.

Specifically, the objectives of the case study were as follows:

(1) To show that previously established theoretical results could be obtained by Blachman's method; to obtain an expression for the spectrum of the output from a sampled quantiser; to establish necessary and sufficient conditions for the quantising noise to be white; and to obtain an expression corresponding to Richards' formula for the NSR.

(2) To obtain some experimental evidence that predictions derived from Blachman's method correspond to measured performance for a pcm system operating under both normal and abnormal conditions.

(3) To find an explanation for the observation that quantising noise spectra are white under conditions drastically different from those established under item (1) above.

(4) To examine the consequences of applying asymptotic approximation techniques to economise the computation of the noise spectrum, checking the accuracy by repeating Robertson's work.

(5) Above all, to reconcile the apparent discrepancies between the various approaches outlined above.

11.2 MATHEMATICAL MODEL OF A SAMPLED QUANTISER

It is easy to show that the model shown in Fig. 11.1 is an adequate representation, for the purposes of quantisation distortion analysis of cascaded ADC/DAC operations, as for instance in pcm.

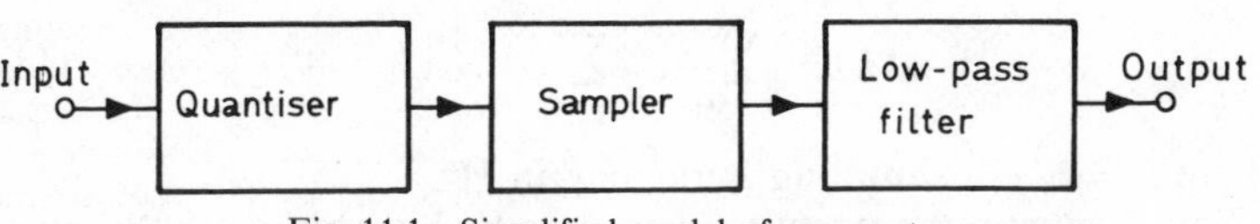

Fig. 11.1 Simplified model of pcm system

The sampling operation can be represented as

$$s(x) = \sum_{m=-\infty}^{\infty} x(mT)\,\delta(t - mT) \tag{11.1}$$

where $x(t)$ is the input signal waveform, T the sampling interval in seconds, and $\delta(t)$ a Dirac delta-functional idealisation of the sampling pulse used in the reconstruction. Denoting the memoryless nonlinear quantisation operation as $y = q(x)$ the sampled quantiser output can be written, using the composition operation $\bullet$, as

$$s \bullet q(x) = \sum_{m=-\infty}^{\infty} q \bullet x(mT)\,\delta(t - mT) \tag{11.2}$$

It follows straightforwardly from Blachman's decomposition that the autocorrelation of the quantiser output can be expressed in terms of input autocorrelation $R_x(\tau)$, in a series of the form

$$R_q(\tau) = \sum_{n=0}^{\infty} a_n R_x^n(\tau) \tag{11.3}$$

Then the autocorrelation of the sampled output is

$$R_{s\bullet q}(\tau) = \frac{1}{T} \sum_{m=-\infty}^{\infty} R_q(\tau - mT) = \frac{1}{T} \sum_{m=-\infty}^{\infty} \sum_{n=0}^{\infty} a_n R_x^n(mT) R_\delta(\tau - nT) \tag{11.4}$$

Here the intermodulation coefficients are related in the usual way to the scalar products q_n in the Fourier–Hermite decomposition of q

$$q_n = \langle q, He_n \rangle = \int_{-\infty}^{\infty} q(x) He_n(x) p(x)\,dx \tag{11.5}$$

$$a_n = q_n^2 / c_n; \quad c_n = \langle He_n^2 \rangle = n! \tag{11.6}$$

and $p(x) = \exp(-x^2/2)/\sqrt{2\pi}$ = pdf of the $N(0, 1)$ Gaussian input signal.

Fourier transformation of eqn. (11.4) gives the structure of the output spectrum in terms of aliased intermodulation products $W_x^{(n)}(f)$, multiple self-convolutions of the input spectrum $W_x(f)$

$$W_{s\bullet q}(f) = \sum_{n=0}^{\infty} a_n \sum_{m=-\infty}^{\infty} W_x^{(n)}(f - mf_s) \tag{11.7}$$

where $f_s = 1/T$ = sampling frequency in Hz.

This is exactly the structure discussed by Velichkin and Shimbo *et al.* It follows that if (and only if) the input spectrum is white and bandlimited to f_b Hz

$$W_x(f) = \begin{cases} 1/2f_b & |f| < f_b \\ 0 & |f| > f_b \end{cases}; \quad R_x(\tau) = \frac{\sin 2\pi f_b \tau}{2\pi f_b \tau} \tag{11.8}$$

and if (and only if) the signal is Nyquist-sampled ($f_s = 2f_b$), the autocorrelation of eqn. (11.4) is impulsive, so that the output spectrum contains a distortion component that is exactly flat. This explains the observed flatness of the quantising noise when the input signal is white, as in the broadband white noise test. Notice that no specific properties of the quantiser have been used to establish the above necessary and sufficient conditions: the result is true for an *arbitrary* nonlinearity. It also follows that we cannot use this result to explain the whiteness of quantising noise spectra when the input signal is non-white or for hyper-Nyquist sampling, as in the case of the narrowband test of a single speech channel.

A further deduction is that under the stated conditions the *entire* quantising noise power aliases back into the inband region. Since both signal and distortion spectra are flat the NSR, calculated as a power ratio, is an adequate summary measure of performance. Separating the signal ($n = 1$) and distortion ($n \geqslant 2$) terms in eqn. (11.7), and ignoring the dc ($n = 0$) term, we find

$$\frac{N}{S} = \sum_{n=2}^{\infty} \frac{a_n}{a_1} \tag{11.9}$$

But, by Parseval's equality for the nonlinearity

$$\sum_{n=2}^{\infty} a_n = \langle q^2 \rangle - a_0 - a_1 = \langle q^2 \rangle - \langle q \rangle^2 - \langle xq \rangle^2 \tag{11.10}$$

leading to the equation

$$\frac{N}{S} = \frac{\langle q^2 \rangle - \langle q \rangle^2 - \langle xq \rangle^2}{\langle xq \rangle^2} = \frac{\langle q^2 \rangle - \langle q \rangle^2}{\langle xq \rangle^2} - 1 \tag{11.11}$$

in complete accord with Richards' equation for the case of a single channel pcm system [9]. Clearly we have achieved the objectives stated in item (1).

11.3 COMPARISON BETWEEN THEORETICAL AND EXPERIMENTAL RESULTS

It was thought to be extremely important to determine whether the elaborate analysis outlined above was capable of producing numerical predictions in accord with experimental measurements. Equation (11.11) can be used successfully, as Richards had shown, to predict the NSR performance for the case of a single channel speech pcm, and there would have been little point in repeating the measurements for that case. Instead, it was decided to test the theory for the case of a (then) state-of-the-art 120 mb/s pcm system [16] designed to carry an fdm telephony 900-channel hypergroup (or alternatively, a single 625-line colour TV signal). NSR measurements on the system were routinely made by means of a broadband (4.028 MHz) white noise test set originally designed for testing for fdm systems [17]. Despite the structural differences in the spectra of the signal processed by the low- and high-speed systems, it was thought that eqn. (11.11) should apply equally to both cases, for Blachman's expansion is always a decomposition into dc, wanted signal and unwanted distortion components.

As a more stringent test it was decided to compare theory and experiment for a quantiser having a globally nonlinear characteristic, a situation where the conventional distortion measure, mean-squared-error (MSE), is known to break down [18, 19]. Thus the experiment took the form of a white noise test of the nonlinearity shown in Fig. 11.2.

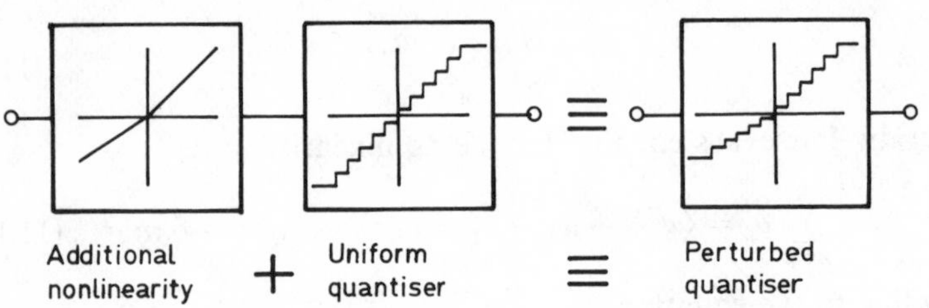

Fig. 11.2 Perturbed quantiser

The predistorting nonlinearity deteriorates the uniform quantiser characteristic so that the quantiser staircase no longer approximates to the equation $y = x$, but to the equation

$$y = \begin{cases} x & x > 0 \\ \lambda x & x < 0 \end{cases} \tag{11.12}$$

The gain for negative signals λ, was chosen to produce distortion from the nonlinearity commensurate with that from the quantiser. Choosing $N = 16$ levels for the quantiser and $\lambda = 0.736$ satisfies this criterion, so that the perturbed quantiser should exhibit a clear deterioration when compared with the unperturbed case.

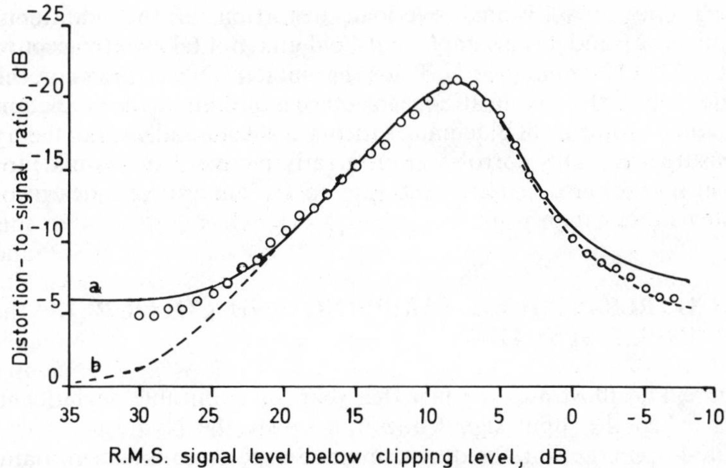

Fig. 11.3a Theoretical and experimental results for the unperturbed quantiser ($\lambda = 1$). (a) Noise/signal ratio ——, (b) mean squared error – – –, (c) experimental points ∘∘∘

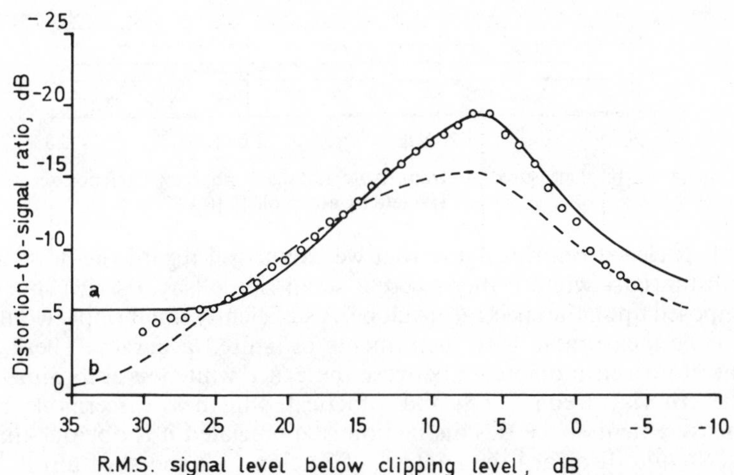

Fig. 11.3b Theoretical and experimental results for the perturbed quantiser ($\lambda = 0.736$). (a) Noise/signal ratio ———, (b) mean squared error – – –, (c) experimental points ∘∘∘

As shown in Fig. 11.3, there is very close agreement between predictions of NSR and measurement in both cases, with and without the predistorting nonlinearity. Note, however, the inadequacy of MSE for the perturbed quantiser. The apparent agreement between MSE and the measurement at high signal levels is in fact spurious, being due to additional overload distortion in the additional nonlinearity and input sample-and-hold unit, not taken into account in the MSE formula (nor NSR for that matter). Nevertheless, in the region where the quantisation mechanism is dominant, the prediction obtained by means of Blachman's theory is substantially validated by measurement. This corroboration greatly increases our confidence when it is applied, *mutatis mutandis*, to the analysis and design of systems other than pcm.

11.4 HYPER-NYQUIST SAMPLING AND COLOURED INPUT SPECTRA

Figure 11.4 illustrates the fact that (barring explicable deviations), even when the input signal does not satisfy the Nyquist-sampled white-noise criterion, the quantising noise spectrum is substantially white.

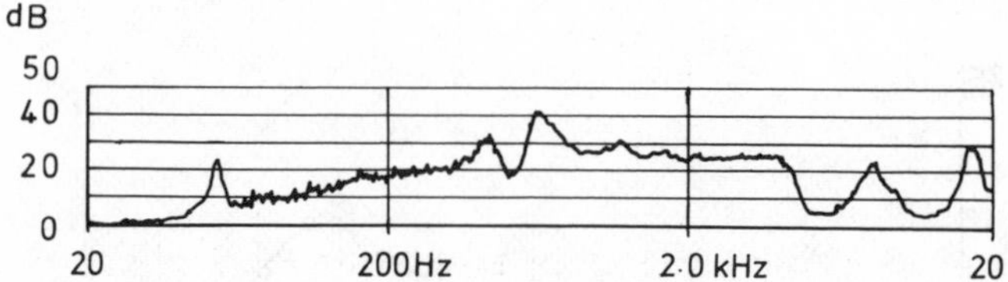

Fig. 11.4 Typical measured spectrum. (Note: test signal is narrow-band noise centred on 500 Hz) (after Cattermole [1])

It is clear from this figure that we would still regard the noise as substantially white if there were a small ripple (say, 0.5 dB) superimposed upon the spectral envelope. A sufficiently small ripple would not be measurable with instruments of limited accuracy. There is therefore a clear difference between the 'exact' whiteness according to the criteria of eqn. (11.8) and 'practical' whiteness discernable by measurement. Once this distinction is appreciated it is obvious that we should disregard the analysis of Section 11.2 and start afresh.

Let us assume that the sampling rate f_s is K (>1) times the Nyquist rate. Equation (11.7) gives the structure of the aliased quantising noise. Apparently the combination of high-order intermodulation products (n-index) with high-order aliasing (m-index) results in a white noise spectrum. The periodicity of the aliased intermodulation spectra

$$AW_x^{(n)}(f) = \sum_{m=-\infty}^{\infty} W_x^{(n)}(f - mf_s)$$

lends itself to Fourier analysis. Equation (11.7) can be written as

$$W_{s\bullet q}(f) = \frac{1}{T}\sum_{n=0}^{\infty} a_n \sum_{m=-\infty}^{\infty} R_x^n(mT)\exp(-j2\pi fmT) \quad (11.13)$$

a form that is much more suitable for the description of a small ripple superimposed upon a constant level. For the sake of simplicity we assume there is no dc line component in the output ($a_0 = 0$). We may then decompose the output spectrum into signal and distortion terms as follows, keeping the more natural aliasing representation for the signal component (and putting $R_x(\tau) = \rho(\tau)$ for notational simplicity)

$$W_{s\bullet q}(f) = a_1 A W_x(f) + \frac{1}{T}\sum_{n=2}^{\infty} a_n \sum_{m=-\infty}^{\infty} \rho^n(mT)\exp(-j2\pi fmT) \quad (11.14)$$

The second term in this equation is the aliased quantising noise (expressed as a Fourier series), which may also be written as

$$N(f) = \frac{1}{T}\sum_{n=2}^{\infty} a_n\left[1 + 2\sum_{m=1}^{\infty} \rho^n(mT)\cos(2\pi fmT)\right] \quad (11.15)$$

a form that is slightly more convenient to handle. A simple argument (see reference [3]) shows that not all the terms in eqn. (11.15) can contribute, if $N(f)$ is to be practically white. Consider the aliased intermodulation products of increasing order: low-order products cannot alias to a white resultant as their bandwidths are too small. But eventually the multiple self-convolution feature will widen the bandwidths of higher-order products so as to ensure the substantial overlapping required for whiteness, as in Fig. 11.5.

Thus, there will be a threshold value L for the intermodulation index n beyond which the aliased intermodulation spectra will be practically white. It follows that if the entire output spectrum is to be white then the early terms in the series with $n < L$ must be swamped by the later terms. Furthermore, since the ripple is assumed to be shallow, we may assume that it is sinusoidal ($m = 1$) and discard the terms $n \leqslant L - 1$, $m \geqslant 2$ in eqn. (11.15). It follows that L can be related to the hyper-Nyquist factor K by the following simple formula (see reference [3] for details)

$$L \simeq cK^2/\alpha^2 \quad (11.16)$$

where α = normalised rms bandwidth of the input spectrum, and c is a function of the spectral ripple

$$c = \frac{2}{\pi^2}\log_e\left[\frac{2(r+1)}{r-1}\right] \quad (11.17)$$

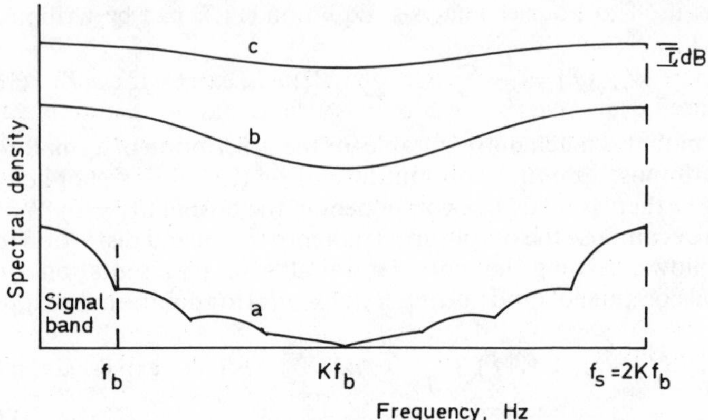

Fig. 11.5 Masking of nonunitary spectral components. (a) Intermodulation series summed to K terms, (b) intermodulation series summed to L terms, (c) total noise spectrum

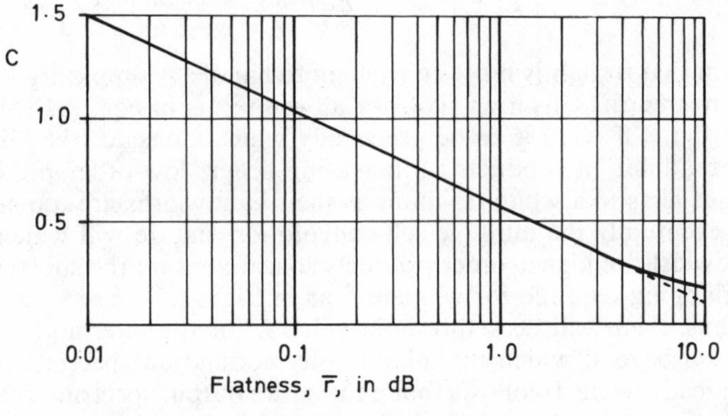

Fig. 11.6 The c–$\bar{r}$ relationship

As Fig. 11.6 shows, the value of c is extremely insensitive to the value of r. For r up to 2.5 (i.e., $\bar{r} < 4$) c is closely approximated by

$$c \simeq 0.58 + 0.47 \log_{10} \bar{r} \qquad (11.18)$$

where $\bar{r}$ is the relative ripple amplitude in dB.

By experimenting numerically with various spectral shapes we found that eqn. (11.16) is always valid provided the sampling rate is greater than five times the Nyquist rate. Again no properties specific to the quantiser are invoked: provided the majority of the intermodulation power resides in components of higher order than $L = cK^2/\alpha^2$, then the distortion by an *arbitrary* nonlinearity of a coloured

Gaussian input signal, sampled at K (>5) times the Nyquist rate, will be white to within $\bar{r}$ dB.

There are several aspects of the above analysis that are counter intuitive. The usual nonlinearities familiar to communications engineers (saturating amplifiers, limiters, hard clippers, etc.), have intermodulation series that converge rapidly, the early terms dominating the later ones. Whereas, the opposite is required to be true for the above flatness criterion. The only way for this to happen is that the quantiser intermodulation series must be so *slowly* convergent that the first $L = cK^2/\alpha^2$ terms can be safely neglected. It is a weird sort of nonlinearity that has an enormously long intermodulation series and *still* manages to produce a close approximation to $y = x$, as required for communication fidelity. It turns out, on account of its discontinuous nature, that the quantiser possesses this unlikely combination of properties. Since the behaviour of large-index intermodulation terms seems to be the crux of the matter is not unnatural to consider exploring the problem by the use of asymptotic approximations.

11.5 ASYMPTOTIC ANALYSIS OF QUANTISER INTERMODULATION TERMS

For convenience we restrict out attention to the odd quantiser considered by Bennett, discarding even terms in the Fourier–Hermite expansion. The distortion component can be written as

$$n(x) = \sum_{n=3}^{\infty} \frac{q_n}{c_n} He_n(x), \quad q_n = \langle q, He_n \rangle \tag{11.19}$$

The odd Hermite polynomials have the following asymptotic approximation [20]

$$He_n(x) \sim \gamma_n \exp(x^2/4) \sin[\sqrt{(n + \tfrac{1}{2})}x] \tag{11.20}$$

where

$$\gamma_n = (-1)^{(n-1)/2} 2^{-n/2} n!/\Gamma(n/2 + 1) \tag{11.21}$$

For large n, the scalar products become

$$q_n \sim \frac{\gamma_n}{\sqrt{2\pi}} \int_{-\infty}^{\infty} n(x) \exp(-x^2/4) \sin(\sqrt{n}\, x)\, dx \tag{11.22}$$

This integral closely resembles a conventional Fourier transform of a product, suggesting the following approach. For a uniform quantiser, having constant step-size Q, $n(x)$ is a periodic sawtooth with period Q

(provided centre or peak clipping effects are negligible). Thus $n(x)$ has a *line* spectrum with Fourier coefficients

$$t_p = 2/p\omega_0 \tag{11.23}$$

where $\omega_0 = 2\pi/Q$. The appearance of the series q_n is therefore a series of discrete lines corresponding to $p = 1, 2, 3, \ldots$ broadened convolutionally by the Fourier transform of $\exp(-x^2/4)$. The lines are spaced quadratically by virtue of the $\sqrt{n}$ argument

$$q_n \sim \frac{\gamma_n}{\sqrt{2}} \sum_{p=1}^{\infty} t_p \exp[-(\sqrt{n} - p\omega_0)^2] \tag{11.24}$$

The negligibility of centre clipping ensures that the lines in the series are disjoint so that

$$a_n = \frac{q_n^2}{c_n} \sim \delta_n \sum_{p=1}^{\infty} t_p^2 \exp[-2(\sqrt{n} - p\omega_0)^2] \tag{11.25}$$

where $\delta_n = \gamma_n^2/2n! \sim 1/\sqrt{2\pi n}$ by Stirling's formula.

This formula was checked by comparison with the exact approach discussed by Robertson [8]. The approximation is accurate to within 1% over all significant regions of the series. The line structure is confirmed by Figs. 11.7 and 11.8, obtained by repeating Robertson's calculation.

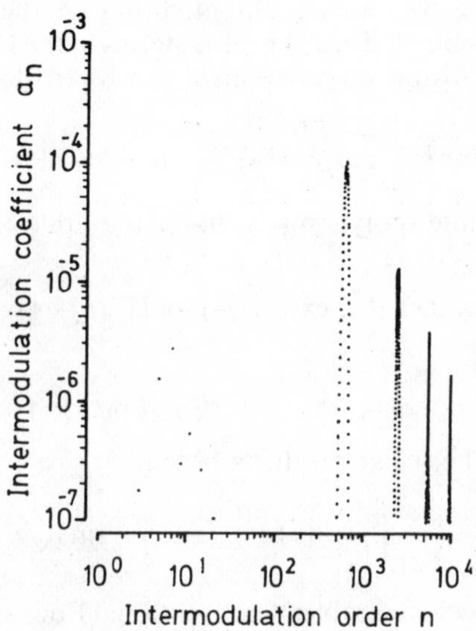

Fig. 11.7 Intermodulation coefficients for a 5 bit uniform quantiser RMS signal level 12 dB below the clipping level

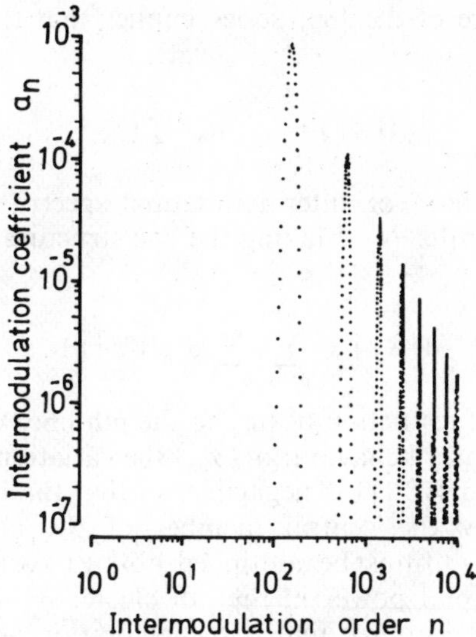

Fig. 11.8 Intermodulation coefficients for a 5 bit uniform quantiser RMS signal level 18 dB below the clipping level

Note the presence of lines centred on $N_p = p^2\omega_0^2 = 4\pi^2p^2/Q^2$ having peak values

$$a_{N_p} = \frac{t_p^2}{\sqrt{(2\pi N_p)}} = 2\sqrt{\frac{2}{\pi}}\,N_p^{-3/2}$$

and exhibiting a Gaussian profile (parabolic on the log scale). So we see that the quantiser intermodulation series is dominated by high-order rather than low-order terms. In Fig. 11.7 the series starts at roughly $L = N_1 = 521$. But $N_1 = 4\pi^2/Q^2$, so this value of L corresponds to $K = \alpha\sqrt{L/c} = 2\pi\alpha/Q\sqrt{c}$ indicating, for instance, that the quantising noise for a white input signal ($\alpha = 1/\sqrt{3}$) would be white for sampling rates up to $K = 38$ times the Nyquist rate. It turns out that eqn. (11.25) is much more economical than the approach adopted by Velichkin and Robertson: for an N-level quantiser the computational work-loads are in the ratio $N:N^3$.

We now consider the companion problem of computing the intermodulation spectra, the multiple self-convolutions of the input spectrum. Knowing that the important components are of very high order, it is natural to use the central-limit approximation

$$W_x^{(n)}(f) \sim \exp\left[-\frac{1}{2}\frac{1}{n}\left(\frac{f}{f_b\alpha}\right)^2\right]\Big/\left[\sqrt{(2\pi n)}\,f_b\alpha\right] \qquad (11.26)$$

The line structure of the $\{a_n\}$ series implies that the output noise spectrum

$$W_n(f) = \sum_{n=3}^{\infty} a_n W_x^{(n)}(f) \tag{11.27}$$

consists of 'clusters' of intermodulation spectra whose central members are of order N_p. Making the line structure explicit we can write

$$W_n(f) = \sum_{p=1}^{\infty} \sum_{n=3}^{\infty} \alpha_{pn} W_x^{(n)}(f) \tag{11.28}$$

where $\{\alpha_{pn}\}$ is the restriction of $\{a_n\}$ to the pth line. We now assume that during the spiky behaviour of $\{\alpha_{pn}\}$ the variation in the shape of the corresponding $W_x^{(n)}(f)$ is negligible, so that the latter can all be approximated by the central member of the cluster $W_x^{(N_p)}(f)$. Accordingly $W_x^{(N_p)}(f)$ must be multiplied, not by its original coefficient a_{N_p}, but by the total power in the pth cluster $\beta_p = \sum_{n=3}^{\infty} \alpha_{pn}$. This summation can be approximated by integration, allowing for the absence of even terms

$$\beta_p \sim \frac{1}{2} \int_0^{\infty} \alpha_{pn}\, dn = \frac{1}{2} \frac{t_p^2}{\sqrt{2\pi}} \int_0^{\infty} \frac{\exp\left[-2(\sqrt{n} - \sqrt{N_p})^2\right]}{\sqrt{n}}\, dn \tag{11.29}$$

A change of variable $y = \sqrt{2}(\sqrt{n} - \sqrt{N_p})$ results in a lower limit $-\sqrt{2N_p}$ which can be approximated by $-\infty$ provided $Q < p\pi$ and we have a standard integral that yields $\beta_p \sim t_p^2/2$. This result is intuitively correct, since all we have done is to remove the convolutional blurring of the sawtooth line spectrum: the factor of 2 derives from the use of two-sided spectra. This can be checked by noting that

$$\begin{aligned}
\langle n^2 \rangle = \int_{-\infty}^{\infty} W_n(f)\, df &= \sum_{p=1}^{\infty} \beta_p = \sum_{p=1}^{\infty} t_p^2/2 \\
&= \sum_{p=1}^{\infty} \frac{Q^2}{2\pi^2 p^2} = \frac{Q^2}{2\pi^2} \sum_{p=1}^{\infty} \frac{1}{p^2} \\
&= \frac{Q^2}{2\pi^2} \frac{\pi^2}{6} = \frac{Q^2}{12}
\end{aligned}$$

as expected.

Combining all these equations gives the following expression for the output spectrum

$$W_n(f) \sim \sum_{p=1}^{\infty} \frac{Q^3}{4\pi^3} \frac{1}{\sqrt{(2\pi f_b \alpha)}} \frac{1}{p^3} \exp\left[-\frac{Q^2}{8\pi^2 p^2} \left(\frac{f}{f_b \alpha} \right)^2 \right] \quad (11.30)$$

First, for the case of a white input spectrum ($\alpha = 1/\sqrt{3}, f_b = 1$) considered by Bennett, this equation is identical to Bennett's B-series (Bennett's k = our Q^2, and the extra factor of 2 derives from Bennett's use of single-sided spectra). The implication is that Bennett's formula is now recognisable as an asymptotic form of a genuine intermodulation series

$$W_n(f) \sim \sum_{p=1}^{\infty} \left(\frac{2}{N_p} \right)^2 \exp\left[-\frac{1}{2} \left(\frac{f}{N_p f_b \alpha} \right)^2 \right] \Big/ [\sqrt{2\pi} N_p f_b \alpha] \quad (11.31)$$

Fourier series for quantiser sawtooth | Central limit form of intermodulation spectrum

—a fact which smartly refutes Velichkin's criticism quoted previously. In fairness to Velichkin, however, we have to say that Bennett was not aware of the intermodulation character of his B-series.

Secondly, we see that we have not only clarified and justified Bennett's original computation, but have also established it for the case of an *arbitrary* input spectrum. Notice that no fine details of the input spectrum are manifest, since for high-order spectra these are obliterated by the multiple self-convolution mechanism: only the rms bandwidth αf_b appears in the formula.

Thirdly, we see that Robertson's assertion that 10,000 terms are needed for convergence is both right and wrong. About 10,000 terms are certainly needed for accurate computation—but the vast majority of these are negligible. In Fig. 11.7, for instance, the total number of terms greater than 1% of the largest is, in each of the four lines 74, 110, 118 and 94—totalling 396. Robertson has clearly overlooked the line structure of $\{a_n\}$ and the computational economies that accrue.

In making these three points we have reconciled the disagreements between the various approaches and objective (5) is achieved.

One final point remains: the effect of companding. In practical speech pcm links it is essential to use nonuniform quantisation to ensure that the designed NSR is maintained for the wide variety of input signal levels encountered. As Figs. 11.9 and 11.10 show, the use of companding does not fill up the initial vacancy of the series of intermodulation coefficients, so that the flatness criterion of Section 11.4 is maintained for the companded case.

It is possible that the quasi-line structure for the practical case of linear-segment companding can be exploited to obtain an

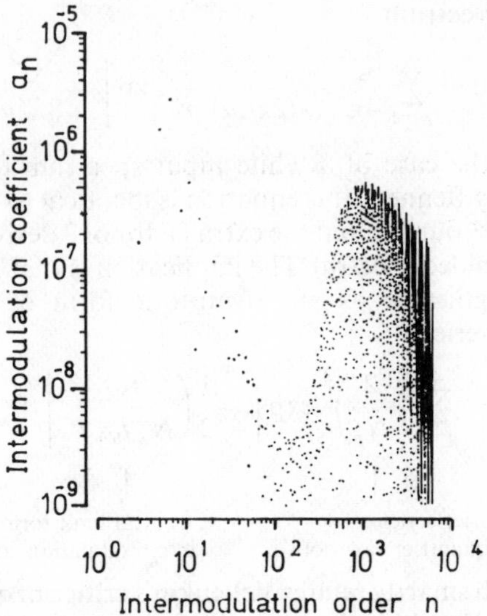

Fig. 11.9 Intermodulation coefficients for a 7 bit quantiser with smooth A-law companding (A = 87.6). RMS signal level 12 dB below the clipping level

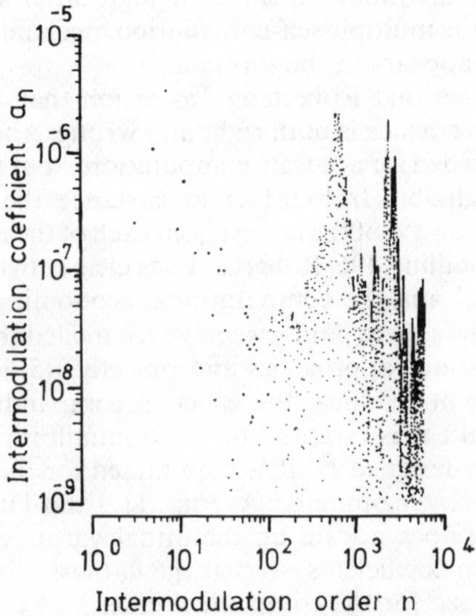

Fig. 11.10 Intermodulation coefficients for a 7 bit quantiser with 13-segment A-law companding (A = 87.6). RMS signal level 12 dB below the clipping level

economised approximation similar to Bennett's B-series. The approach suggested by Gurevich and Driatsky [21], to use the series in its unapproximated form, in the same fashion as Velichkin, is excessively and probably unnecessarily expensive.

11.6 SUMMARY AND CONCLUSIONS

The five objectives outlined in Section 11.1 have all been achieved. The result is a profound clarification of Bennett's work and its relationship with that of Velichkin, Shimbo *et al.*, Robertson and Richards. The various approaches have been unified by the use of Blachman's diagonal expansion technique as a starting point, augmented by asymptotic numerical methods to solve the computational problem on the one hand, and experimental evidence to validate the theory on the other. The general conclusion of this and other studies is that for conditions under which pcm systems are normally operated—and also considerably beyond—the spectral distribution of quantising noise is substantially white. The implication is that we need not bother with the computation of quantising noise spectra at all, and the NSR (and in some cases MSE) is a sufficiently accurate measure of system performance. In view of the cost of spectral computations (even when Bennett's approach is adopted) this conclusion is most welcome.

One point remains: the entire case study was devoted to the case of a Gaussian input signal. There is a wealth of practical experience to suggest that similar conclusions would apply to non-Gaussian signals. Apart from the fact that the estimate $Q^2/12$ is known to be distribution free, there appears to be no similarly-detailed justification at the theoretical level undertaken in this study. It is extremely likely, however, that similar analyses could be carried through for all diagonal processes whose intermodulation spectra approach flatness with increasing order. Since the expansion functions are orthogonal, in the continuous case they would possess Sturm–Liouville-type zero-interlacing properties, and this would appear to imply the asymptotic quasi-sinusoidal behaviour necessary for the line structure of the intermodulation series for a uniform quantiser, or the dominance of high-order terms for a nonuniform one.

Even more challenging is the case when the high-order spectra do *not* approach flatness, as for the sinusoidal process. High-order terms, being merely harmonics of the input, do not become more and more broadband. Nevertheless, for the asynchronously sampled case, experimental evidence shows the quantising noise spectrum to be white. Presumably, the multiple aliasing mechanism plays a greater part under these conditions than in the Gaussian case, where the situation is dominated by central-limit spectral flatness of the output from high-order nonlinearities. It may be that the conditions for which spectral flatness is maintained are quite different in the two cases. Since quantising a sinewave is of no particular practical significance there is no clear incentive to pursue this problem further. It is sufficient that one case of practical importance, the Gaussian, has been fully analysed and understood from both a theoretical and experimental viewpoint. For other cases the weight of experimental evidence alone is sufficiently persuasive.

Acknowledgements

This case study was generously supported by facilities and personnel within the Telecommunications Research Laboratories and Data Processing Department of the GEC Hirst Research Centre. Specifically, F. M. Clayton, J. C. Barlow and M. Carter of the Systems Theory Group; P. Wells, P. Bylanski and D. G. W. Ingram of the Switching and Transmission Group; and F. W. Parker, Miss M. E. Doo and J. G. Wynne of Data Processing all provided invaluable assistance.

Figure 11.4 is taken from Fig. 3.52 of reference [1]. Figure 11.6 is taken from reference [2] and the remainder from reference [3].

REFERENCES

1. Cattermole, K. W., *Principles of Pulse Code Modulation*, Iliffe, p. 125 (1969)
2. Lever, K. V., 'Quantization noise in pcm systems', MSc Thesis, University of Essex (1972)
3. Lever, K. V. and Cattermole, K. W., 'Quantising noise spectra', *Proc. IEE*, **121**, No. 9, 945–954 (1974)
4. Bennett, W. R., 'Spectra of quantized signals', *Bell Syst. Tech. J.*, **27**, No. 4, 446–472 (1948)
5. Rice, S. O., 'Mathematical analysis of random noise', *Bell Syst. Tech. J.*, **24**, No. 46, Part IV (1945)
6. Velichkin, A. I., 'Correlation function and spectral density of a quantised process', *Telecommun. Radio Eng. Pt. 2*, **17**, 70–77 (1962)
7. Shimbo, O., Ohira, T. and Nitadori, K., 'A generalised formula for the power spectrum of a pulse-modulated signal and its application to pcm transmission problems', *Electron. Commun. Jap.*, **46**, 9–17 (1963)
8. Robertson, G. H., 'Computer study of quantizer output spectra', *Bell Syst. Tech. J.*, **48**, 2393–2403 (1969)
9. Richards, D. L., 'Distortion of speech by quantising', *Electron. Lett.*, **3**, 230–231 (1967)

10. Barrett, J. F. and Lampard, D. G., 'An expansion for some second-order probability distributions', *IRE Trans.*, **IT-1**, 10–15 (1955)
11. McGraw, D. K. and Wagner, J. F., 'Elliptically symmetric distributions', *IEEE Trans. Inf. Th.*, **IT-14**, No. 1, 110–120 (1968)
12. McFadden, J. A., 'A diagonal expansion in Gegenbauer polynomials for a class of second-order probability densities', *SIAM J. Appl. Math.*, **14**, No. 6, 1433–1436 (1966)
13. Wong, E. and Thomas, J. B., 'On polynomial expansions of second-order distributions', *SIAM J. Appl. Math.*, **10**, No. 3, 507–516 (1962)
14. Blachman, N. M., 'The signal × signal, noise × noise and signal × noise output of a nonlinearity', *IEEE Trans. Inf. Th.*, **IT-14**, 21–27 (1968)
15. Blachman, N. M., 'The uncorrelated output components of a nonlinearity', *IEEE Trans. Inf. Th.*, **IT-14**, 250–255 (1968)
16. Ingram, D. G. W. and Bylanski, P., 'Digital coding of broadband signals', *J. Sci. Technol.*, **38**, 151–156 (1971)
17. Easter, B., *et al.*, 'Improved equipment for the measurement of intermodulation distortion using a noise spectrum', GEC Telecommunications Research Laboratories Report No. 12, 590 (1956)
18. Cattermole, K. W., *Principles of Pulse Code Modulation*, Iliffe, p. 241 (1969)
19. 'Compatibility of pcm encoders and decoders', CCITT 4th Plenary Assembly, Mar del Plata, 23rd Sept.–25th Oct. 1968 in White Book Vol. 3, 'Line Transmission', Appendix L (to Annexe 4), p. 12 (1968)
20. Szegö, G., 'Orthogonal polynomials', *Proc. Am. Math. Soc.*, **23**, 194, Formula 8.22.8 (1939)
21. Gurevich, V. E. and Driatsky, I. N., 'Correlation function of noise of nonuniform quantization', *Rad. Eng. Electron Phys.*, 1379–1382 (1972)

12

Numerical Calculation of Intermodulation Products Produced by a Memoryless Nonlinearity

F. M. Clayton

12.1 INTRODUCTION

In many situations of practical interest, it is necessary to analyse the response of a nonlinear device to an input signal which consists of a number of independent sinusoidal components. Such circumstances arise, for example, in the multicarrier operation of microwave amplifiers in staellite communication systems, in the detection of angle modulated signals in the presence of co-channel interference, and, if one considers the limiting case of an infinite number of randomly phased tones, in any problem which involves nonlinear processing of Gaussian noise.

Performance calculations usually begin by splitting the frequency components present in the output waveform into three distinct classes: the input frequencies themselves; their various harmonics; and the intermodulation products which occur at frequencies obtained from sums and differences of multiples of the fundamentals. In certain circumstances, these intermodulation products can fall close to or coincide with the input frequencies, and it then becomes important to establish the amount of energy which they contain. This is particularly relevant when the nonlinearity is a distorting imperfection in a nominally linear system—in such cases a useful measure of signal degradation is often obtained from the carrier to intermodulation ratio (C/I)

$$(C/I) = \frac{\text{Power in a wanted carrier}}{\text{Total intermodulation power in a neighbourhood of specified bandwidth}}$$

Direct measurement of this parameter becomes very difficult when the input signal contains more than two or three tones [1], and consequently there is strong motivation for the development of theoretical analysis procedures which can be used to predict system behaviour from the results of those experimental investigations which *are* feasible.

This topic has a history which extends over many years [1–17], reaching back to the time when electronic devices were first introduced into telecommunication systems. However, in spite of the considerable efforts devoted to the development of suitable analysis methods, the subject is still very much alive, and there are a number of unsolved problems providing scope for further investigations. The purpose of this chapter is to describe a particular computational technique which has proved successful in various applications, to show its relationship to methods devised by other workers in the field, and to indicate some possible directions for future research.

The method was originally conceived by Medhurst [6] and is based on the following notion. If a nonlinear device is subjected to experimental investigation, during which certain characteristics of its performance are measured, then together these provide a picture of particular aspects of its behaviour. From this picture it should be possible to deduce various other performance characteristics: the required information is often contained within the basic data, although considerable processing may be necessary in order to extract it. The technique to be described aims to achieve this processing without recourse to an explicit model of the underlying input/output nonlinearity, thereby simplifying its application to practical situations.

12.2 THE CALCULATION PROCEDURE AND ASSOCIATED MATHEMATICS

Suppose that $F_1(u)$ is given as a known characteristic of a nonlinear device, where u is some variable input parameter. For example, $F_1(u)$ could be the amplitude of the fundamental component of the output waveform which results from a single tone input of amplitude u. The problem is to deduce the form of another characteristic of the device, $F_2(u)$, say. To continue the example, this might be the amplitude of the '$2\omega_1 - \omega_2$' intermodulation product appearing when the input consists of two sinewaves at incommensurate frequencies, each with amplitude u.

At this stage, a further constraint is imposed, namely that, if possible, the approximation shall take the form of a weighted linear sum

$$F_2(u) \simeq b_1 F_1(c_1 u) + b_2 F_1(c_2 u) + \cdots + b_n F_1(c_n u) \qquad (12.1)$$

Here

- n is the *order* of the approximation—determined by the accuracy required
- $\{b_i\}$ is the set of *sample weights*
- $\{c_i\}$ is the set of *sample multipliers*

The task is then to find the b's and c's which, while being specific to a particular problem (i.e. to the *nature* of the behaviour represented by the characteristics F_1 and F_2) and to the approximation order n, will be independent of the *shape* of the particular F_1 curve used as data.

One possible approach is to assume that $F_1(u)$ can be expressed as a power series in u

$$F_1(u) = \sum_{n=0}^{\infty} \alpha_n u^n \qquad (12.2)$$

It is then usually possible to represent $F_2(u)$ by another power series, which, for convenience, is written as a modification of (12.2)

$$F_2(u) = \sum_{n=0}^{\infty} A_n \alpha_n u^n \qquad (12.3)$$

The apparent simplicity of this equation is somewhat illusory—in many cases the analysis and computation needed to determine the set $\{A_n\}$ may be formidable.

However, the analysis proceeds by the substitution of eqns. (12.2) and (12.3) into (12.1), and the supposition that $\{b_i\}$ and $\{c_i\}$ are chosen to make the latter equation valid up to the maximum possible degree. A comparison of the coefficients of successive powers of n gives the following set of equations

$$\begin{aligned}
b_1 &+ b_2 &+ \cdots + b_n &= A_0 \\
b_1 c_1 &+ b_2 c_2 &+ \cdots + b_n c_n &= A_1 \\
b_1 c_1^2 &+ b_2 c_2^2 &+ \cdots + b_n c_n^2 &= A_2 \\
\vdots & \quad \vdots & \vdots & \quad \vdots \\
b_1 c_1^{2n-1} &+ b_2 c_2^{2n-1} &+ \cdots + b_n c_n^{2n-1} &= A_{2n-1}
\end{aligned} \qquad (12.4)$$

Methods for solving these equations are known; that given in the Appendix produces the c's as roots of a polynomial equation and the b's subsequently from a set of linear equations (the first n of (12.4)).

Provided the solution is physically meaningful (see Section 12.3.1 below), eqn. (12.1) then provides a simple means of obtaining $F_2(u)$ in terms of values of $F_1(u)$. Furthermore, the required data can be read from a measured characteristic without the necessity for any intermediate curve fitting.

Since approximation with high-order polynomials can be a very critical process, it might be expected that the expression (12.1), derived from an assumed polynomial fit, would show a high sensitivity to changes in the F_1 values. In practice, provided the original problem is itself well posed, the calculation method turns out to be quite stable, a feature which will be examined further in Sections 12.3 and 12.4.

12.3 APPLICATIONS OF THE TECHNIQUE

12.3.1 Two equal amplitude sinewaves through a nonlinear device

Here the basic problem is: given a curve of single tone output amplitude as a function of input amplitude, to deduce the behaviour of the same nonlinear device when the input signal consists of two equal amplitude tones having incommensurate frequencies. The particular output characteristics of interest are

(1) the amplitude of the fundamental component at frequency ω_1 (or ω_2)
(2) the amplitude of the 3rd-order intermodulation product at frequency $2\omega_1 - \omega_2$ (or $2\omega_2 - \omega_1$)

For the analysis, write the single tone as

$$v(t) = B \cos t\ \omega t \qquad (12.5)$$

and suppose that the nonlinearity has a 'static' input/output characteristic

$$V(v) = \sum_{n=1}^{\infty} a_{2n-1} v^{2n-1} \qquad (12.6)$$

(Notice that even-order terms are omitted, since they make no contribution to the outputs being considered.)

Then, the output amplitude at frequency ω is

$$F_1(B) = \sum_{n=1}^{\infty} \frac{1}{2^{2n-2}} \binom{2n-1}{n-1} a_{2n-1} B^{2n-1} \qquad (12.7)$$

If the two-tone input is written as

$$v(t) = B \cos \omega_1 t + B \cos \omega_2 t \qquad (12.8)$$

The corresponding output is

$$V(t) = \sum_{n=1}^{\infty} a_{2n-1}B^{2n-1}(\cos \omega_1 t + \cos \omega_2 t)^{2n-1}$$

$$= \sum_{n=1}^{\infty} a_{2n-1}B^{2n-1}2^{2n-1} \cos^{2n-1} \omega_S t \cos^{2n-1} \omega_D t \quad (12.9)$$

with

$$\omega_S = (\omega_1 + \omega_2)/2; \quad \omega_D = (\omega_1 - \omega_2)/2$$

Expanding the two cosine terms as harmonic series yields the amplitude of the ω_1 fundamental (from the product of the fundamentals of each expansion) as

$$F_2(B) = \sum_{n=1}^{\infty} \frac{1}{2^{2n-2}} \binom{2n-1}{n-1}^2 a_{2n-1}B^{2n-1} \quad (12.10)$$

and the amplitude of the $2\omega_1 - \omega_2$ product as

$$F_3(B) = \sum_{n=2}^{\infty} \frac{1}{2^{2n-2}} \binom{2n-1}{n-1}\binom{2n-1}{n-2} a_{2n-1}B^{2n-1} \quad (12.11)$$

The corresponding A values to be inserted in eqn. (12.4) thus take the particularly simple forms

$$A_n^{(2)} = \binom{2n-1}{n-1} \quad [n \geqslant 1] \quad (12.12)$$

$$A_n^{(3)} = \binom{2n-1}{n-2} \quad [n \geqslant 2] \quad (12.13)$$

The corresponding sample weights and multipliers for three term approximations are given in Tables 12.1 and 12.2.

It should be noted that in both cases the c's are distributed across

Table 12.1. WEIGHTS AND MULTIPLIERS TO DERIVE ω_1 FUNDAMENTAL OUTPUT FOR TWO-TONE INPUT FROM SAMPLES OF SINGLE-TONE OUTPUT

$$F_2(B) = \sum_{i=1}^{3} b_i F_1(c_i B)$$

i	b_i	c_i
1	0.1240	0.8678
2	0.2234	1.5673
3	0.2786	1.9499

Table 12.2. WEIGHTS AND MULTIPLIERS TO DERIVE $2\omega_1 - \omega_2$ INTERMODULATION PRODUCT WITH TWO-TONE INPUT FROM SAMPLES OF SINGLE-TONE OUTPUT

$$F_3(B) = \sum_{i=1}^{3} b_i F_1(c_i B)$$

i	b_i	c_i
1	−0.19245	0.68404
2	−0.19245	1.28558
3	0.19245	1.96962

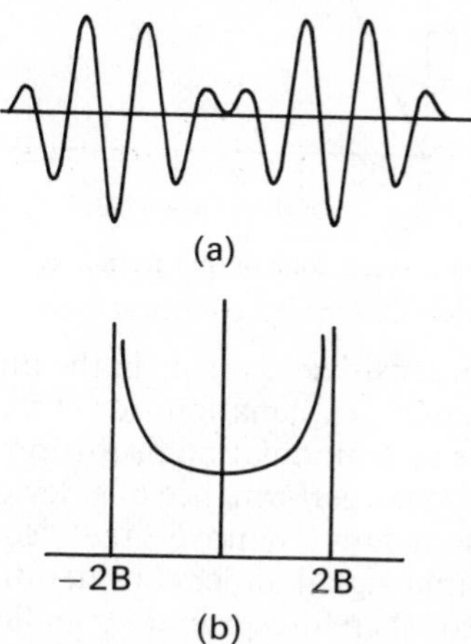

Fig. 12.1 Test signal; (a) time domain structure, (b) pdf of instantaneous amplitude

the interval (0, 2), which is to be expected, since this corresponds to the range of instantaneous amplitude of the composite input signal. The fact that samples are drawn preferentially from the upper end of the interval reflects the fact that the statistical distribution of this instantaneous amplitude corresponds to that of a sinusoid of magnitude $2B$ (Fig. 12.1). Consequently the behaviour of $F_1(B)$ at the extreme end of the range will be most important in determining the specified output characteristics.

A further point of considerable practical interest appears in a comparison of the behaviour of the sample weights $\{b_i\}$ in the two formulae. In the approximation for $F_2(B)$, these are all positive, so that with a positive F_1, calculations will be free from any loss of numerical accuracy through cancellation. However, negative weights occur in the approximation for $F_3(B)$ and there is then a danger that

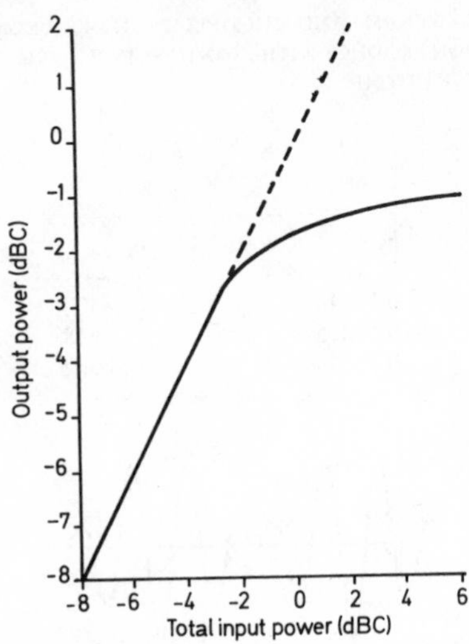

Fig. 12.2 Fundamental single tone output from a symmetric ideal clipper

this formula could be sensitive to errors in the input data. There is a straightforward physical explanation for this, namely that, in attempting to deduce intermodulation distortion from any measurement of the effective linear performance of a device, the computation procedure must somehow remove the 'equivalent linearity' appropriate to the input signal under consideration. Small measurement inaccuracies can therefore produce significant changes in the residual device characteristics [8]. This point will be considered further in Sections 12.3.2 and 12.4.1.

To illustrate the numerical results which can be obtained from these formulae, Figs. 12.3 and 12.4 provide comparisons between exact and approximate calculations performed for the ideal peak clipper, whose single tone response is given in Fig. 12.2. These examples make it clear that even a relatively low-order formula can produce accurate predictions, this being due mainly to the fact that the basic data curve is approximated only over the range required for the particular output point being calculated. For example, if the two-tone signal excites only the linear portion of the clipper characteristic, the formulae predict unit gain and zero intermodulation distortion, and are precisely correct. Any attempt at *global* curve fitting for the basic characteristics (either static or single-tone input/output) is, in contrast, unlikely to be this successful, since the behaviour of the approximant in the region of interest will be affected (through continuity restrictions, for example) by data values in other regions.

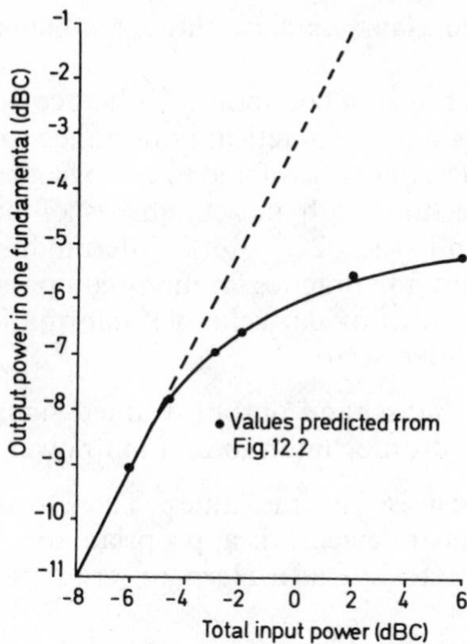

Fig. 12.3 Fundamental output power in one tone. Two equal tones input to a symmetric ideal clipper

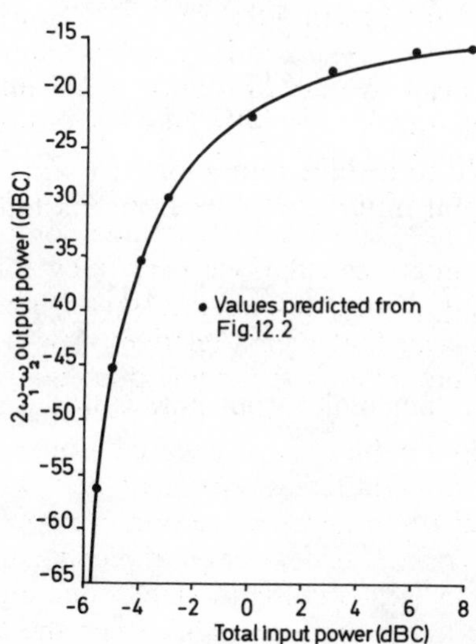

Fig. 12.4 Output power at frequency $2\omega_1-\omega_2$. Two equal tones input to a symmetric ideal clipper

12.3.2 Band-limited Gaussian noise through a nonlinear device

This situation has much in common with that considered in Section 12.3.1, but it is of sufficient practical importance to merit a separate discussion. The basic data is assumed to come from a two-tone test of the device in question, with measurements of both fundamental output power and the '$2\omega_1 - \omega_2$' intermodulation distortion product. This is common practice for many component specifications nowadays, and can provide valuable information. The output characteristics of interest are:

(1) the coherent noiseband output (wanted signal) power G_1^2
(2) the total third-order intermodulation power G_3^2.

Suppose that the noise has rms value σ. Then, to analyse its passage through the nonlinear device, it is appropriate to represent the static input/output characteristic as a Hermite series

$$V(v) = \sum_{n=1}^{\infty} \frac{g_{2n-1}}{(2n-1)!} He_{2n-1}(v/\sigma) \tag{12.14}$$

(As before, only odd degree terms are included.) The coefficients $\{g_i\}$ may be written in terms of the $\{a_i\}$ of eqn. (12.6), and in particular the two terms of interest are given as

$$g_1(\sigma) = \sum_{n=1}^{\infty} \frac{(2n-1)!}{(n-1)!\, 2^{n-1}} a_{2n-1}\sigma^{2n-1} \tag{12.15}$$

$$g_3(\sigma) = \sum_{n=2}^{\infty} \frac{(2n-1)!}{(n-2)! 2^{n-2}} a_{2n-1}\sigma^{2n-1} \tag{12.16}$$

These may be used to deduce values for G_1^2 $(=g_1^2)$ and G_3^2 $(=g_3^2/6)$.

Keeping the total input power fixed means that the basic data curves are

$$\Phi_1(\sigma) \equiv \frac{F_2(\sigma)}{\sqrt{2}} = \frac{1}{\sqrt{2}} \sum_{n=1}^{\infty} \frac{[(2n)!]^2}{(n!)^4 2^{2n-4}} a_{2n-1}\sigma^{2n-1} \tag{12.17}$$

$[\Phi_1(\sigma)]^2$ is the fundamental output power produced by a two-tone input with total power σ^2.

$$\Phi_3(\sigma) \equiv \frac{F_3(\sigma)}{\sqrt{2}}$$

$$= \frac{1}{\sqrt{2}} \sum_{n=2}^{\infty} \frac{[(2n-1)!]^2}{(n+1)!\, n!\, (n-1)!\, (n-2)!\, 2^{2n-2}} a_{2n-1}\sigma^{2n-1} \tag{12.18}$$

Table 12.3. WEIGHTS AND MULTIPLIERS TO DERIVE NOISEBAND FUNDAMENTAL OUTPUT POWER FROM SAMPLES OF TWO-TONE FUNDAMENTAL OUTPUT

$$[G_1(\sigma)]^2 = \left[\sum_{i=1}^{3} b_i\Phi_1(c_i\sigma)\right]^2$$

i	b_i	c_i
1	0.93723	0.83703
2	0.42170	1.41272
3	0.01633	2.08016

Table 12.4. WEIGHTS AND MULTIPLIERS TO DERIVE NOISEBAND THIRD-ORDER INTERMODULATION POWER FROM SAMPLES OF TWO-TONE $2\omega_1 - \omega_2$ OUTPUT

$$[G_3(\sigma)]^2 = \left[\sum_{i=1}^{3} b_i\Phi_3(c_i\sigma)\right]^2$$

i	b_i	c_i
1	1.28488	1.19605
2	0.41391	1.75942
3	0.01191	2.40666

$[\Phi_3(\sigma)]^2$ is the $2\omega_1 - \omega_2$ output power produced by (a two-tone) input with total power σ^2.

Tables 12.3 and 12.4 give the sample weights and multipliers which correspond to the assumption of three term approximating formulae. Again the statistical properties of the input signal are mirrored in the behaviour of the coefficients $\{b_i\}$ and $\{c_i\}$. However, it is significant that because the third-order intermodulation behaviour is in this case deduced from a third-order intermodulation measurement, albeit of a quite different nature, the appropriate prediction formula has positive weights and is therefore relatively insensitive to perturbations of the input data.

Figure 12.6 shows the result of applying the formulae to the data given in Fig. 12.5, which corresponds to the measured performance of a typical present-day microwave FET amplifier. The predicted performance is deduced on the assumption that the amplifier behaves essentially as a resistive nonlinearity and, provided that the noise is sufficiently narroband to allow effective zonal separation of harmonic components in the amplifier output, the results may be used to deduce the NPR curve shown in Fig. 12.7.

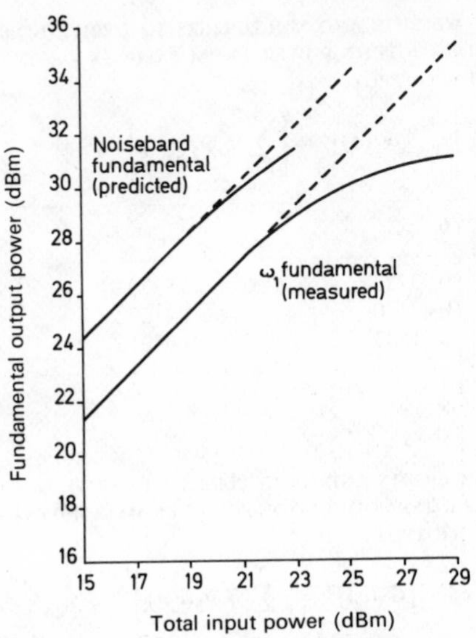

Fig. 12.5 Fundamental characteristic for a microwave power amplifier (noiseband or two tone input)

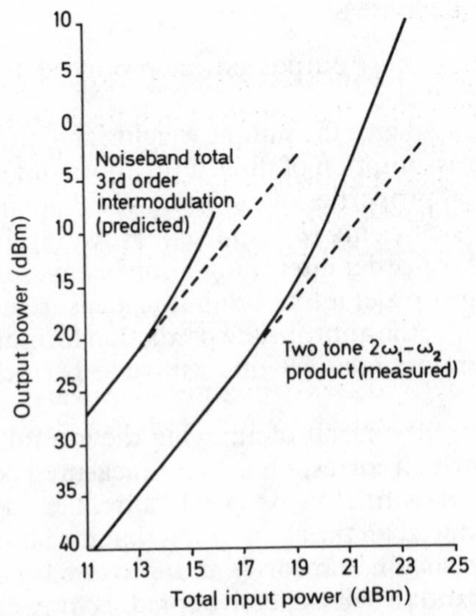

Fig. 12.6 Intermodulation characteristics of power amplifier (noiseband or two tone input)

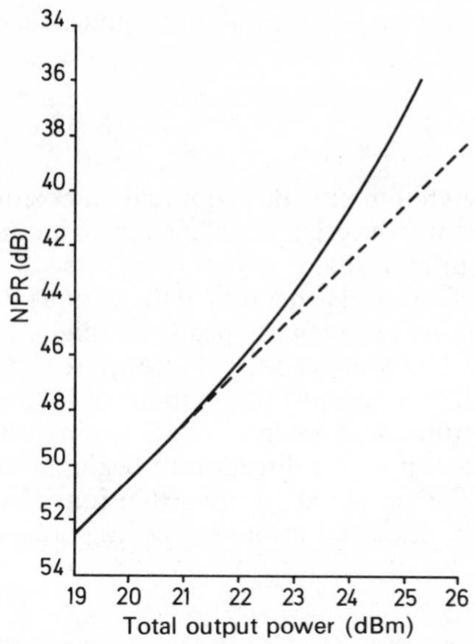

Fig. 12.7 NPR values for narrow band noise input to microwave power amplifier

12.4 RELATIONSHIPS WITH OTHER METHODS AND POSSIBLE DIRECTIONS FOR FUTURE WORK

12.4.1 Gaussian quadrature

From the structure of eqn. (12.1), which forms the basis for the approximation procedure, it is apparent that the method has much in common with the familiar numerical analysis technique of Gaussian quadrature [19]. The parallel can be made explicit in several cases—for example, it is shown in reference [6] how the relationship between $F_2(B)$ and $F_1(B)$ of Section 12.3 can be written as an integral

$$F_2(B) = \frac{1}{\pi} \int_0^2 \frac{x}{\sqrt{(4 - x^2)}} F_1(Bx)\, dx \tag{12.19}$$

Consequently, the derived approximation is equivalent to the use of a Gauss formula for integration on (0, 2), with weight function

$$w(x) = \frac{1}{\pi} \frac{x}{\sqrt{(4 - x^2)}}$$

The fact that this weight function is positive ensures [19] that the coefficients $\{b_i\}$ and $\{c_i\}$ will exist for approximation formulae of all orders, and that the $\{b_i\}$ will also be positive. This substantiates mathematically the earlier empirical observations.

The corresponding integral formulation applicable to the function F_3 is

$$F_3(B) = \frac{1}{\pi} \int_0^2 \frac{(x^2 - 2)}{\sqrt{(4 - x^2)}} F_1(Bx)\, dx \tag{12.20}$$

In this case the weight function does not remain positive throughout (0, 2) and consequently the existence of an approximation formula no longer follows automatically.

Considering the powerful generality of the various theorems which relate to Gauss quadrature with positive weight functions, it is interesting to speculate whether or not those approximation formulae which are successful numerically (G_3 in terms of F_3, for example) can always be transformed into integral relationships which make this property manifest. A possible direction to begin an investigation of this point could be provided by an extension of the describing function approach, discussed in the following paragraph.

12.4.2 Describing functions and Chebyshev transforms

For a memoryless nonlinearity with sinusoidal input (12.5), the mth-order Chebyshev transform of $V(v)$ is defined [17] as

$$C_m(B) = \frac{2}{\pi} \int_{-B}^{B} \frac{T_m(v/B)}{\sqrt{(B^2 - v^2)}} V(v)\, dv \tag{12.21}$$

where $T_m(x) = \cos(m \arccos x)$ is the Chebyshev polynomial of degree m. Physically, $C_m(B)$ is the amplitude of the mth harmonic in the output waveform (except for $m = 0$, when the dc component is $\frac{1}{2}C_0$). In control theory, $C_m(B)/B$ is often referred to as the mth describing function [14].

Since eqn. (12.21) relates the harmonic output (as a function of input amplitude) to the underlying device characteristic, various authors [14–17] considered possible ways to derive an inverse transform which would produce properties of the device from data obtained by observing its output. This was finally achieved by Blachman [17].

In only odd (quâ function) nonlinearities are considered, then [17] shows that inversion of $C_{2n-1}(B)$ defines V up to an arbitrary (odd) polynomial of degree $2n - 3$. Thus $C_3(B)$, for example, should contain sufficient information to enable the characterisation of any third-order intermodulation behaviour for any specified input waveform. If proof of this statement is forthcoming, it is conceivable that it might be further extended to show that different functions of the same order (e.g. F_1, F_2 and G_1; F_3, G_3 and C_3) are in some sense equivalent—that is to say, any can be deduced from any other. This might then lead to a proof of the earlier postulate concerning the stability of approximations.

12.4.3 The problem of quadrature distortion in bandpass nonlinearities

Once again, this is a phenomenon which has been recognised for some time, and various modelling techniques have been proposed for its solution [6, 9–13]. Of these, the quadrature summation introduced by Kaye, George and Eric [10] (Fig. 12.8) seems to have proved

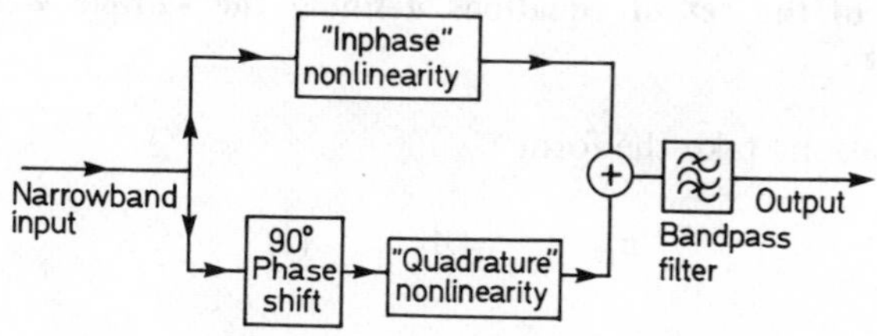

Fig. 12.8 I-Q model for bandpass nonlinearities

successful, and it is quite straightforward to apply the present technique to their formulation, provided that information about both the in-phase and quadrature nonlinearities is available. The basic difficulty which arises is in the interpretation of measurement data, particularly the disentangling of output components which may have arisen in either nonlinearity. Although recognised as a problem, particularly for the understanding of experimental $2\omega_1 - \omega_2$ intermodulation product behaviour, it is only recently that investigations in the area have been reported [12, 13]. There is scope for both theoretical and practical work here, with the ultimate goal a 'vector spectrum analyser' which could provide a more complete characterisation of bandpass nonlinearities.

12.5 CONCLUSIONS

The computation method described above allows distortion products of a specified type to be calculated quickly and easily from measured characteristics, using simple approximation formulae. Certain numerical constants must be determined in advance, but, once these are known, the formulae apply without modification to any shape of characteristic.

Acknowledgements

My interest in this topic was first aroused in conversations with Dr. R. G. Medhurst and R. A. Harris. J. A. Gillespie collaborated with me to extend their work and derive numerical values for various approximation formulae.

APPENDIX

Solution of the set of equations defining the sample weights and multipliers

The equations take the form

$$\begin{aligned} \sum_{i=1}^{n} b_i &= A_0 \qquad (1) \\ \sum_{i=1}^{n} b_i c_i &= A_1 \qquad (2) \\ \vdots \qquad & \qquad \vdots \\ \sum_{i=1}^{n} b_i c_i^{2n-1} &= A_{2n-1} \qquad (2n-1) \end{aligned} \tag{A.1}$$

Suppose that the c_i's are the roots of the polynomial equation

$$z^n + \sum_{k=0}^{n-1} q_k z^k = 0 \tag{A.2}$$

The coefficients q_k are obtained by the following procedures

Multiply equation (1) by q_0

(2) by q_1

$(n-2)$ by q_{n-1}

$(n-1)$ by 1

and add

$$A_n + \sum_{k=0}^{n-1} q_k A_k = 0 \tag{A.3}$$

By treating eqns.(2) to (n) in the same way, followed by (3) to $(n+1)$, etc., a set of linear equations for the q_k is derived

$$\sum_{k=0}^{n-1} q_k A_{k+i} = -A_{n+i} \quad [i = 0, 1, 2, \ldots, n-1] \tag{A.4}$$

Once (A.2) is solved for the values of c_i, substitution in the first n eqns. (A.1) gives a set of linear equations for the b_i.

REFERENCES

1. Westcott, R. J., 'Investigation of multiple fm/fdm carriers through a satellite TWT operating near to saturation, *Proc. IEE*, **114**, 726–740 (1967)
2. Espley, D. C., 'Harmonic analysis by the method of central differences', *Phil. Mag.*, **28**, 338–352 (1939)
3. Bloch, A., 'The calculation of intermodulation products by means of a difference table', *J. IEE*, **93**, Pt. III, 211–216 (1946)
4. Bennett, W. R., 'The biased ideal rectifier', *Bell Syst. Tech. J.*, **26**, 139–169 (1947)
5. Brockbank, R. A. and Wass, C. A. A., 'Nonlinear distortion in transmission systems, *J. IEE*, **92**, Pt. III, 45–56 (1945)
6. Medhurst, R. G. and Harris, R. A., 'Calculation of nonlinear distortion products', *Proc. IEE*, **115**, 909–917 (1968)
7. Blachman, N. M., 'The uncorrelated output components of a nonlinearity', *IEE Trans. Inf. Theory*, **IT-14**, 250–255 (1968)
8. Sette, G. R., 'Calculation of intermodulation from a single carrier amplitude characteristic', *IEEE Trans. Commun.*, **COM-22** (3), 319–323 (1974)
9. Berman, A. L. and Mahle, L. E., 'Nonlinear phase shift in travelling wave tubes as applied to multiple access communications satellite', *IEEE Trans. Commun.*, **COM-18**, 37–48 (1970)
10. Kaye, A. R., George, D. A. and Eric, M. J., 'Analysis and compensation of bandpass nonlinearities for communications', *IEEE Trans. Commun.*, **COM-20**, 965–972 (1972)
11. Blachman, N. M., 'The output signals and noise from a nonlinearity with amplitude dependent phase shift', *IEEE Trans. Inf. Theory*, **IT-25**, 77–79 (1979)
12. Maseng, T., 'On the characterisation of a bandpass nonlinearity by two-tone measurements', *IEEE Trans. Commun.*, **COM-26**, 746–754 (1978)
13. Chie, C. M., 'A modified Barrett–Lampard expansion and its application to bandpass nonlinearities with both am–am and am–pm conversion', *IEEE Trans. Commun.*, **COM-28**, 1859–1866 (1980)
14. Gibson, J. E. and di Tada, E. S., 'On the inverse describing function problem', *Proc. IFAC 1963*, Butterworth, London, pp. 29–34 (1964)
15. Schmideg, I., 'Note on the evaluation of Fourier coefficients of power-law devices', *Proc. IEEE* (*Letters*), **56**, 1383–1384 (1968)
16. Schmideg, I., 'Effects of limiting on angle and amplitude modulated signal', *Proc. IEEE* (*Letters*), **57**, 1302–1303 (1969)
17. Blachman, N. M., 'Detectors, bandpass nonlinearities and their optimisation: inversion of the Chebyshev transform', *IEEE Trans. Inf. Theory*, **IT-17**, 398–404 (1971)
18. Kopal, Z., *Numerical Analysis*, Chapman and Hall (1955)
19. Stroud, A. M. and Secrest, D., *Gaussian Quadrature Formulas*, Prentice-Hall (1966)

13

Intermodulation in Single-Channel per Carrier Satellite Communication Systems

B. G. Evans

13.1 INTRODUCTION

This chapter deals with the problems of non-linearities and stochastic variations as applied to analogue satellite communication systems. It represents a summary of some of the work on intermodulation and interference analysis and simulation applied to single-channel-per-carrier (SCPC) satellite systems carried out by the Essex group over the last five years. In this chapter we will concentrate particularly on the methods used to calculate baseband intermodulation noise degradation due to multi-carrier operation through a non-linear satellite transponder. The latter is a specific example of the more general mathematical modelling of non-linearities covered in Chapter 9. However, the associated problems of interference between radio/satellite systems parallels this so closely that we have been able to apply our results in this area as well. The treatment of non-linearities presented here is not new, but its application to the sort of systems considered is new. The specific problems associated with the signal-processing of SCPC signals will be mentioned only briefly.

Let us first set the problem in context; we have a multi-access satellite communication system with many carriers simultaneously being amplified by a non-linear satellite transponder (see Fig. 13.1). The overall quality of the communication over any link in the system will be given by its signal to noise or probability of error via

$$\left\{\frac{S}{N} \text{ or } P_e\right\} = f_n\left\{\frac{C}{N_{\mathrm{TOT}}} \times m\right\} \tag{13.1}$$

where m is the modulation advantage and C/N_{TOT} is the carrier to total noise. The latter term is obviously crucial in satellite systems

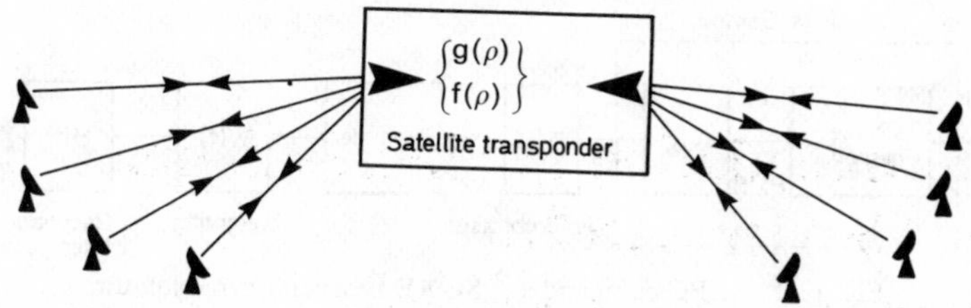

Fig. 13.1 Multiple access satellite system

design, which really amounts to the optimum distribution of the noise components. These can be summarised as

$$\text{Noise total} = \begin{Bmatrix}\text{Up-path and down-path}\\ \text{thermal noise}\end{Bmatrix} + \begin{Bmatrix}\text{Equipment}\\ \text{noise}\end{Bmatrix} + \begin{Bmatrix}\text{Intermodulation}\\ \text{noise}\end{Bmatrix} + \begin{Bmatrix}\text{Interference}\\ \text{noise}\end{Bmatrix}$$

Herein we are concerned with the latter two components. Notice that once the rf intermodulation power spectral density (PSD) has been calculated from non-linearity considerations, both of the latter can be treated similarly as far as derivation of equivalent baseband noise is concerned.

The problem can then be summarised as

(1) rf intermodulation analysis
—both levels and PSD's of IMP's

(2) baseband distortion produced from either
(i) intermodulation
or (ii) interference } unwanted rf components

Many studies have dealt with the determination of intermodulation product amplitudes and PSD's [1–5]. Perhaps the most complete and unified is that due to Fuenzalida, Shimbo and Cook [6] which accounts for both amplitude and phase non-linearities and hence AM/PM conversion generated IMP's as well as non-linear amplitude IMP's. The method uses the time-domain representation of the output of the non-linear device and calculation in this domain of the resultant baseband intermodulation noise spectra. This technique allows an exact calculation of the baseband distortion since the statistical dependence between the intermodulation noise and the carriers is accounted for. We have used the technique in our studies as, unlike others, it can be applied for carriers and interference with more general spectral shapes.

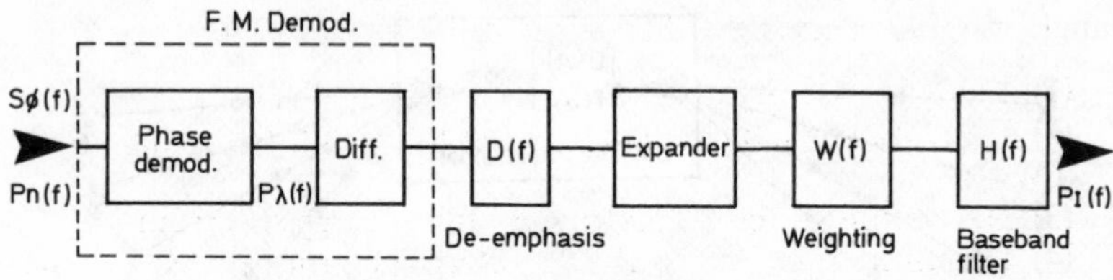

Fig. 13.2 SCPC companded FM, RF-baseband demodulator

The particular system which we have investigated is companded fm SCPC which is in use with many domestic and regional satellites. The intermodulation analysis problem then amounts to a simulation of the system shown in Fig. 13.2. Unlike the well-used and well-documented fm/fdma systems this system has several important differences:

(1) The spectrum of the SCPC carrier cannot be approximated by a simple Gaussian and is not stationary.
(2) The number of carriers, and hence IMP's, is increased markedly and the voice switching means that the number is not constant.
(3) All carriers are not necessarily equal in amplitude or spectrum.
(4) In general, energy dispersion is not used.
(5) The use of an additional non-linear device, the syllabic compandor affects both the spectrum and statistics of the IMP's.

The above have forced us into looking at statistical variations of the parameters and the use of stochastic descriptions of both the spectra and the operations [7]. We will mention these only briefly in Section 13.3 but essentially they are the major problems in dealing with intermodulation SCPC systems, as, once solved, the general method of Fuenzalida may be applied.

There is one other important area to which these calculations are relevant and that is in the Earth-station transmitter power requirements. The transmitters are either Klystrons or TWT's and both need to be backed-off to limit the effects of intermodulation products. In choosing the power capacity of Earth-station transmitters, which may simultaneously be required to amplify fdm and SCPC carriers, it is of prime importance that intermodulation levels are accurately assessed.

13.2 MATHEMATICAL MODELLING FOR INTERMODULATION CALCULATIONS

We consider here intermodulation noise resulting from the amplification by a non-linear TWT of an input signal composed of a sum of angle modulated carriers. The steps in the analysis are

(1) Modelling of the non-linearity of the TWT.
(2) Derivation of a time domain representation of the output of the TWT.
(3) Derivation of the rf intermodulation power spectrum.
(4) Determination of baseband intermodulation noise by demodulation.

13.2.1 Modelling of non-linearity

Following the mathematics of Shimbo [4] we characterise the non-linear device in terms of the input and output envelopes. Besides being physically more meaningful than the instantaneous voltage characterisation, it lends itself ideally to simple laboratory measurements of the amplitude and phase non-linearities.

The general representation of an input bandpass signal is

$$\begin{aligned} e_i(t) &= R_e\{V(t)\exp(j\omega_0 t)\} \\ &= R_e\{\rho(t)\exp(j\omega_0 t + j\theta(t))\} \end{aligned} \tag{13.2}$$

where

$$V(t) = \rho(t)\exp[j\theta(t)] = \text{complex input envelope}$$

$$\rho(t) = \text{amplitude of } V(t)$$

$$\theta(t) = \text{phase of } V(t)$$

Two real non-linear functions are necessary to describe the properties of the non-linearity:

$$g(\rho) = \text{amplitude non-linearity}$$

$$f(\rho) = \text{phase non-linearity}$$

For the so-called 'memoryless' case, that is, both of the above independent of frequency, the bandpass output from the device is

$$e_0(t) = R_e\{g[\rho(t)]\exp(j\omega_0 t + j\theta(t) + jf[\rho(t)]\} \tag{13.3}$$

which may be written in terms of a complex envelope gain function

$$e_0(t) = R_e\{G[\rho(t)]e_i(t)\} \tag{13.4}$$

where $G[\rho(t)] = g(\rho)\exp\{jf(\rho)\}/\rho(t)$.

In Fig. 13.3 we show typical measured amplitude and phase characteristics of a TWT (in practice they do not vary much) using envelope measurements. Although a number of series expansions of

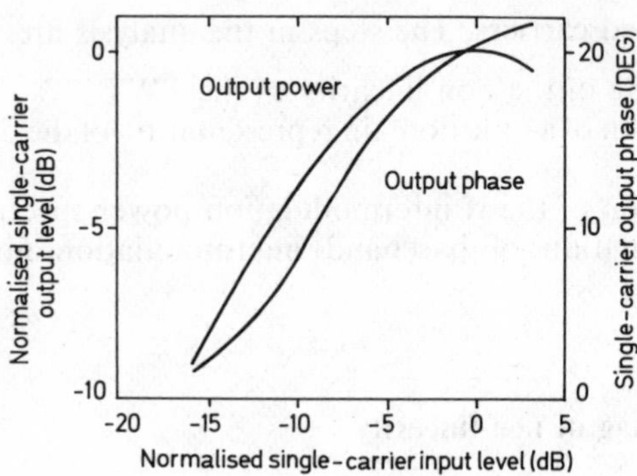

Fig. 13.3 Measured nonlinear TWT characteristics

$g(\rho)\exp\{jf(\rho)\}$ are possible [2], the Bessel function expansion proves particularly suitable for the analogue input signals considered in this work.

$$g(\rho)\exp\{jf(\rho)\} = \sum_{s=1}^{L} b_s J_1(\alpha s \rho) \tag{13.5}$$

where b_s are complex coefficients to be determined from the fitting procedure and α is a scaling factor [6].

13.2.2 Determination of the time-domain representation of the output

Treating the simplest case in which the input is formed by the sum of m narrowband bandpass signals,

$$\begin{aligned} e_i(t) &= R_e \sum_{j=1}^{m} A_j(t)\exp[j\omega_0 t + j\theta_j t)] \\ &= R_e\{\rho(t)\exp j[\omega_0 t + \theta(t)]\} \end{aligned} \tag{13.6}$$

and the bandpass output is represented by eqn. (13.8).

The fundamental component of the output can be expressed as

$$\hat{e}_0(t) = \mathscr{H}[\hat{e}_i(t)] \tag{13.7}$$

and the method used to obtain the inband output is very similar to the 'characteristic function method'. The transfer function $\mathscr{H}$ is expressed as the inverse Fourier transform of its Fourier transform

$$\mathscr{H}(z) = \mathscr{F}^{-1}[\mathscr{F}(\mathscr{H}(z))] \tag{13.8}$$

which is, in fact, a double Fourier transform as z is complex. This formulation does lead to a simplification as it allows one to separate a multiple integral, and to calculate each part separately. After some manipulation this results in [6, 7],

$$e_0(t) = R_e\left\{\exp\left[j\omega_0 t\right] \sum_{k_1 \cdots k_m = -\infty}^{\infty} \exp\left[j \sum_{i=1}^{m} k_i \theta_i(t)\right] \times M(k_1, k_2, \ldots, k_m)\right\} \quad (13.9)$$

and the complex amplitude $M(k_1, \ldots, k_m)$ which involves a double integral over Bessel function products may be further simplified to

$$M(k_1, k_2, \ldots, k_m) = \int_0^{\infty} \gamma \left\{\prod_{i=1}^{m} J_{k_i}[A_i(t)\gamma]\right\} d\gamma \times \int_0^{\infty} \rho g(\rho) \exp\left[jf(\rho)\right] J_1(\gamma\rho)\, d\rho \quad (13.10)$$

whence using the Bessel expansion of (13.5) yields the simple result

$$M(k_1, k_2, \ldots, k_m) = \sum_{s=1}^{L} b_s \prod_{i=1}^{m} J_{k_i}[\alpha s A_i(t)] \quad (13.11)$$

So for this case the bandpass output from the non-linearity is also formed by angle modulated signals. Each component (carrier as well as IMP) is described by a set of integers $k_1, \ldots, k_m$. The corresponding amplitude is $M(k_1, \ldots, k_m)$, and its angle modulation is given by $\sum_{i=1}^{m} k_i \theta_i(t)$.

Using this formulation one can deal with stationary signals such as fdm blocks, but not SCPC carriers or mixtures of the latter which involve time dependence. The latter cases have been solved [7] but are more complex.

13.2.3 rf intermodulation power spectrum

The PSD of the bandpass output $e_0(t)$ is obtained by first calculating the autocorrelation function of $e_0(t)$ and then taking the Fourier transform term by term. The general method is given in reference [4] and the particular case for the Bessel function expansion of the non-linearity in reference [6]. The resulting expression for the output power spectrum is

$$P(f) = \frac{1}{4} \sum_{k_1, \ldots, k_m = -\infty}^{+\infty} |M(k_1, k_2, \ldots, k_m)|^2 \times \left[\mathscr{P}_{\Delta\phi}\left(f - f_0 - \sum_{i=1}^{m} k_i f_i\right) + \mathscr{P}_{\Delta\phi}\left(-f - f_0 - \sum_{i=1}^{m} k_i f_i\right)\right] \quad (13.12)$$

The above has been extended to identical fdm and SCPC carriers where

$$f_0 + f_i = \frac{\omega_0 + \omega_i}{2\pi} = \text{frequency of the } i\text{th carrier}$$

$$\mathscr{P}_{\Delta\phi}(f) = \mathscr{F}\left\{E\left[\exp j \sum_{i=1}^{m} k_i(\phi_\rho(t+\tau) - \phi_\rho(t))\right]\right\}$$

$$\Delta\phi = k_1\phi_1 + \cdots + k_m\phi_m \tag{13.13}$$

Now $\mathscr{P}_{\Delta\phi}(f)$ represents the low pass equivalent power spectral density, normalised to unit power, of a carrier angle modulated by the signal $\sum_{i=1}^{m} k_i\phi_\rho(t)$. For signals $\phi_\rho(t)$ which are statistically independent, one can show [8].

$$\mathscr{P}_{\Delta\phi}(f) = S_{k1}\phi_1(f) * S_{k2}\phi_2(f) * \cdots * S_{km}\phi_m(f)$$

where

$$S_{ki}\phi_i = \mathscr{F}\{E[\exp jk_i(\phi_i(t-\tau) - \phi_i(t))]\} \tag{13.14}$$

represents the low-pass equivalent PSD (normalised to unit power) of a carrier angle modulated by a signal $k_i\phi_i(t)$. It will be noticed from the above that many convolutions $*$ will be required for large numbers of carriers such as in SCPC.

Except for cases employing linearised TWT's the intermodulation level is mainly dominated by the level of third-order products [1]. These can be defined as type A; $-(2f_1 - f_2)$ and type B; $-(f_1 + f_2 - f_3)$, and eqns. (13.12)–(13.14) evaluated [7] to produce the PSD of third-order intermodulation products.

$$\begin{aligned}
\mathscr{P}_3(f) = {} & \frac{1}{4}\sum_{\{A\}} |M(k_1, k_2, \ldots, k_m)|^2 \\
& \times [S_{2\theta_i} * S_{\theta_j}(f - f_0 - 2f_i + f_j) \\
& \qquad + S_{2\theta_i} * S_{\theta_j}(-f - f_0 - 2f_i + f_i)] \\
& + \frac{1}{4}\sum_{\{B\}} |M(k_1', k_2', \ldots, k_m')|^2 \\
& \times [S_{\phi_i} * S_{\phi_j} * S_{\phi_k}(f - f_0 - f_i - f_j + f_k) \\
& \qquad + S_{\phi_i} * S_{\phi_j} * S_{\phi_k}(-f - f_0 - f_i - f_j + f_k)]
\end{aligned} \tag{13.15}$$

The above has been extended to identical FDM and SCPC carriers as well as the case of mixed FDM and SCPC blocks in reference [7].

13.2.4 Baseband intermodulation spectrum

We have seen that the PSD associated with an IMP can be regarded as the spectrum of an angle modulated signal, the modulating signal of which is determined by the modulating signals of the carriers generating the product. Then the problem of demodulation of a wanted carrier in the presence of IMP's can be accomplished successively and the sum taken to form the baseband noise (see Fig. 13.2). The derivation of the baseband noise spectrum is classical if somewhat involved [7]; the method being to express the temporal output of the demodulator,

$$S(t) = \phi(t) + \lambda_i(t)$$

where $\lambda_i(t)$ is the phase modulation due to a particular IMP. To take the auto-covariance function of $\lambda_i(t)$ and thence the Fourier transform of this gives the PSD, $P_{\lambda_i}(f)$ as

$$P_{\lambda_i}(f) = \frac{1}{A^2}\left[P_n(f+f_c) + P_n(f-f_c)\right] * S_\phi(f) \qquad (13.16)$$

where $P_n(f)$ is the PSD of $n(t)$ and $S_\phi(f)$ is the PSD of the carrier.

The treatment is quite general and any input signal may be considered providing its spectrum can be modelled. The COMSAT work restricts the PSD's of inputs to be Gaussian or rectangular for numerical evaluation. This constraint is imposed by computer time, as many convolutions are required and with the latter these are closed form. With SCPC carriers the rf spectrum is very peaky, with a large proportion of the power located near to the carrier and this causes problems with spectrum definition and ultimate noise determination. The noise due to IMP's in the baseband with reference to Fig. 13.2 will be given by

$$N = \sum_i \int_{-\infty}^{\infty} \mathscr{P}_{\lambda_i}(f) f^2 D(f) W(f) H(f)\, df \qquad (13.17)$$

and the noise power ratio (NPR) as

$$\mathrm{NPR} = \sigma + 10\log(p) - 10\log\left\{\int_{-\infty}^{\infty} \frac{P_\lambda(f)}{W(f)}\, df\right\} \qquad (13.18)$$

where p and σ represent the activity ratio and speech volume in dBmO in the SCPC channel.

13.3 PARTICULAR PROBLEMS RELATED TO SCPC COMPANDED fm

13.3.1 Spectrum simulation

Unlike the fdm/fm carrier whose spectrum can be approximated well by a Gaussian, the SCPC carrier spectrum depends on the dynamic statistics of speech and the talker variations which produce a peaky spectrum with large components close-in to the carrier. The PSD must therefore be considered as a function of both the frequency and the modulating level σ; $S(\sigma, f)$. We have measured typical companded and non-companded spectra and attempted to fit the mean spectra with (1) a 'delta function plus Gaussian' model and (2) a 'delta function plus gamma function' model, viz

$$\hat{S}(\sigma, f) = k\delta(f) + (1-k)\sqrt{\left[\frac{b(\sigma)}{\pi}\right]}\exp\{-b(\sigma)f^2\} \quad (13.19)$$

$$\hat{S}(\sigma, f) = k\delta(f) + \frac{1-k}{2}\sqrt{\left[\frac{b(\sigma)}{\pi|f|}\right]}\exp\{-b(\sigma)|f|\} \quad (13.20)$$

The results obtained [7] indicate that the gamma function fit is better, but is more time consuming when involved in multiple convolutions. This latter restriction may not be so important, as recently we have succeeded in evaluating intermodulation noise power directly in the time domain.

13.3.2 Voice switching of the carrier

Most SCPC systems employ some form of voice switching of the carrier to conserve power. Continuous switching on and off of the carriers causes the amplitude of both wanted carriers and intermodulation products to vary in a random manner. In general, this will improve the intermodulation noise performance. Equation (13.11) gives the complex amplitude at the output of the TWT of either a wanted carrier or an IMP. For an input of n, SCPC/fm carriers, the output amplitude of the rth carrier is

$$\begin{aligned} M(r,t) &= M(0, \ldots, 1, \ldots, 0) \\ &= \sum_{s=1}^{L} b_s J_1(\alpha_s A_r(t)) J_0^{\hat{X}_r(t)}(\alpha_s A) \prod_{\substack{\rho=n+1 \\ \rho \neq r}}^{m} J_0(\alpha_s A_\rho) \end{aligned} \quad (13.21)$$

where $\hat{X}_r(t)$ represents the total number of active SCPC other than the rth.

If we define a voice activation factor as

$$\text{Probability}\,[A_\rho(t) = A] = p \tag{13.22}$$

then $A_\rho(t) = AX_\rho(t)$ where $X_\rho(t)$ is a Bernouilli random variable and

$$\hat{X}_r(t) = \sum_{\substack{i=1 \\ \neq r}}^{n} X_i(t) \tag{13.23}$$

where $\hat{X}_r(t)$, being the sum of Bernouilli random variables, is a binomial random variable. It is possible to show [7] that if n is large then the expression for $M(r, t)$ can be approximated by

$$\hat{M}(r, t) = \sum_{s=1}^{L} b_s J_1(\alpha_s A_r(t)) \prod_{\rho=n+1}^{m} J_0(\alpha_s A_\rho)[pJ_0(\alpha_s A) + 1 - p]^{n-1} \tag{13.24}$$

The same technique can be used to express the amplitude of the IMP's. Hence the PSD of the carrier and IMP's may be obtained in a similar manner to that of Section 13.2 except that the process is considerably more complex. The result is

$$P_{\lambda 1}(f)_{\text{ON}} = \frac{1}{|\hat{M}(r)|^2}[P_n(f+f_r) + P_n(f-f_r)] * \bar{S}_{r\text{ON}}(f) \tag{13.25}$$

and the mathematical model for $\bar{S}_{r\text{ON}}(f)$, the average PSD associated with a voice activated, speech modulated SCPC/fm carrier, has been derived [7] from measurements.

13.3.3 Effects of syllabic companding

The use of a syllabic compandor complicates the analysis of the SCPC system somewhat. The compressor will modify the rf spectrum of the speech modulated carrier as well as the IMP's. Using a model for the compressor action it has been possible to verify a simple relationship linking the companded and uncompanded spectra. The action of the expander is more difficult to model as, during speech pauses, its gain is controlled by the noise power at its input which will reduce the ultimate noise output level markedly. An ideal way of taking this into account would be to consider the gain of the expandor as a random process whose statistics are dependent on the speech statistics. However, this treatment has proved too complex and we have simplified it by dealing with the long-term average gain of the expandor for the active and non-active conditions. In addition a stochastic process $X(t)$ is defined as

$$\begin{aligned} X(t) &= 1, \quad \text{wanted signal active at time } t \\ &= 0, \quad \text{for silence periods} \end{aligned} \qquad (13.26)$$

which has been modelled as a two-state continuous-time Markov random process. Using these two concepts we have formulated the PSD of the demodulated noise for IMP's as previously for both voice-activated and non-voice-activated cases. The PSD of the demodulated speech signal remains essentially unchanged.

13.4 INTERMODULATION NOISE IN VOICE ACTIVATED SCPC COMPANDED FM CHANNELS

The theory presented in Section 13.2 together with the extensions to the SCPC case, listed briefly in Section 13.3, allow us to calculate the baseband intermodulation noise in a wanted SCPC channel due to combinations of SCPC and fdm/fm block loadings of a transponder. In a similar manner to the interference reduction factor method, we specify the noise as

$$\text{Noise power in baseband channel} = \{10 \log [I/C] + X(\sigma, \Delta f)\} \qquad (13.27)$$

where C/I is the rf carrier to intermodulation ratio, σ the wanted channel modulating level and Δf the frequency separation between the wanted carrier and IMP considered. Thus the problem is to evaluate $X(\sigma, \Delta f)$. Let us consider an $(f_a + f_b + f_c)$-type third-order IMP, its spectrum $P_3(f)$ is given from eqn. (13.15) as

$$\mathscr{P}_3(f) = \frac{p^i}{4} |\hat{M}(a, b, -c)|^2 [\bar{S}_{abc}(f - f_p) + \bar{S}_{abc}(-f - f_p)] \qquad (13.28)$$

where p is the activity ratio of the i voice activated SCPC carriers, f_p and f_c are the frequencies of the IMP and carrier respectively. If $\bar{S}_{\text{ON}}(\sigma, f)$ is the average normalised PSD of the demodulated wanted carrier, then the demodulated noise spectrum is given from eqn. (13.25) as

$$\begin{aligned} \mathscr{P}_{\lambda_1}(\sigma, f, \Delta f)_{\text{ON}} &= \frac{p^i}{4} \frac{|\hat{M}(a, b, -c)|^2}{|\hat{M}|^2} \\ &\quad \times [\bar{S}_{abc}(f - \Delta f) + \bar{S}_{abc}(-f - \Delta f)] * \bar{S}_{\text{ON}}(\sigma, f) \end{aligned} \qquad (13.29)$$

where $\hat{M}$ is the complex average amplitude of the wanted carrier.

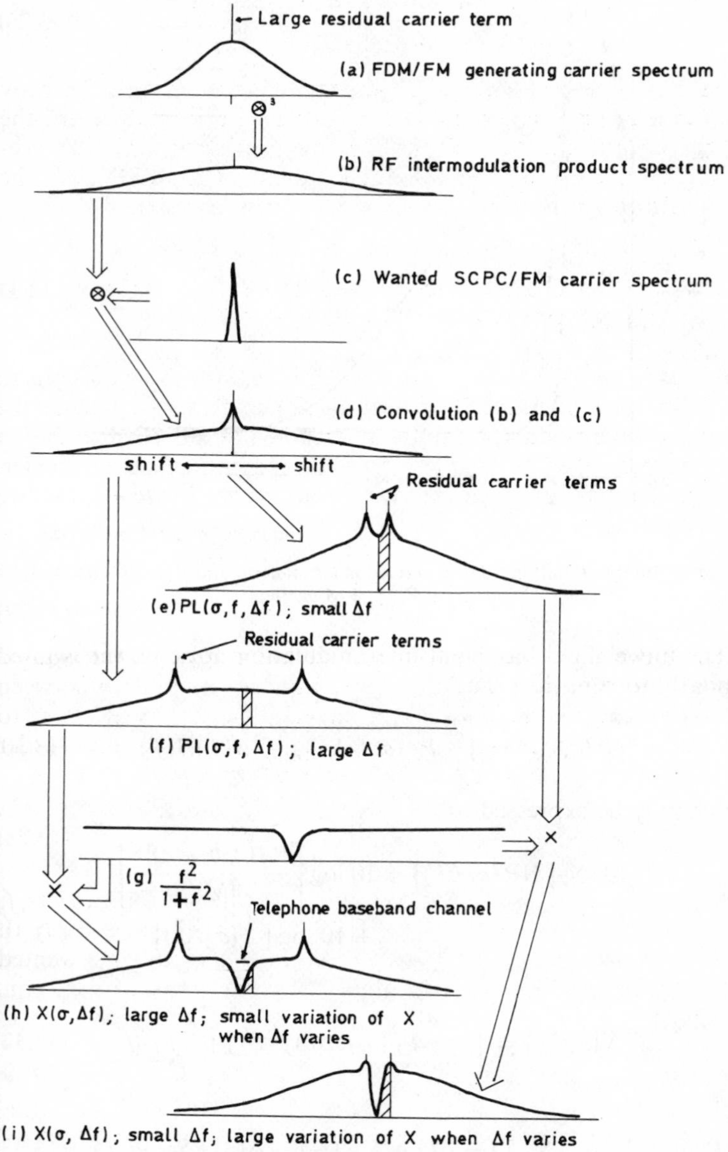

Large residual carrier term

(a) FDM/FM generating carrier spectrum

(b) RF intermodulation product spectrum

(c) Wanted SCPC/FM carrier spectrum

(d) Convolution (b) and (c)

shift shift

Residual carrier terms

(e) $PL(\sigma, f, \Delta f)$; small Δf

Residual carrier terms

(f) $PL(\sigma, f, \Delta f)$; large Δf

(g) $\frac{f^2}{1+f^2}$

Telephone baseband channel

(h) $X(\sigma, \Delta f)$; large Δf; small variation of X when Δf varies

(i) $X(\sigma, \Delta f)$; small Δf; large variation of X when Δf varies

Fig. 13.4 Illustration of the calculation of $X(\sigma, \Delta f)$

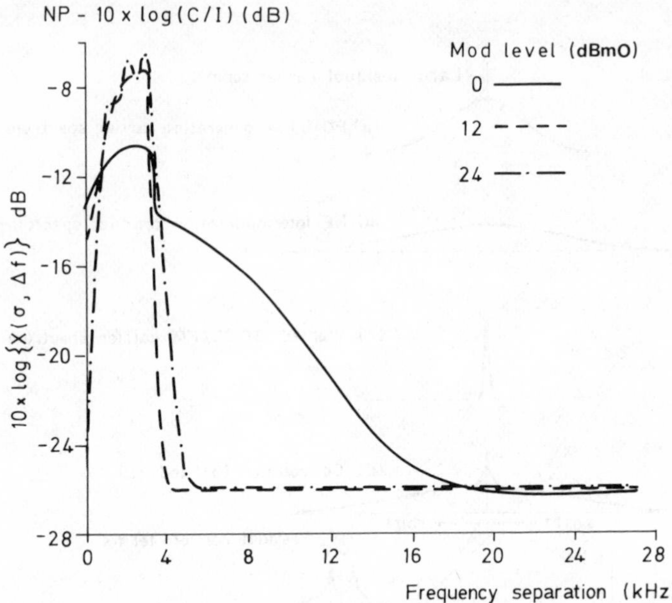

Fig. 13.5 Intermodulation noise generated in a non-companded SCPC channel by 3 identical FDM/FM carriers

The unweighted baseband intermodulation noise power is given similarly to eqn. (13.17) as

$$\mathrm{NP}\,(\sigma, \Delta f) = \int_{\infty}^{\infty} P_{\lambda_1}(\sigma, f, \Delta f)_{\mathrm{ON}}\, f^2 D(f) H(f)\, df \qquad (13.30)$$

which may be expressed as

$$10 \log\left[\mathrm{NP}\,(\sigma, \Delta f)\right] = 10 \log\left[\frac{|\hat{M}(a, b, -c)|^2}{|\hat{M}|^2}\right] + 10 \log\left[X(\sigma, \Delta f)\right] \qquad (13.31)$$

where

$$X(\sigma, \Delta f) = \int_{\infty}^{\infty} \frac{p^i}{4} P_L(\sigma, f, \Delta f) f^2 D(f) H(f)\, df \qquad (13.32)$$

and

$$P_L(\sigma, f, \Delta f) = [\bar{S}_{abc}(f - \Delta f) + \bar{S}_{abc}(-f - \Delta f)] * \bar{S}_{\mathrm{ON}}(\sigma, f) \quad (13.33)$$

In Fig. 13.4 we summarise the calculation of $X(\sigma, \Delta f)$.

As examples of the calculations, results for three fdm/fm blocks are

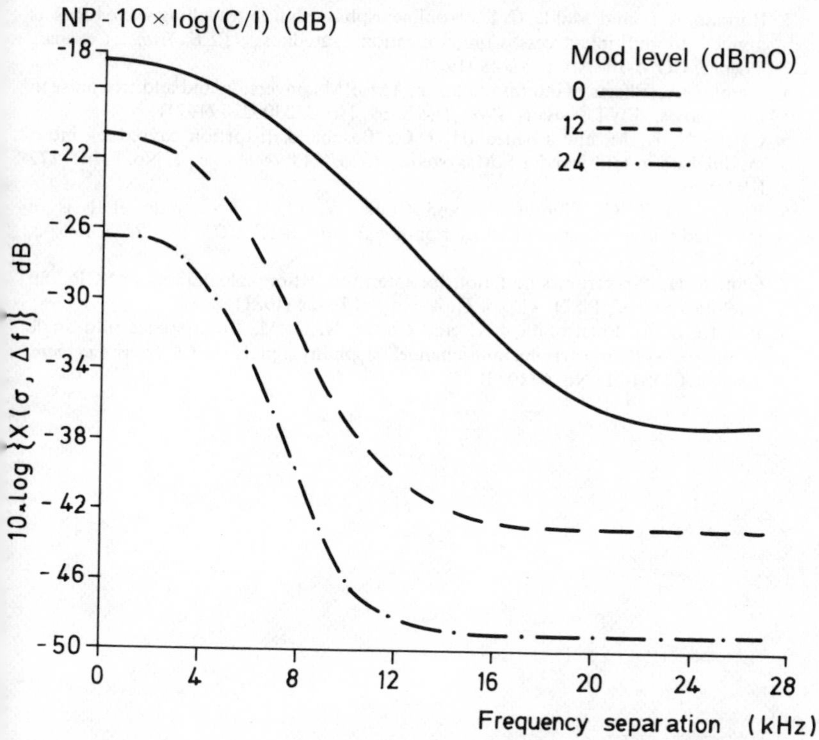

Fig. 13.6 Intermodulation noise generated in a companded SCPC channel by the combination of one FDM/FM carrier and two identical SCPC/FM carriers

shown in Figs. 13.5 and 13.6, the case for one fdm block and two identical voice-switched companded fm carriers.

In conclusion, we have used the above technique to investigate intermodulation noise in various SCPC systems, as once spectral information is known the technique given can be applied. We have found that in order to optimise practical satellite transponder frequency plans, the computation time can still be excessive. A new method in which $X(\sigma, \Delta f)$ is evaluated in the time-domain shows promise in reducing this time significantly.

REFERENCES

1 .Westcott, R. J., 'Investigation of multiple FM/FDM carriers through a satellite TWT operating near to saturation', *Proc. IEE*, **114**, No. 6, 726–750 (1967)

2. Berman, A. L. and Podracky, E., 'Experimental determination of intermodulation distortion produced in wide band communication repeaters', *IEEE Int. Con. Rec.*, **15**, Pt. 2, 69–88 (1967)

3. Borman, A. L. and Mahle, C. E., 'Nonlinear phase shift in travelling wave tubes as applied to multiple-access Communications satellites', *IEEE Trans. Commun. Theory*, **COM-18**, No. 1, 37–48 (1970)
4. Shimbo, O., 'Effects of intermodulation, AM–PM conversion and additive noise in multi-carrier TWT systems, *Proc. IEEE*, **59**, No. 2, 230–238 (1971)
5. Chitre, N. K. M. and Fuenzalida, J. C., 'Baseband distortion caused by intermodulation in multicarrier F.M. systems', *COMSAT Tech. Rev.*, **2**, No. 1, 147–172 (1972)
6. Fuenzalida, J. C., Shimbo, O. and Cook, W. L., 'Time-domain analysis of intermodulation effects caused by non-linear amplifiers', *COMSAT Tech. Rev.*, **3**, No. 1, 89–127 (1973)
7. Ganem, H., 'Spectrum simulation for intermodulation calculations in SCPC/fm satellite systems', Ph.D. Thesis, University of Essex (1981)
8. Ponato, B. A., Fuenzalida, J. C. and Chitre, N. K. M., 'Interference into angle modulated systems carrying multichannel telephony signals', *IEEE Trans. Commun. Theory*, **COM-21**, No. 6 (1973)

14

Representation and Analysis of Digital Line Codes

K. W. Cattermole

14.1 BLOCK CODES

There is a large class of line codes in which a block of N binary digits from the source is translated to a block of M digits, each with L levels, for line transmission. On electrical lines, ternary codes are common, i.e. $L = 3$; in which case $M \leqslant N$, subject to the obvious constraint $L^M \geqslant 2^N$. On optical fibres, binary codes are common, in which case $M \geqslant N$, The treatment here is applicable to both cases, and indeed more generally. The normal purposes of a line code require some redundancy, so that the strict inequality $L^M > 2^N$ usually applies.

The line code must be uniquely decodable, i.e. each block of M line digits corresponds to one and only one block of N information digits. There may, however, be more than one line code block corresponding to one information code block. Many line codes of interest are *alternative codes*: that is, some or all of the information blocks have two or more alternative translations into line code. These translations will differ in some property which is significant in line transmission, and the choice between them is made on each block occurrence so as to preserve an accumulated property of the complete sequence. Usually, the digital sum (or disparity) of the block is chosen so as to minimise fluctuations in accumulated digital sum. Sometimes the parity (oddness or evenness) of a binary line block is chosen so as to preserve a specific accumulated parity.

The statistical properties of a communication system are derived by modelling the source information as a stationary random process; usually, the simplest source model with independent equiprobable binary digits is employed. A non-redundant line code would then give rise to a stationary random line signal with independent digits and simple statistical properties. Block codes complicate this picture in two ways.

Firstly, the smallest potentially independent unit is the block. If we suppose for the moment a unique translation, then independent equiprobable source blocks will translate to independent equiprobable line blocks. The line digits will, however, be neither independent nor, in general, equiprobable. The probability distributions at each digit position in the block, and the joint distribution of digit pairs, will depend on the code structure. The line signal statistics will in general be cyclo-stationary rather than stationary: that is, statistics such as digit probability distribution, correlation between adjacent pairs, etc. will vary periodically, with a period equal to the duration of a block. For some purposes we can derive stationary statistics by averaging over all phases of the periodic variation.

Secondly, in alternative codes not even the blocks are independent. For the translation of any specific source block depends upon past history. This history is adequately described by a single variable, such as accumulated digital sum or accumulated parity, whose value determines the choice among a finite set of translations. So it turns out that each variable associated with the line signal can be modelled as a Markov process. In the next section we develop this model.

14.2 THE SEQUENTIAL MACHINE MODEL

We consider the encoder as a sequential machine, which in response to input signals can change its internal state and which gives output signals dependent both on the current input and the current state. The ideal sequential machine is a well-known concept in the fields of computation and automation theory, and is also well suited to many coding problems. Its application to line code analysis is primarily due to Cariolaro and Tronca [1,2].

The sequential machine is defined in terms of three sets (the inputs, the outputs and the states) and two functions (determining the next output and the next state). In our present context, these properties can be described as follows.

(1) The input set B of N-digit binary code words. We assume the source code to be non-redundant, so that $|B| = 2^N \equiv K$. A code word considered in the abstract will be denoted by the symbol β_u ($u = 1, 2, \ldots, K$). The nth codeword in an input sequence will be denoted by b_n (n any integer).

(2) The set C of states. We assume this to be finite, with $|C| = J$. A state considered in the abstract will be denoted by γ_i ($i = 1, 2, \ldots, J$). The state at the time the nth codeword is translated will be denoted by c_n.

(3) The output set A of M-digit L-level codewords. It is convenient

to identify output codewords by a double index: the word α_{iu} is the translation of the input word β_u which is delivered when the encoder is in state γ_i. Consequently $|A| = JK$: but not all these words are distinct, since the same word may be used in more than one state. The foregoing notation applies to codewords in the abstract: the nth codeword in an output sequence will be denoted by a_n.

(4) The output function h has domain $B \times C$ and range A. It specifies the translation $\alpha_{iu} = h(\gamma_i, \beta_u)$. In operation,

$$a_n = h(c_n, b_n) \tag{14.1}$$

In our context, the function must be chosen so that it has a special form of inverse $h^{-1}(\alpha_{iu}) = \beta_u$: that is, knowledge of α must identify β without reference to γ, to ensure unique decodability.

(5) The state transition function g has domain $B \times C$ and range C. It specifies the state in which the encoder is left after translating one codeword; so that $c_{n+1} = g(c_n, b_n)$. This definition accords with the usual sequential-machine convention, but in our present context a useful state transition function may be defined with the output, rather than the input, as an argument. This function f has domain $A \times C$ and range C, and defines the next state as

$$c_{n+1} = f(c_n, a_n) = g\{c_n, h^{-1}(a_n)\} \tag{14.2}$$

Given the property of unique decodability, these approaches are equivalent, but since the states of a line encoder derive directly from the history of the output, the second version may be simpler.

The great majority of useful line codes constrain the accumulated digital sum of the line signal. Let the digit values be a balanced set of integers, e.g. $\{-1, 0, +1\}$ for ternary: and let the value of the ith digit of the nth word be a_{ni}. Then the digital sum of the nth word is

$$d(a_n) = \sum_{i=1}^{M} a_{ni} \tag{14.3}$$

and the state may be defined as the prior accumulated digital sum

$$c_n = \sum_{j=0}^{n-1} d(a_j) \tag{14.4}$$

The state transition function then takes the form

$$c_{n+1} = c_n + d(a_n) \tag{14.5}$$

For the simplest codes of this class, the output function (14.1) is chosen so that, for any β_u, $d(\alpha_{iu})$ is either opposite in sign to γ_i or is zero: then the sum (14.5) obviously cannot build up to either a positive or negative extreme. In particular, AMI [3, 4] has two states: 4B3T [5, 6] and 10B7T-2M [7, 8] have six. Other codes have somewhat more complex rules which may limit the number of states even further: MS43 [9] and 10B7T-3M [7, 8] have only four states. These examples show that the sequential-machine encoding model is a feasible and realistic one.

The sequence of states occasioned by the transmission of a random signal is a discrete random process. Moreover, it is a Markov chain. This is defined as a sequence of discrete random variables $X_0, X_1, X_2, \ldots$ having the property that given the value X_m for any time instant m, then for any later time instant $m + n$ the probability distribution of X_{m+n} is completely determined, and the values $X_{m-1}, X_{m-2}, \ldots$ at times earlier than m are irrelevant to its determination†. A stationary Markov chain is characterised by a matrix $\mathbf{p}$ of transition probabilities p_{ij}

$$p_{ij} = \text{prob}\,\{X_{m+1} = j \mid X_m = i\} \tag{14.6}$$

The probability distribution of the states at time m is a row vector $\mathbf{p}^{(m)}$ with component probabilities $p_i^{(m)}$

$$p_i^{(m)} = \text{prob}\,\{X_m = i\} \tag{14.7}$$

The change in probability distribution with time is expressed by the matrix equation

$$\mathbf{p}^{(m+n)} = \mathbf{p}^{(m)}\mathbf{P}^n \tag{14.8‡}$$

Under suitable conditions, always fulfilled if X is the state of a block encoder, the state distribution converges towards a limiting distribution $\boldsymbol{\pi}$ such that

$$\boldsymbol{\pi} = \boldsymbol{\pi}\mathbf{P} \tag{14.9}$$

We now show that the sequence of states, in an encoder driven by a source of independent random codewords with stationary distribution, is a stationary Markov chain. Firstly, by (14.2) the state c_{n+1} depends only on the last state c_n and on the current input b_n which by definition is independent of previous states or inputs: so the history is adequately defined by c_n, as required by the definition. Secondly, the transition $c_n \rightarrow c_{n+1}$ is defined by the input, which has a stationary distribution, so we can establish a matrix $\mathbf{P}$ of stationary transition probabilities, from which subsequent statistical properties follow.

†The theory of Markov chains is discussed in many textbooks and in Chapter 3 of this volume. The quotation here is from Cox and Miller [10], page 84.

‡Some authorities defined $p^{(m)}$ as a column vector, premultiplied by a stochastic matrix which is the transpose of P. There is no essential difference of principle. Our notation here follows reference [10].

14.3 FIRST-ORDER STATISTICS

To find the transition probability p_{ij} we must identify the subset of codewords which, on their occurrence in the presence of state i, cause a transition to state j. In the notation of Section 14.2, this is

$$B_{ij} \equiv \{\beta_u \mid g(\gamma_i, \beta_u) = \gamma_j\} \tag{14.10}$$

If the state transition is defined by the line word disparity as in eqn. (14.5), then the subset is most directly identified by

$$B_{ij} = \{\beta_u \mid d(\alpha_{iu}) = j - i\} \tag{14.11}$$

The transition probabilities (14.6) are then

$$p_{ij} = \sum_{\beta_u \in B_{ij}} \text{prob}\,(\beta_u) \tag{14.12a}$$

$$= \frac{|B_{ij}|}{|B|} \quad \text{for equiprobable words} \tag{14.12b}$$

Analysis is usually conducted on the 'equiprobable' assumption, in which case p_{ij} is a property solely of the code structure; the examples in this chapter are mainly of this type. However, it has been shown by Bates [1] that outputs from practical pcm transmitters differ significantly in their statistics from ideal independent equiprobable bit streams. Insofar as the effects can be modelled by a probability distribution of independent words, the present analysis can be extended to take some account of them. Cariolaro and Tronca [2] give examples in which the source bit probability is varied, and show that there can be significant changes in the line signal statistics as a result: also, one simple example appears in Section 14.5 below.

Reverting to the equiprobable case, the transition probability defined by (14.12b) and (14.11) is just the proportion of words with the appropriate disparity. We take as example 4B3T. The 16 binary words are translated into (1) six ternary words of zero disparity, used in all states (2) 10 pairs of ternary words; each pair contains words of opposite disparity, the choice depending on the state. With equiprobable source words we do not need to consider details of the mapping: the numbers of ternary words in each disparity class give us the probabilities. The words used in each of the six states are as in Table 14.1 (for brevity we write simply + and − for +1 and −1 digit values):

Table 14.1

Ternary word disparity	*States −2, −1, 0*	*States 1, 2, 3*	*Proportion*
0	0+− +0− +−0 0−+ −0+ −+0		6/16
±1	−++ +−+ ++− +00 0+0 00+	+−− −+− −−+ −00 0−0 00−	6/16
±2	0++ +0+ ++0	0−− −0− −−0	3/16
±3	+++	−−−	1/16

The corresponding transition matrix is

$$\mathbf{P} = \frac{1}{16}\begin{bmatrix} 6 & 6 & 3 & 1 & 0 & 0 \\ 0 & 6 & 6 & 3 & 1 & 0 \\ 0 & 0 & 6 & 6 & 3 & 1 \\ 1 & 3 & 6 & 6 & 0 & 0 \\ 0 & 1 & 3 & 6 & 6 & 0 \\ 0 & 0 & 1 & 3 & 6 & 6 \end{bmatrix} \tag{14.13}$$

The process can be displayed on a state transition diagram (Fig. 14.1) in which nodes represent states, arcs represent possible transitions, and the numbers against the arcs are transition probabilities. (Precisely the same set of nodes and arcs serves for certain other 6-state codes such as 10B7T-2M, though with different probabilities.) The equilibrium distribution $\boldsymbol{\pi}$ follows on solving eqn. (14.9). By the symmetry of the structure, the six state probabilities must comprise three equal pairs, and the solution is easily found to be

$$\boldsymbol{\pi} = \frac{1}{30}[1 \quad 4 \quad 10 \quad 10 \quad 4 \quad 1] \tag{14.14}$$

Given the distribution of states, and the distribution of codewords in each state, it is possible to find the distribution of digit values. For an arbitrary code, the various digits could have different distributions. In 4B3T (and some though not all other practical codes) the symmetry in

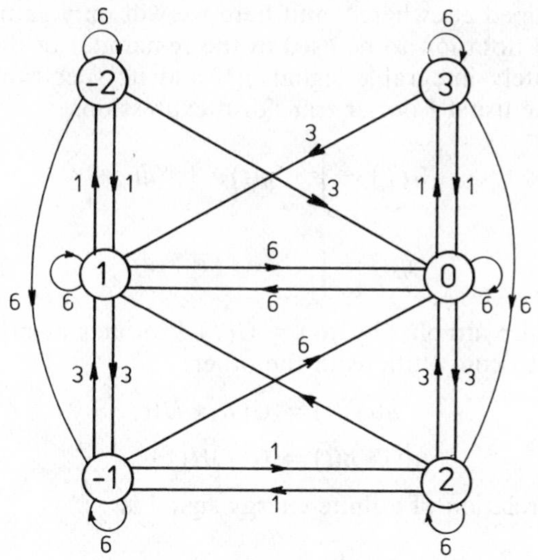

Fig. 14.1 State transition diagram of 4B3T. Transition probability: divide numbers against arcs by 16

the code structure ensures that all digit distributions are alike; the probabilities of 0, + and − respectively are 10/32, 11/32, 11/32.

Let the coded message be transmitted on the line with an elementary pulse waveform $s(t)$ for each digit and an element period T. The signal waveform is a random process

$$X(t) = \sum_{m} a_m s(t - mT) \tag{14.15}$$

where the index m ranges over all digits. Its mean value is (taking a time average)

$$E(X) = E(a_m)\frac{1}{T}\int_{-\infty}^{\infty} s(t)\,dt \tag{14.16}$$

A signal to be conveyed over electrical paths normally has a zero mean. This may be attained either by annulling the integral of $s(t)$, for example by using dipulses, or with greater bandwidth economy by annulling $E(a_n)$, as in 4B3T and many other line codes.

14.4 REVIEW OF SPECTRAL THEORY

The most significant statistic of a communication signal is probably its power spectrum. Fourier analysis of random signals is a topic amply developed elsewhere*, and here we will only summarise the concepts and notation to be used in the remainder of the paper.

An absolutely integrable signal $g(t)$ and its spectrum $G(f)$ are related by the usual Fourier transform expressions

$$G(f) = \int_{-\infty}^{\infty} g(t)\, e^{-j2\pi ft}\, dt \tag{14.17a}$$

$$g(t) = \int_{-\infty}^{\infty} G(f)\, e^{j2\pi ft}\, df \tag{14.17b}$$

which we shall symbolise by $g(t) \Rightarrow G(f)$. Products in either domain transform into convolutions in the other:

$$g(t)h(t) \Rightarrow G(f) * H(t) \tag{14.18a}$$

$$g(t) * h(t) \Rightarrow G(f)H(f) \tag{14.18b}$$

The autocorrelation of a finite-energy signal is

$$R_g(\tau) = \int_{-\infty}^{\infty} g(x)g(x+\tau)\, dx \tag{14.19a}$$

$$= g(t) * g(-t) \tag{14.19b}$$

the second version following on comparison with the usual convolution integral. The Fourier transform of the finite autocorrelation function is, from (14.18b) and (14.19b)

$$R_g(\tau) \Rightarrow G(f)G(-f) = |G(f)|^2 \tag{14.20}$$

namely the energy spectrum. To extend spectral theory to the realisations of a stationary random process, which of course are not integrable, it is necessary to perform a limiting operation on the time average, so that the autocorrelation function of an infinitely extended waveform is

$$R_h(=) = \operatorname*{limit}_{T\to\infty} T^{-1} \int_{-\frac{1}{2}T}^{\frac{1}{2}T} h(x)h(x+\tau)\, dx \tag{14.21}$$

A corresponding operation in the frequency domain gives the power spectrum

$$P_h(f) = \operatorname*{limit}_{T\to\infty} T^{-1} |H(f) * \operatorname{sinc} fT|^2 \tag{14.22}$$

and an extension of the reasoning in eqn. (14.20) gives

$$R_h(\tau) \Rightarrow P_h(f) \tag{14.23}$$

* In many textbooks of communication theory, in some textbooks of probability, including [10] and Chapter 2 of this volume.

For an ergodic random process the autocorrelation

$$R_X(\tau) = E[X(t)X(t+\tau)] \tag{14.24}$$

found by time-averaging as in eqn. (14.21) is by definition equal to the ensemble average. Then the autocorrelation transforms to a power spectrum

$$R_X(\tau) \Rightarrow P_X(f) \tag{14.25}$$

which is the Wiener–Khintchine theorem.

Passage through a linear filter is treated readily in spectral terms. Let a covariance-stationary process $X(t)$ be applied to a filter of impulse response $g(t) \Rightarrow G(f)$. The output process $Y(t)$ has a power spectrum

$$P_Y(f) = P_X(f)|G(f)|^2 \tag{14.26a}$$

and so, by Fourier transformation, an autocorrelation function

$$R_Y(\tau) = R_X(\tau) * R_g(\tau) \tag{14.26b}$$

We shall encounter many random processes which are cyclo-stationary rather than stationary: that is to say, the statistics are periodic with some period T. A useful technique for some purposes is to define an equivalent stationary process whose statistics are those of the cyclo-stationary process averaged over the period T. We shall use this method to define and derive a power spectrum for block-coded line signals. The physical justification is that, if measurements are made without a time reference, these are the statistics which will normally be observed. A practical spectrum analyser comprises a narrow-band filter, tunable over the range of interest, whose output is averaged over a period much longer than the periodic times of the signal. We can use such an instrument to measure the power spectrum of a coded line signal, and the result will be relevant to such problems as estimating the mutual interference between two relatively asynchronous transmission systems*.

The distribution between stationary and cyclo-stationary processes is sometimes glossed over in the technical literature, but it is an important one, and in the following sections we shall indicate clearly where some phase-averaging is required. For a full discussion of this topic, see the book by Franks [12], or the pioneering paper by Bennett [13], who first derived the key results of Section 14.5.

* It is important to remember that there are some problems for which a fuller description and analysis is necessary, e.g. estimating mutual interference between systems with some synchronous relationship, or analysing clock recovery circuits. See also Chapters 2 and 10.

14.5 DIGITAL SIGNALS WITHOUT BLOCK STRUCTURE

We begin by considering the relatively simple case of digital sequences with no block structure, and stationary statistics. This is both a valid model for some practical signals, and a good foundation for the more complex theory in later sections. Let the line signal comprise elementary pulses of waveform $s(t)$, with amplitudes in successive time slots according to a stationary stream of digit values $\{a_n\}$. This can be written as

$$X(t) = \sum_n a_n s(t - nT) \tag{14.27a}$$

$$= s(t) * \sum_n a_n \delta(t - nT) \tag{14.27b}$$

$$\equiv s(t) * Y(t) \tag{14.27c}$$

Now the first term of the convolution is just a fixed linear filter. The second term is a random process $Y(t)$ which conveys the information; if we can define its statistics, then these can be appropriately modified to deal with the filtering operation. The process $Y(t)$ is cyclostationary, but phase-averaging is in this case particularly easy. The phase-averaged mean is an impulse of moment $E(a_n)$ averaged over a time interval T: and this is multiplied by the zero-frequency filter response to obtain

$$E(X) = T^{-1}E(a_n)S(0) \tag{14.28a}$$

$$= T^{-1}E(a_n)\int_{-\infty}^{\infty} s(t)\,dt \tag{14.28b}$$

confirming the time average (14.16). The autocorrelation of $Y(t)$, being the convolution of two impulse trains of interval T, is itself an impulse train of period T, independent of the time origin. So the phase-averaged version is

$$R_Y(\tau) = T^{-1}\sum_k E(a_n a_{n+k})\delta(\tau - kT) \tag{14.29a}$$

$$= T^{-1}\sum_k R_a(k)\delta(\tau - kT) \tag{14.29b}$$

where in the second version we have used the notation $R_a(k)$ for the autocorrelation of a discrete sequence. Fourier transformation gives the power spectrum

$$P_Y(f) = T^{-1}\sum_k R_a(k)\, e^{-j2\pi fkT} \tag{14.30}$$

The statistics of the signal process are, from eqn. (14.26)

$$R_X(\tau) = R_Y(\tau) * R_s(\tau) \tag{14.31a}$$

$$= T^{-1} \sum_k R_a(k) R_s(\tau - kT) \tag{14.31b}$$

and

$$P_X(f) = P_Y(f)|S(f)|^2 \tag{14.32a}$$

$$= |S(f)|^2 T^{-1} \sum_k R_a(k) e^{-j2\pi fkT} \tag{14.32b}$$

The mean power is

$$\int_{-\infty}^{\infty} P_X(f)\, df = R_X(0) \tag{14.33}$$

In the absence of word structure, these functions may take simple forms. We give three examples.

(1) Independent digits. Only $R_a(0)$ is non-zero; the spectrum is simply that of the element waveform, and the mean power is the pulse energy divided by T.
(2) Autocorrelation decreases exponentially with time displacement, say $R_a(k) = \gamma^{|k|}$. This will happen if the a_n are derived from a binary Markov process: discrete Markov process sources of a wider class can produce somewhat similar phenomena with a sum of several exponentially decreasing terms. Substituting in (14.30) gives

$$TP_Y(f) = 1 + \sum_{k=1}^{\infty} (\gamma\, e^{j2\pi fT})^k + \sum_{k=1}^{\infty} (\gamma\, e^{-j2\pi fT})^k$$

$$= \frac{1 - \gamma^2}{1 - 2\gamma \cos 2\pi fT + \gamma^2} \tag{14.34}$$

This is periodic, as follows generally from the fact that such spectra are the Fourier transforms of a row of equally spaced impulses. A low-pass filter function $S(f)$ will select the peak in the baseband region. Some examples are drawn in Fig. 14.2a. The filter power spectrum shown is

$$|S(f)|^2 = \text{sinc}^2 fT$$

appropriate to rectangular pulses of width T. Note that positive correlation, which tends to produce bursts of similar digits, enhances the low-frequency components: negative correlation, which tends to produce alternations, enhances the higher end of the

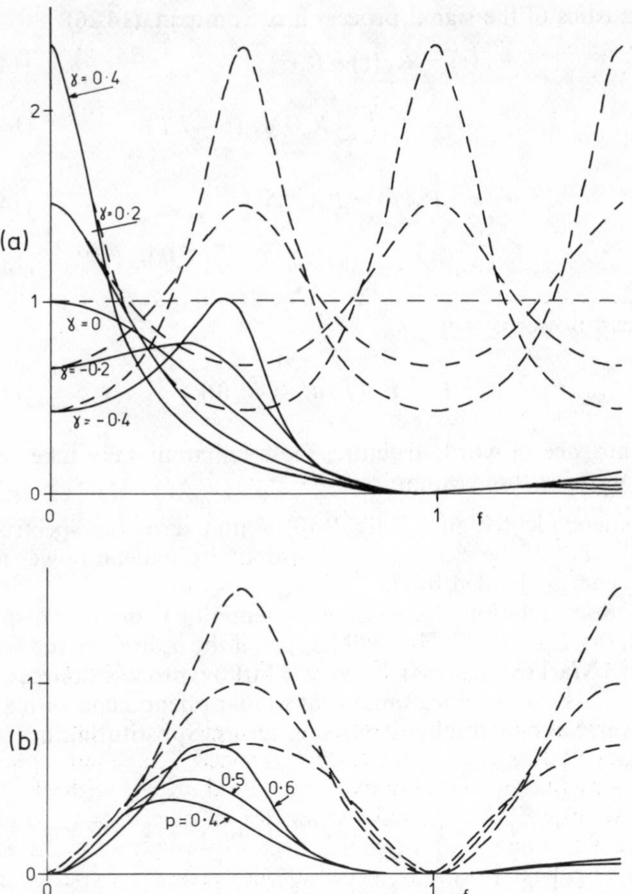

Fig. 14.2 Power spectra of some digital signals: (a) binary, (b) AMI. – – – Py(f), ——— Px(f) with $|S(f)|^2 = \text{sinc}^2 f$

baseband, with a peak somewhwere near the Nyquist frequency $1/2T$.

(3) The well-known alternative code AMI fits our general description of block codes in Sections 1 and 2, but having a block length of unity it can be analysed by the simple technique of this section. Binary digits $\{0, 1\}$ are translated by ternary digits $\{-1, 0, +1\}$: binary 0 is represented as ternary 0, and binary 1 by ternary $+1$ and -1 alternatively. Assuming the binary digits to be independent and equiprobable, the autocorrelation $R_a(k)$ of the ternary sequence can be found as follows. Only ± 1 values contribute, and

$R_a(0) = \frac{1}{2}$. If a_n and a_{n+1} are both of modulus 1, they must be of opposite sign, and $R_a(1) = -\frac{1}{4}$. If a_n and a_{n+k} $(k > 1)$ are both of modulus 1, their product is $+1$ or -1 according as there is an odd or even number of ± 1 digits intervening: these conditions are equiprobable, so $R_a(k) = 0$. Consequently

$$TP_Y(f) = \tfrac{1}{2} - \tfrac{1}{4}\,e^{j2\pi fT} - \tfrac{1}{4}\,e^{-j2\pi fT}$$

$$= \tfrac{1}{2}\{1 - \cos 2\pi fT\} \tag{14.35}$$

If the binary digits are not equiprobable, then $R_a(k)$ does not vanish but decreases exponentially. It can be shown that, for '1' probability p, the power spectrum is

$$TP_Y(f) = \frac{2p(1-p)\{1 - \cos 2\pi fT\}}{1 - 2(1-2p)\cos 2\pi fT + (1-2p)^2} \tag{14.36}$$

The negative correlation shifts the power towards the upper end of the baseband, as in the previous example: but the enforced balance of positive and negative digits ensures that the sero-frequency spectral density is zero, for all source probabilities. This is one of the normal aims of an alternative code.

14.6 BLOCK STRUCTURE: GENERAL THEORY (I)

We have seen in the last section that a digital line signal is cyclo-stationary with period T, but can be assigned equivalent stationary statistics by phase-averaging over the period. Codes with line block length M may have a second periodicity, of period MT; to obtain equivalent stationary statistics it is necessary to phase-average over this longer period. Formally, a continuous average over period MT is all that is required to eliminate both periodicities. Given the previous treatment, however, it is simplest to make use of the known results of continuous averaging over the digit period T, and to supplement this with a further discrete averaging over the M digits of the word.

We revert to the notation of Section 14.2, and denote the nth word in the sequence by a row vector with digits as components

$$\mathbf{a}_n = (a_{n1}, a_{n2}, \ldots, a_{nm}) \tag{14.37}$$

(It is legitimate to think of the a_n in Section 14.5 as one-digit blocks, a concept which we used explicitly in describing AMI.) The general expression for a line signal process is then

$$X(t) = \sum_n \sum_i a_{ni} s[t - (nM + i - 1)T] \tag{14.38a}$$

$$= s(t) * \sum_n \sum_i a_{ni}\delta[t - (nM + i - 1)T] \quad (14.38b)$$

$$= s(t) * Y(t) \quad (14.38c)$$

which reduce to eqn. (14.27) when $M = 1$. Digit means and correlation must be defined separately for each position within the block. For example, the mean of the ith digit is $E(a_{ni})$, and we average these to obtain the overall mean

$$E(X) = (MT)^{-1} \sum_{i=1}^{M} E(a_{ni})S(0) \quad (14.39a)$$

$$= (MT)^{-1} \sum_{i=1}^{M} E(a_{ni}) \int_{-\infty}^{\infty} s(t)\, dt \quad (14.39b)$$

This reduces to eqn. (14.28) when $M = 1$.

The autocorrelation must take into account the block interval, and the position of both elements within a block. We define

$$R_a^{ij}(k) = E(a_{ni}a_{n+k,j}) \quad (14.40)$$

Then the sequence $Y(t)$ has autocorrelation and power spectrum

$$R_Y(\tau) = (MT)^{-1} \sum_k \sum_i \sum_j R_a^{ij}(k)\delta[\tau - (kM + j - i)T] \quad (14.41a)$$

$$P_Y(f) = (MT)^{-1} \sum_k \sum_i \sum_j R_a^{ij}(k) \exp[-j2\pi f T(kM + j - i)] \quad (14.41b)$$

Note that each distinct term in the series is the average of M contributions deriving from different digit positions, in accordance with the phase-averaging principle: and that these equations reduce to (14.29b) and (14.30) respectively, when $M = 1$.

Cariolaro and Tronca [2] use a matrix formulation for the correlations, as follows. Let the $R_a^{ij}(k)$ in eqn. (14.40) be elements of a matrix

$$\mathbf{R}_k = E\{\mathbf{a}_a^T \mathbf{a}_{n+k}\} \quad (14.42)$$

where the superscript T signifies the transpose: and let $\mathbf{V}$ be a row vector

$$\mathbf{V} = (e^{j2\pi fT}, e^{j4\pi fT}, \ldots, e^{j2M\pi fT}) \quad (14.43)$$

with transposed conjugate $\mathbf{V}^*$. Then

$$P_Y(f) = (MT)^{-1} \sum_k \mathbf{V}\mathbf{R}_k\mathbf{V}^* e^{-j2\pi fkMT} \quad (14.44)$$

It can be shown by multiplying out the matrix product that this is identical with (14.41b).

Now it is a conclusion from the general theory of stationary Markov processes that sequences of distributions will normally converge towards a limiting distribution. Let us suppose, therefore, that as k increases $\mathbf{R}_k$ tends to a limit $\mathbf{R}_\infty$ (which may, or may not, be zero). We can separate the sequence $\mathbf{R}_k$ into a constant term $\mathbf{R}_\infty$ and a sequence $\mathbf{R}_k - \mathbf{R}_\infty$ converging to zero. In this notation the power spectrum (14.44) can be written as

$$P_Y(f) = P_Y^c(f) + P_Y^d(f) \tag{14.45a}$$

where

$$\begin{aligned} MTP_Y^c(f) &= \sum_{k=-\infty}^{\infty} \mathbf{V}(\mathbf{R}_k - \mathbf{R}_\infty)\mathbf{V}^* \, e^{-j2\pi fkMT} \\ &= \mathbf{V}(\mathbf{R}_0 - \mathbf{R}_\infty)\mathbf{V}^* \\ &\quad + 2Re \sum_{k=1}^{\infty} \mathbf{V}(\mathbf{R}_k - \mathbf{R}_\infty)\mathbf{V}^* \, e^{-j2\pi fkMT} \end{aligned} \tag{14.45b}$$

and

$$\begin{aligned} MTP_Y^d(f) &= \mathbf{V}\mathbf{R}_\infty\mathbf{V}^* \sum_{k=-\infty}^{\infty} e^{-j2\pi fkMT} \\ &= \mathbf{V}\mathbf{R}_\infty\mathbf{V}^* \sum_{k=-\infty}^{\infty} \delta(f - kMT) \end{aligned} \tag{14.45c}$$

These two components of the spectrum have distinct physical significance. $P_Y^c(f)$ is a continuous spectral density. Our $M = 1$ examples, eqns. (14.34–14.36), are of this class; in each case the correlations converge and the series is summable. Cariolaro and Tronca [2] have shown that the series in (14.45b) converges for all coded digital processes of interest. There is, therefore, always a continuous spectrum, which like our examples is periodic. The other component $P_Y^d(f)$ comprises a series of discrete lines at harmonics of the block frequency $1/MT$: note that in deriving the second form of (14.45c) we have used the theorem that a row of impulses in the time domain transforms into a row of impulses in the frequency domain. These lines will vanish if $\mathbf{R}_\infty = 0$, as in our $M = 1$ examples: their presence implies some persistent feature in the codewords.

It is clear that the remaining problem is to derive the $\mathbf{R}_k$ in a form convenient for evaluation of the spectral functions (14.45). We shall first present an example, practical but with some simplifying features: and then revert to the general theory.

14.7 BLOCK STRUCTURE: AN INTRODUCTORY EXAMPLE

The 4B3T code described in Section 14.3 is a good practical example of block structure, but has some simplifying features. We will exhibit some of its properties in the special case of equiprobable source words b_n. The line words a_n are then drawn equiprobably from the alphabet appropriate to the encoder state. Table 14.1 in Section 14.3 shows that there are two alphabets. Let these be denoted by A_1 and A_2 respectively. The three digit columns for each alphabet have the same proportions of 0, + and − element values, as follows (Table 14.2).

Table 14.2

Proportion of:	A_1	A_2
0	5/16	5/16
+	8/16	3/16
−	3/16	8/16

The expected digit products (14.40) are therefore the same for all pairs of digit positions, so that for $k > 0$ the matrix $\mathbf{R}_k$ consists of M^2 identical entries which we will denote by R_k. Now it is easy to evaluate the expected digit product of random words from given alphabets. Let the proportion of + and − digits in the alphabet from which a_n is drawn be $p_+(n)$ and $p_-(n)$ respectively. Then

$$\begin{aligned} R_k \text{ (given alphabets)} = {} & p_+(n)p_+(n+k) + p_-(n)p_-(n+k) \\ & - p_+(n)p_-(n+k) - p_-(n)p_+(n+k) \end{aligned} \tag{14.46}$$

There are four possible combinations of alphabets in 4B3T, and the corresponding conditional values of R_k are as follows (Table 14.3):

Table 14.3

	$a_n \in A_1$	$a_n \in A_2$
$a_{n+k} \in A_1$	25/256	−25/256
$a_{n+k} \in A_2$	−25/256	25/256

The unconditional R_k is the weighted sum of the conditional values,

$$R_k = \frac{25}{256}\{\text{prob (same alphabets)} - \text{prob (different alphabets)}\}$$

$$\equiv \frac{25}{256}\{P_s^{(k)} - P_D^{(k)}\} \tag{14.47a}$$

Since the alphabets are associated with states—A_1 with the 'negative' states $\{-2, -1, 0\}$ and A_2 with the 'positive' states $\{+1, +2, +3\}$—these probabilities derive from the state transition probabilities. Let us define these sets of states as C_1, C_2 respectively. Let the (i, j) element of the k-step transition matrix $\mathbf{P}^k$ be denoted by $p_{ij}^{(k)}$. Let the system be in equilibrium, so that $\mathbf{p}_n = \boldsymbol{\pi}$. Then the probability that a_n and a_{n+k} are both in the same alphabet is

$$P_s^{(k)} = \sum_{i \in c_1} \sum_{j \in c_1} \pi_i p_{ij}^{(k)} + \sum_{i \in c_2} \sum_{j \in c_2} \pi_i p_{ij}^{(k)} \tag{14.47b}$$

The probability that they are in different alphabets is

$$P_D^{(k)} = \sum_{i \in c_1} \sum_{j \in c_2} \pi_i p_{ij}^{(k)} + \sum_{i \in c_2} \sum_{j \in c_1} \pi_i p_{ij}^{(k)} \tag{14.47c}$$

There is no need to calculate both of these, since clearly their sum is unity. Moreover, in the case of 4B3T the symmetries of $\boldsymbol{\pi}$ and $\mathbf{P}$ ensure that the two summations in each equation are equal. Consequently

$$R_k = \frac{25}{256}\left\{4 \sum_{i \in c_1} \sum_{j \in c_1} \pi_i p_{ij}^{(k)} - 1\right\} \tag{14.48}$$

So the correlations follow directly if we know the powers of the transition matrix, $\mathbf{P}^k$. It is easy to calculate the first few directly: for $k = 1, 2, 3$ the bracketed probability in (14.48) takes the values 1/40, $-3/80$, $-19/1024$ respectively. Moreover, as we shall sketch in Section 14.8, matrix theory gives a general solution for the powers of stochastic matrices. However, we will leave this for the moment, and continue to extract as much as possible by easy methods.

The theory of Section 14.6 makes use of the limiting correlation R_∞, which implies that we need to find the limit of the infinite product $\mathbf{P}^\infty$. Now the physical significance of $\mathbf{P}^\infty$ is simple: it is the transition matrix which turns an arbitrary initial distribution $\mathbf{p}_0$ into the equilibrium distribution $\boldsymbol{\pi}$. So it is a matrix each of whose rows is $\boldsymbol{\pi}$, in this case

$$\mathbf{P}^{\infty} = \frac{1}{30}\begin{bmatrix} 1 & 4 & 10 & 10 & 4 & 1 \\ 1 & 4 & 10 & 10 & 4 & 1 \\ 1 & 4 & 10 & 10 & 4 & 1 \\ 1 & 4 & 10 & 10 & 4 & 1 \\ 1 & 4 & 10 & 10 & 4 & 1 \\ 1 & 4 & 10 & 10 & 4 & 1 \end{bmatrix} \quad (14.49)$$

Application of eqns. (14.47) or (14.48) to this case soon shows that $R_{\infty} = 0$. There are two important conclusions. Firstly, it is verified that the $\mathbf{P}^k$ converge to a limit, which is necessary if the theory of Section 14.6 is to hold. Secondly, it appears that the discrete spectrum (14.45c) vanishes in this case: the spectrum of 4B3T is continuous.

The expression for the continuous spectrum, eqn. (14.45b), contains matrix products which can be simplified in this case. We have found that all entries in $\mathbf{R}_k$ are equal, and that $\mathbf{R}_{\infty} = 0$. Consequently,

$$\mathbf{V}(\mathbf{R}_k - \mathbf{R}_{\infty})\mathbf{V}^* = \mathbf{V}\mathbf{R}_k\mathbf{V}^* = R_k\mathbf{V}\mathbf{U}\mathbf{V}^* \quad (14.50)$$

where $\mathbf{U}$ is an $M \times M$ matrix each of whose elements is unity. Using definition (14.43) and writing $e^{j2\pi fT} = \theta$ for convenience, we have

$$\mathbf{V}\mathbf{U}\mathbf{V}^* = [\theta \quad \theta^2 \quad \cdots \quad \theta^m] \begin{bmatrix} 1 & 1 & \cdots & 1 \\ 1 & 1 & \cdots & 1 \\ \vdots & & & \\ 1 & 1 & \cdots & 1 \end{bmatrix} \begin{bmatrix} \theta^{-1} \\ \theta^{-2} \\ \vdots \\ \theta^{-m} \end{bmatrix} \quad (14.51a)$$

$$= \sum_{i=1}^{M} \sum_{j=1}^{M} \theta^{i-j} \quad (14.51b)$$

$$= \sum_{k=-M+1}^{M-1} (M - |k|)\theta^k \quad (14.51c)$$

The form (14.51b) would also follow directly from eqn. (14.41b). The exception to the foregoing is $\mathbf{R}_0$, which for 4B3T is

$$\mathbf{R}_0 = \frac{1}{16}\begin{bmatrix} 11 & -1 & -1 \\ -1 & 11 & -1 \\ -1 & -1 & 11 \end{bmatrix} \quad (14.52)$$

whence

$$M^{-1}\mathbf{V}\mathbf{R}_0\mathbf{V}^* = \frac{11}{16} - \frac{1}{12}\cos 2\pi fT - \frac{1}{24}\cos 4\pi fT \quad (14.53)$$

Let us sum up what we can deduce about the spectrum of 4B3T from this elementary approach. It is a continuous density, with no discrete lines. It is periodic, and can be represented as a cosine series of which we can calculate the first few terms. The structure will ensure that the spectrum vanishes at zero frequency: however, the fact that the short-range correlation R_1 is positive suggests that low-frequency components will be larger than in AMI. Moreover, there is no one dominant correlation as in AMI, but (even on our present limited evidence) several small terms, which would suggest a broader and flatter spectrum. These statements are borne out by examination of the true spectrum (Fig. 14.3). This graph, taken from [8], was calculated by the exact methods to be outlined in Section 14.8. It shows the spectra of the 4B3T code described here, and of two other block codes of the same length but with different coding rules.

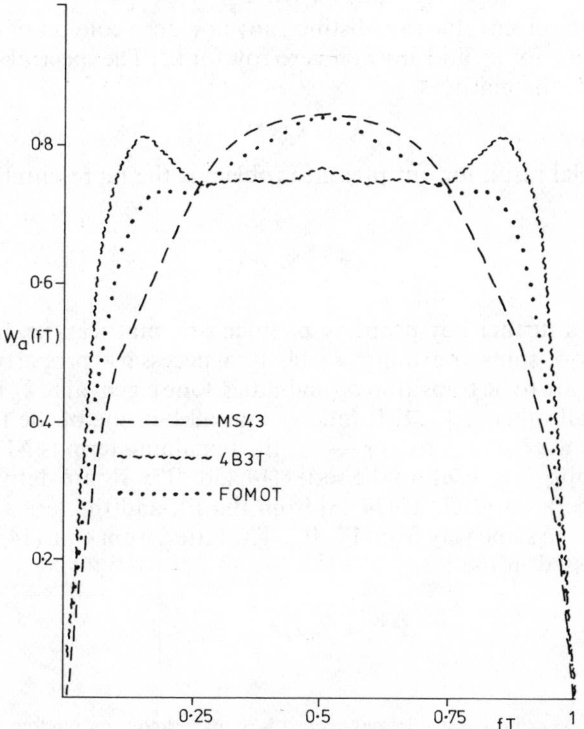

Fig. 14.3 Average power–density spectra for some 4B/3T line codes

14.8 BLOCK STRUCTURE: GENERAL THEORY (II)

It is clear that the matrices $\mathbf{R}_k$ which enter into eqns. (14.45) depend on the codeword table and on the powers $\mathbf{P}^k$ of the state transition matrix. The techniques for calculating the $\mathbf{P}^k$ are our next topic.

The mathematical theory of matrices, and of stochastic matrices in particular, is presented in many textbooks*. The key results are as follows. A non-singular matrix $\mathbf{M}$ can be characterised by its eigenvalues λ_i and eigenvectors $\mathbf{x}_i$, x_i^* which satisfy the equations

$$\mathbf{M}\mathbf{x}_i = \lambda_i \mathbf{x}_i$$
$$\mathbf{x}_i^* \mathbf{M} = \lambda_i \mathbf{x}_i^*$$

The eigenvalues are the roots of the equation

$$|\mathbf{M} - \lambda \mathbf{I}| = 0$$

For any eigenvalue λ_k define the matrix

$$\mathbf{M}^{(k)} = \mathbf{M} - \lambda_k \mathbf{I}$$

Then, if the eigenvalues are distinct, any non-zero column of adj $\mathbf{M}^{(k)}$ is a solution for $\mathbf{x}_k$, and any non-zero row for $\mathbf{x}_k^*$. The spectral set of $\mathbf{M}$ comprises the matrices

$$\mathbf{A}_i = \mathbf{x}_i (\mathbf{x}_i^*)^T$$

The crucial result for our present problem is the representation of a power,

$$\mathbf{M}^r = \sum_i \lambda_i^r \mathbf{A}_i \tag{14.54}$$

There is a further key property of stochastic matrices, i.e. matrices whose row sums are unity which is a necessary property if the elements are to be transition probabilities. One eigenvalue λ_0 is unity: and for all others $|\lambda| < 1$. It follows that all but one of the terms in (14.54) converge to zero as $r \to \infty$: the remaining term is $\mathbf{M}^\infty = \mathbf{A}_0$. Let us apply this result to the series (14.45b). The $\mathbf{R}_k$ are derived by a linear operation such as (14.48) from the $\mathbf{P}^k$, and the terms $\mathbf{R}_k$–$\mathbf{R}_\infty$ derive in the same way from $\mathbf{P}^k$–$\mathbf{P}^\infty$. The latter, from eqn. (14.54), can be expressed in the form

$$\mathbf{P}^k - \mathbf{P}^\infty = \sum_{i \neq 0} \lambda_i^k \mathbf{A}_i, \quad |\lambda_i| < 1 \tag{14.55}$$

* A brief exposition in the present context is in many books on random processes, including [10]. For a more extensive treatment see Broyden [14].

where the $\mathbf{A}_i$ are the spectral set of $\mathbf{P}$. It follows that the series (14.45b) converges. Moreover, it will be the sum of a number of series of the form (14.34), with the λ_i in place of γ. This method therefore gives a complete solution for the power spectrum of a code, in closed form.

The foregoing procedure finds all the $\mathbf{P}^k$, hence all the $\mathbf{R}_k$, whence the series can be summed. Another approach goes directly to the sum of the series. Let $\mathbf{P}$ be a matrix whose powers converge to some $\mathbf{P}^\infty$, as in (14.53). Define the generating function

$$\mathbf{P}(z) = \sum_{k=0}^{\infty} z^k \mathbf{P}^k \tag{14.56}$$

which is clearly a matrix of the same dimensions as $\mathbf{P}$. It is clear that

$$\mathbf{P}(z)(\mathbf{I} - z\mathbf{P}) = \sum_{k=0}^{\infty} z^k \mathbf{P}^k - \sum_{k=1}^{\infty} z^k P^k = \mathbf{I} \tag{14.57a}$$

whence

$$\mathbf{P}(z) = (\mathbf{I} - z\mathbf{P})^{-1} = \frac{\operatorname{adj}(\mathbf{I} - z\mathbf{P})}{|I - z\mathbf{P}|} \tag{14.57b}$$

Cariolaro and Tronca [2] point out that this result can be applied to a series of terms $\mathbf{P}^k$–$\mathbf{P}^\infty$, as required to evaluate eqn. (14.45b). The transition matrix $\mathbf{P}^\infty$ has the property that it converts any initial distribution into π. Consequently, for all k,

$$\mathbf{P}^k\mathbf{P}^\infty = \mathbf{P}^\infty \tag{14.58a}$$

whence

$$(\mathbf{P} - \mathbf{P}^\infty)^k = \mathbf{P}^k\text{–}\mathbf{P}^\infty, \quad k > 0 \tag{14.58b}$$

Using eqns. (14.57b) and (14.58b) gives

$$\sum_{k=0}^{\infty} z^k(\mathbf{P}^k\text{–}\mathbf{P}^\infty) = \sum_{k=0}^{\infty} z^k(\mathbf{P} - \mathbf{P}^\infty)^k - \mathbf{P}^\infty$$
$$= [\mathbf{I} - z(\mathbf{P} - \mathbf{P}^\infty)]^{-1} - \mathbf{P}^\infty \tag{14.59a}$$

Cariolaro and Tronca derive an equivalent form

$$(\mathbf{I} - \mathbf{P}^\infty)[\mathbf{I} - z(\mathbf{P} - \mathbf{P}^\infty)]^{-1} \tag{14.59b}$$

by further application of eqn. (14.58a).

Finally, it remains to be added that the operation of deriving correlations $\mathbf{R}_k$ from transitions $\mathbf{P}^k$ (which we have discussed only in connection with a specific example in Section 14.7) can be given a general formulation in terms of matrix algebra. Cariolaro and Tronca [2] use a matrix $\mathbf{A}$ partitioned into sub-matrices $\mathbf{A}_u$ whose rows are code words α_{iu}, a set of matrices $\mathbf{E}_u$ to define state transitions due to the input words β_u, a diagonal matrix $\mathbf{D}$ whose non-zero elements are the equilibrium probabilities π_i, and probabilities q_u for the input words β_u. The correlation matrices are then

$$\mathbf{R}_0 = \sum_u q_u \mathbf{A}_u^T \mathbf{D} \mathbf{A}_u \tag{14.60a}$$

$$\mathbf{R}_k = \left\{ \sum_u q_u \mathbf{E}_u^T \mathbf{D} \mathbf{A}_u \right\}^T \mathbf{P}^{k-1} \left\{ \sum_u q_u \mathbf{A}_u \right\} \tag{14.60b}$$

For further explanation, the reader should consult the original papers [1, 2] which are the major primary sources for this topic*. Reference [2] gives the more complete theoretical development, while [1] has additional examples including spectral formulae for 4B3T and MS43 codes. Chapter 15 of this volume gives numerous examples including all the line codes in current use.

14.9 OTHER STATISTICS FOR DIGITAL LINE SIGNALS

Our analysis in Sections 14.4–14.8 has concentrated on the power spectrum of digital line signals, this being both an important and a complex problem. There are several other statistical properties which we notice briefly.

(1) Intersymbol interference. Estimation of mean squared isi requires knowledge of the correlation between digits over as wide a range as is necessary to encompass the element waveforms. This emerges in the course of the spectral analysis, which is based on autocorrelation. Munoz [8] has shown that the effect of differences in correlation between various codes is small compared with the effect of equalising tolerances on the pulse waveform.

(2) Timing jitter. The statistics of jitter in a line system carrying independent random digits are well known [15]. The effect of digit correlations is complicated by the passage of signals through a non-linear device (typically squarelaw): the analysis requires the use of statistics up to fourth order, which can be derived by extension of the methods used for spectral analysis [8, 16].

(3) Digit pattern recurrence. Block codes require the receiver to be in block alignment, and if this is lost it must be regained by scanning the received signal for features which indicate misalignment, such as invalid words, states or state sequences. The interval between such events is a random variable whose probability distribution can be calculated: like the other statistics, it depends

* It may be noticed that the equation for $\mathbf{R}_k$ contains a factor $\mathbf{P}^{k-1}$ whereas we have in the present paper associated $\mathbf{R}_k$ with $\mathbf{P}^k$. The explanation is that the matrices $\mathbf{E}_u$ account for one transition.

both on code structure and on state transition probabilities. Similar problems arise in error monitoring, in frame alignment [17], and in the imposition of ancillary channels on a primary channel by means of modified 'signal opportunity' patterns [18].

REFERENCES

1. Cariolaro, G. L. and Tronca, G. P., 'Correlation and spectral properties of multi-level (M, N) coded digital signals with applications to pseudo-ternary (4, 3) codes', *Alta Frequenza*, **43**, 2–15 (1974)
2. Cariolaro, G. L. and Tronca, G. P., 'Spectra of block coded digital signals', *Trans. IEEE*, **COM-22**, 1555–1564 (1974)
3. Barker, R. H., British Patent 706,687
4. Aaron, M. R., 'PCM transmission in the exchange plant', *Bell. Syst. Tech. J.*, **41**, 99–141 (1962)
5. Cattermole, K. W., *Principles of Pulse Code Modulation*, Iliffe, London, pp. 379–381 (1969)
6. Jessop, A. and Waters, D. B., '4B3T, an efficient code for PCM coaxial line', *18th Int. Congress on Electronics*, Rome (1970)
7. Munoz Rodriguez, D. and Cattermole, K. W., 'New codes for digital transmission', *Electron. Lett.*, **13**, 340–342 (1977)
8. Munoz Rodriguez, D., 'Digital communication: relationship between signal design and system performance', Ph.D. Thesis, University of Essex (1979)
9. Franaczek, P. A., 'Sequence state coding for digital transmission', *Bell Syst. Tech. J.*, **47**, 143–157 (1968)
10. Cox, D. R. and Miller, H. D., 'The theory of stochastic processes', Methuen, London (1965)
11. Bates, R. J. S., 'Topics in the design of digital transmission systems', Ph.D. Thesis, University of Cambridge (1980)
12. Franks, L. E., *Signal Theory*, Prentice-Hall (1969)
13. Bennett, W. R., 'Statistics of regenerative digital transmission', *Bell Syst. Tech. J.*, **37**, 1501–1542 (1958)
14. Broyden, C. G., *Basic Matrices*, Macmillan, London (1975)
15. Bylanski, P. and Ingram, D. W. G., *Digital Transmission Systems*, Peter Peregrinus, London, pp. 190–209 (1976)
16. Munoz Rodriguez, D. and Cattermole, K. W., 'Time jitter in self-timed regenerative repeaters with correlated transmitted symbols', *Electron. Circuits Syst.*, **3**, 109–115 (1979)
17. Bylanski, P. and Ingram, D. W. G., *Digital Transmission Systems*, Peter Peregrinus, London, pp. 88–106 (1976)
18. Cattermole, K. W. and Grover, W. D., 'Optical line codes bearing ancillary channels', *IEEE Colloquium on Data Transmission Codes*, November (1980)

15

Power Spectral Density Characterisation of Digital Transmission Codes

G. S. Poo

15.1 INTRODUCTION

Various line codes have been devised for the transmission of binary information over digital line and optical fibre systems, see, for example, [1, 2]. The power spectral density (PSD) of a code is an important property, providing information concerning dc and low frequency content, average power, and bandwidth. Such information is of interest when designing a data system in that it facilitates a comparison of different pulse codes, aiding the selection of an appropriate code for a given system or environment. Broadly speaking, knowledge of the PSD helps in defining amplitude requirements and eases spectrum planning and crosstalk analyses. The calculation of the PSD for a line code is, however, a laborious and lengthy process; the spectrum depends on the coding rules and on the statistics of the input symbol sequence. The primary purpose of the present paper is to provide a detailed characterisation of the spectral properties of a large number of commonly encountered codes. Herein, results relating to some thirteen different codes are presented, a rather more extensive collection than has previously been available.

The code spectra presented are derived following the method of Cariolaro and Tronca [3] with the aid of a suite of computer programs described previously [4]. These programs considerably ease the task of obtaining spectral characterisations. Codes with bounded total disparity (finite digital sum) are considered directly while block codes with unbounded total disparity are studied by way of a successive-approximation method [5].

15.2 COMPUTATIONAL PROCEDURE

A tutorial exposition of the rather complex procedures involved in deriving a spectral characterisation for a line code is provided in Chapter 14 of this volume. Here we will simply provide a summary of the algorithm [3] employed in the computer programs [4] and describe a successive approximation procedure which allows this algorithm to be used to obtain a spectral characterisation for unbounded codes.

15.2.1 Summary of algorithm

The system under consideration is composed of an input source, an encoder and a generated sequence. The input source consists of an infinite string of random binary words. Each binary word of length M is coded into an L-ary codeword of length N. In general the mapping is many-folded, M, $N > 1$.

The normalised power spectral density of the coded signal can be written as

$$\frac{W(f)T}{|G(f)|^2} = \mathrm{PSD}_{\mathrm{cont}} + \mathrm{PSD}_{\mathrm{discrete}} \tag{15.1}$$

where the continuous component gives

$$\mathrm{PSD}_{\mathrm{cont}} = [X_1(f) + X_2(f)]/N \tag{15.2}$$

and the discrete component

$$\mathrm{PSD}_{\mathrm{discrete}}\, T = N^{-2} X_d(f) \sum_{k=-\infty}^{\infty} \delta\left(f - \frac{k}{NT}\right) \tag{15.3}$$

The scalar functions X_1, X_2 and X_d contain all the correlations. These functions can be expressed in an attractive cosine form as follows:

$$X_1(f) = \sum_{k=0}^{N-1} \varepsilon_k v_k \cos k\theta \tag{15.4}$$

$$X_2(f) = \frac{\sum_{k=0}^{(I+1)N} \varepsilon_k N_k \cos k\theta}{\sum_{k=0}^{I-1} \varepsilon_k D_k \cos kN\theta} \tag{15.5}$$

$$X_d(f) = \sum_{k=0}^{N-1} \varepsilon_k \mu_k \cos k\theta \tag{15.6}$$

where

$\varepsilon_k = 1$ for $k = 0$ and $\varepsilon = 2$ for $k > 1$

$$\theta = 2\pi f T$$

$$v_k = \sum_{i=1}^{N-k} [\mathbf{R}_0(i,i+k) - \mathbf{R}_\infty(i,i+k)]$$

$$\mu_k = \sum_{i=1}^{N-k} R_\infty(i, i+k)$$

$$N_k = N_{p+qN} = \sum_{h=0}^{I-q} [d_{I-h} n_{(h+q)N+p} + d_{I-h-q} n_{hN-p}]$$

$$n_s = n_{p+kN} = \sum_{i=1}^{N-p} \mathbf{H}_k(i+p,i) + \sum_{i=1}^{p} \mathbf{H}_{k+1}(i,i+N-p)$$

$$D_k = \sum_{l=1}^{I-k} d_{I-l} d_{I-l-k}$$

The reader is referred to Cariolaro and Tronca [3] for the semantics of the notations. The statistics of the code are contained in the matrices $\mathbf{R}_0$, $\mathbf{R}_\infty$, $\mathbf{d}_k$, $\mathbf{G}_k$ and $\mathbf{H}_k$. In terms of Markov transition matrices π, matrices of encoder function $\mathbf{E}$ and codeword matrices $\mathbf{A}$, the codeword correlation matrices

$$\mathbf{R}_k = \mathbf{E}\{a_n^T a_{n+k}\} \tag{15.7}$$

can be analysed to yield

$$\mathbf{R}_0 = \sum_{u=1}^{K} \mathbf{q}_u \mathbf{A}_u^T \mathbf{D}_\infty \mathbf{A}_u$$

$$\mathbf{R}_k = \mathbf{C}_1^T \pi^{k-1} \mathbf{C}_2 \quad (k > 1)$$

$$\mathbf{R}_\infty = \mathbf{C}_1^T \pi_\infty \mathbf{C}_2$$

$$\mathbf{C}_1 = \sum_{u=1}^{K} \mathbf{q}_u \mathbf{E}_u^T \mathbf{D}_\infty \mathbf{A}_u$$

$$\mathbf{C}_2 = \sum_{u=1}^{K} \mathbf{q}_u \mathbf{A}_u$$

$$\mathbf{K} = 2^M$$

and

$$\mathbf{q}_u = \mathbf{p}^{M_u}(1-p)^{M-M_u},$$
$$= 2^{-M} \quad \text{for random binary input}$$

The limiting matrices π_∞ and $\mathbf{D}_\infty$ are obtained by solving the system of linear equations

$$\mathbf{p}^\infty = \pi^T \mathbf{p}^\infty \tag{15.8}$$

$$\sum_{i=1}^{I} p_i^{\infty} = 1 \tag{15.9}$$

The coefficients d_k and the matrix coefficients $\mathbf{G}_k$ are related by the recursive equations:

$$d_k = -k^{-1}\,\mathrm{tr}[\mathbf{F}\mathbf{G}_{k-1}] \tag{15.10}$$

$$\mathbf{G}_k = \mathbf{F}\mathbf{G}_{k-1} + d_k\mathbf{U} \quad (k = 1,2,\ldots,I), \tag{15.11}$$

where $\mathbf{F} = \boldsymbol{\pi} - \boldsymbol{\pi}_{\infty}$. Consequently, the matrices $\mathbf{H}_k$ can be computed from

$$\mathbf{H}_k = \mathbf{C}_1^T(\mathbf{U} - \boldsymbol{\pi}_{\infty})\mathbf{G}_{I-k-1}C_2 \quad (k = 0, 1, \ldots, I-1) \tag{15.12}$$

The above analysis is based on the Mealy model in which the output codeword is dependent on the input word, and on the state of the encoder. Hence, encoder state is the key of the analysis. Nevertheless, the method is also applicable to adaptive codes where the output codeword is not only dependent on the input word and history, but also requires some degree of foresight. In this case, it is more appropriate to analyse the code by states of the codeword. The above algorithm could be interpreted by setting $k = 1$ and $\mathbf{E}_1 = \boldsymbol{\pi}$ [6].

The above procedure is well suited to implementation on a digital computer if the code disparity is bounded, in which case the encoder state diagram has a finite number of states. This is not the case for all codes of practical interest, however. For example, the 6B4T and 3B2T block codes do not possess deterministically bounded digital sums; the number of possible states is thus unlimited and the problem appears intractable. However, not all possible states are equally probable. Indeed, for an essentially random input data sequence the probability distribution of the total disparity decreases exponentially with the modulus of the digital sum [8]. Such codes are referred to as statistically bounded in that the probability that the modulus of the digital sum will exceed some large value B becomes vanishingly small with increasing B. This property makes it possible to adopt a successive approximation method for the spectral analysis of such codes: the occupancy probability for the outer (large modulus) states is very low and elimination of these states may thus produce only small errors. To apply the method one begins the calculation with a basic number of states, the number being chosen to encapsulate the essential features of the central portion of the encoder state diagram. The transitions at the extreme ends are to be truncated, corresponding to the elimination of the tail portions of the probability distribution of total disparity. The error introduced by this truncation is assessed by performing a sequence of code spectrum calculations

Table 15.1. MAIN CHARACTERISTICS OF BLOCK CODES

Codes	*Alphabets*	*Word disparities*	*Encoder states*	*Authorised words*	*Prohibited words*	*Code efficiency (%)*
5B6B	2	3: −2, 0, +2	2: −2, 0	44: 20 disparity 0 12 disparity +2 12 disparity −2	6: + + + + − − − + + + + − + inverses	83.33
7B8B	2	5: −4, −2, 0, 2, 4	4: −4, −2, 0, 2	189: 67 disparity 0 48 disparity 2 48 disparity −2 13 disparity 4 13 disparity −4	49: 3 disparity 0 8 disparity 2 8 disparity −2 15 disparity 4 15 disparity −4	87.5
CMI (1B2B type)	2	3: −2, 0, 2	2: −2, 0	3: 1 disparity 0 1 disparity 2 1 disparity −2	1: + −	50.00
4B3T-1	2	7: −3, −2, −1, 0, 1, 2, 3	6: −3, −2, −1, 0, 1, 2	26: 6 disparity 0 6 disparity 1 6 disparity −1 3 disparity 2 3 disparity −2 1 disparity 3 1 disparity −3	1: 000	84.12
4B3T-2	2	5: −2, −1, 0, 1, 2	4: −2, −1, 0, 1	25: 7 disparity 0 6 disparity 1 6 disparity −1 3 disparity 2 3 disparity −2	2: + + + − − −	84.12

AMI (1B1T)	2	3: −1, 0, 1	2: −1, 0	3: 1 disparity 0 1 disparity 1 1 disparity −1			63.09
MS43	3	5: −2, −1, 0, 1, 2	4: −2, −1, 0, 1	26: 6 disparity 0 6 disparity 1 6 disparity −1 3 disparity 2 3 disparity −2 1 disparity 3 1 disparity −3	1: 000		84.12
FOMOT	4	5: −2, −1, 0, 1, 2	4: −2, −1, 0, 1	26: 6 disparity 0 6 disparity 1 6 disparity −1 3 disparity 2 3 disparity −2 1 disparity 3 1 disparity −3	1: 000		84.12
3B2T		5: −2, −1, 0, 1, 2	Unlimited				
6B4T		7: −3, −2, −1, 0 1, 2, 3	Unlimited				

increasing the number of states at each step and comparing the values obtained from successive computations. The procedure is terminated when convergence occurs and the successive values differ by less than some acceptably small amount. Three points deserve special mention:

(1) The results obtained have been found to be insensitive to the actual form adopted for truncation of the encoder states.
(2) Rapid convergence is obtained in the high frequency region of the spectrum but in the important low frequency region convergence can be slow for some codes. Much depends, of course, on the spread of the tails of the probability distribution of total disparity.
(3) Truncation of the state diagram does not constitute an approximation if we are interested in a practical encoder rather than an abstract encoding rule. Any practical encoder will be a finite state machine and the number of possible encoder states will be limited by the capacity of the counter used to record the accumulated digital sum. If the truncation adopted in the analysis corresponds to that of the physical encoder no approximation is involved. Indeed, given that spectrum control can be one of the motives for line encoding the successive approximation procedure can be used to determine the complexity required in a practical realisation of the coder if the requisite spectral characteristics are to be obtained to some desired precision.

15.3 ANALYSIS OF SPECIFIC LINE CODES

Some 13 different block codes have been investigated; their main characteristics are summarised in Table 15.1. The codes fall into four main groups, as noted below.

15.3.1 Zero mean disparity codes

This category covers a large number of balanced codes which possess a common property, i.e. 'zero mean disparity'. The disparity of a codeword is defined as the algebraic digital sum at the end of a word. Common coding rules apply. The code consists of codewords with zero and finite disparities arranged in two alphabets. The zero disparity words are uniquely allocated to input words. Other codewords are allocated in pairs of equal and opposite disparity. During transmission, the code converter selects whichever word brings the cumulative line disparity nearest to zero. This ensures that

Table 15.2. TRANSLATION TABLE 5B6B

Input binary	*Transmitted binary*	
11000	110001	
10100	101001	
10010	100101	
10001	100011	
01100	011001	
01010	010101	
01001	010011	
00110	001101	
00101	001011	
00011	000111	
00111	001110	
01011	010110	
01101	011010	
01110	011100	
10011	100110	
10101	101010	
10110	101100	
11001	110010	
11010	110100	
11111	111010	000101
11110	110110	001001
11101	101110	010001
11011	111001	000110
10111	110101	001010
01111	101101	010010
10000	011101	100010
01000	110011	001100
00100	101011	010100
00010	011011	100100
00001	100111	011000
00000	010111	101000
	Used alternately	

the transmitted code has zero dc content. The total line disparity (i.e. the digital sum variation) is bounded.

Two classes of line codes are included in this category: Binary/Binary and Binary/Ternary codes. Three examples of each class are examined. These are 5B6B, 7B8B, CMI and 4B3T-1, 4B3T-2 and AMI. The codeword dictionaries obtained from the literature [1, 9, 10] are reproduced in Tables 15.2–15.7. The matrices of encoder function are determined by the codeword dictionary and the disparity values. The codeword matrices are obtained by arranging the codewords in appropriate states. The encoder state diagrams are illustrated in Fig. 15.1.

Table 15.3. CODEWORD DICTIONARY FOR 7B8B CODE

	Input binary	*Codeword*
1	0000111	00010111
2	0001011	00011011
3	0001101	00011101
4	0001110	00011110
5	0010101	00100111
6	0011001	00101011
7	0011011	00101101
8	0011100	00101110
9	0011111	00110011
10	0100001	00110101
11	0100010	00110110
12	0100100	00111001
13	0100101	00111010
14	0100110	00111100
15	0101100	01000111
16	0110000	01001011
17	0110010	01001101
18	0110011	01001110
19	0110110	01010011
20	0000000	01010101
21	0111000	01010110
22	0111010	01011001
23	0111011	01011010
24	0111100	01011100
25	0111111	01100011
26	1000001	01100101
27	1000010	01100110
28	1000100	01101001
29	1000101	01101010
30	1000110	01101100
31	1000111	01110001
32	1001000	01110010
33	1001001	01110100
34	1001010	01111000
35	1001111	10000111
36	1010011	10001011
37	1010101	10001101
38	1010110	10001110
39	1011001	10010011
40	1011011	10010101
41	1011100	10010110
42	1011110	10011001
43	1011111	10011010
44	1100000	10011100
45	1100011	10100011
46	1100101	10100101
47	1100110	10100110
48	1101000	10101001
49	1101001	10101100

Table 15.3 *(continued)*

	Input binary	*Codeword*	
50	1101010	10110001	
51	1101011	10110010	
52	1101100	10110100	
53	1101101	10111000	
54	1110000	11000011	
55	1110010	11000101	
56	1110011	11000110	
57	1110101	11001001	
58	1110110	11001010	
59	1110111	11001100	
60	1111000	11010001	
61	1111001	11010010	
62	1111010	11010100	
63	1111011	11011000	
64	1111100	11100001	
65	1111101	11100010	
66	1111110	11100100	
67	1111111	11101000	
68	0000011	11101100	00010011
69	0000101	11101010	00010101
70	0000110	11101001	00010110
71	0001001	11100110	00011001
72	0001010	11100101	00011010
73	0001100	11100011	00011100
74	0010001	11011100	00100011
75	0010011	11011010	00100101
76	0010100	11011001	00100110
77	0010111	11010110	00101001
78	0011000	11010101	00101010
79	0011010	11010011	00101100
80	0011101	11001110	00110001
81	0011110	11001101	00110010
82	0100000	11001011	00110100
83	0100011	11000111	00111000
84	0101000	10111100	01000011
85	0101010	10111010	01000101
86	0101011	10111001	01000110
87	0101110	10110110	01001001
88	0101111	10110101	01001010
89	0110001	10110011	01001100
90	0110100	10101110	01010001
91	0110101	10101101	01010010
92	0110111	10101011	01010100
93	0111001	10100111	01011000
94	0111101	10011110	01100001
95	0111110	10011101	01100010
96	1000000	10011011	01100100
97	1000011	10010111	01101000
98	1001011	01111100	10000011

Table 15.3 *(continued)*

	Input binary	*Codeword*	
99	1001101	01111010	10000101
100	1001110	01111001	10000110
101	1010001	01110110	10001001
102	1010010	01110101	10001010
103	1010100	01110011	10001100
104	1010111	01101110	10010001
105	1011000	01101101	10010010
106	1011010	01101011	10010100
107	1011101	01100111	10011000
108	1100001	01011110	10100001
109	1100010	01011101	10100010
110	1100100	01011011	10100100
111	1100111	01010111	10101000
112	1101110	00111110	11000001
113	1101111	00111101	11000010
114	1110001	00111011	11000100
115	1110100	00110111	11001000
116	0000001	11101110	00010001
117	0001111	11011110	00100001
118	0000010	11101101	00010010
119	0010000	11011101	00100010
120	0100111	10111101	01000010
121	0000100	11101011	00010100
122	0010010	11011011	00100100
123	0101001	10111011	01000100
124	1001100	01111011	10000100
125	0001000	11100111	00011000
126	0010110	11010111	00101000
127	0101101	10110111	01001000
128	1010000	01110111	10001000

Table 15.4. TRANSLATION TABLE FOR CMI

	Codeword	
Binary	*−2*	*0*
0	− +	− +
1	+ +	− −

Table 15.5. TRANSLATION TABLE FOR 4B3T-1

	Ternary transmitted when total disparity is:	
Binary	*Negative*	*Positive*
0000	+ 0 −	+ 0 −
0001	− + 0	− + 0
0010	0 − +	0 − +
0011	+ − 0	+ − 0
0100	+ + 0	− − 0
0101	0 + +	0 − −
0110	+ 0 +	− 0 −
0111	+ + +	− − −
1000	+ + −	− − +
1001	− + +	+ − −
1010	+ − +	− + −
1011	+ 0 0	− 0 0
1100	0 + 0	0 − 0
1101	0 0 +	0 0 −
1110	0 + −	0 + −
1111	− 0 +	− 0 +

Table 15.6. TRANSLATION TABLE FOR 4B3T-2

	Ternary transmitted when total disparity is:	
Binary	*Negative*	*Positive*
0000	+ 0 −	+ 0 −
0001	− + 0	+ 0 −
0010	0 − +	0 − +
0011	+ − 0	+ − 0
0100	+ + 0	− − 0
0101	0 + +	0 − −
0110	+ 0 +	− 0 −
0111	0 0 0	0 0 0
1000	+ + −	− − +
1001	− + +	+ − −
1010	+ − +	− + −
1011	+ 0 0	− 0 0
1100	0 + 0	0 − 0
1101	0 0 +	0 0 −
1110	0 + −	0 + −
1111	− 0 +	− 0 +

Table 15.7. TRANSLATION TABLE FOR AMI

	Codeword	
Binary	*−1*	*0*
0	0	0
1	1	−1

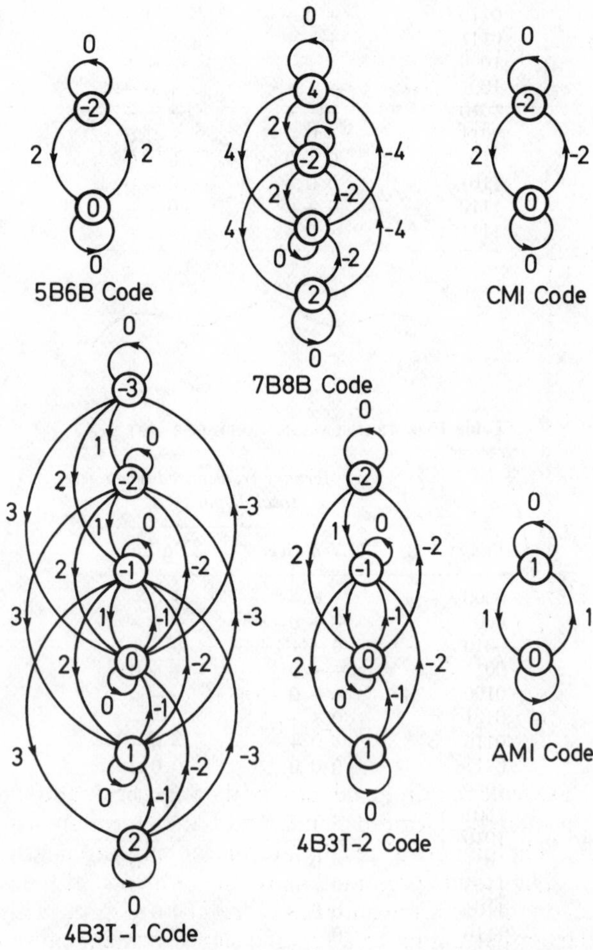

Fig. 15.1 Encoder state diagram of 2-alphabet zero mean disparity codes

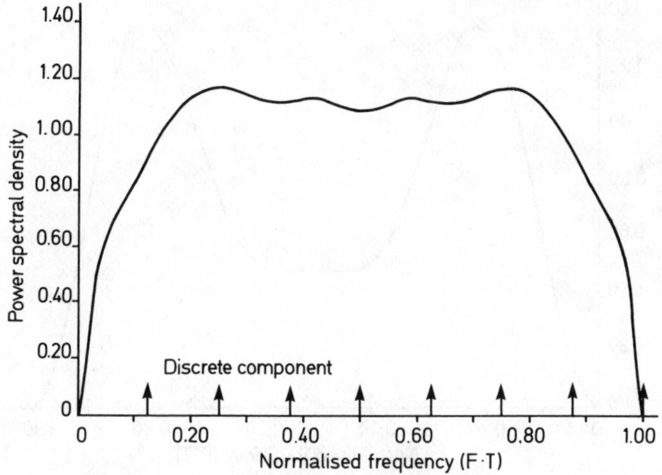

Fig. 15.2 Power spectral density for 7B8B code

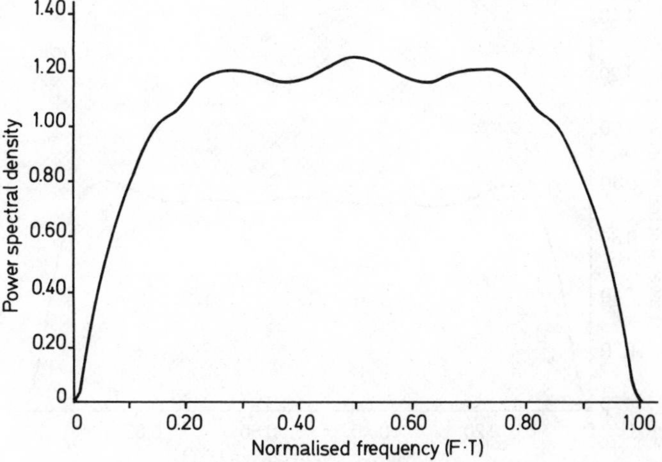

Fig. 15.3 Power spectral density for 5B6B code

The PSD calculation produces both analytical and numerical results, the analytical formulae in closed cosine form are listed in the Appendix. The numerical results are presented graphically in Figs. 15.2–15.7. The results presented show the influence of the encoding operation on the spectrum and thus correspond to impulse signalling. The spectrum pertaining to a specific signal element pulse shape is obtained by multiplying the spectra presented by the energy

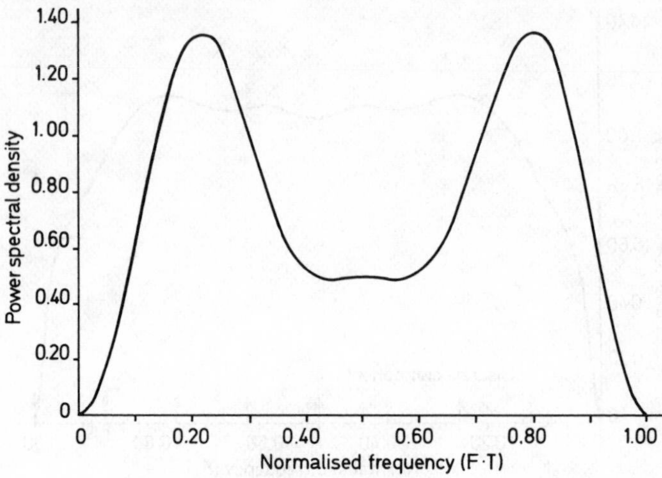

Fig. 15.4 Power spectral density for CMI code

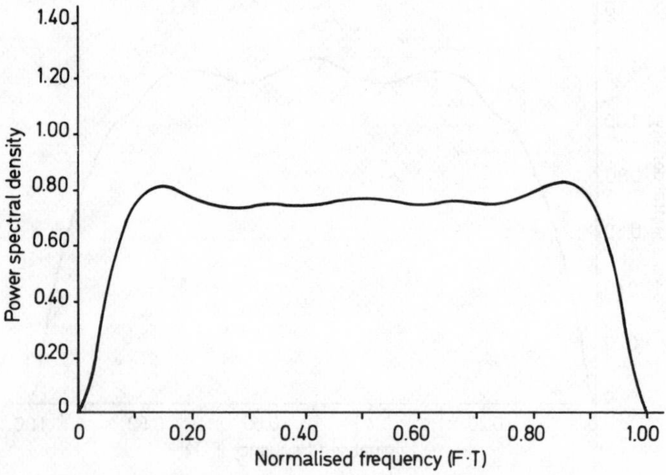

Fig. 15.5 Power spectral density for 4B3T-1 code

density spectrum of the signal element. As one would expect, these curves are dc free. The line spectrum vanishes for most of the codes, although a small discrete component is detected for the 7B8B code. This is probably caused by the non-standard selection of codewords for the translation table [8] which in this case does not retain the complete cyclosymmetry.

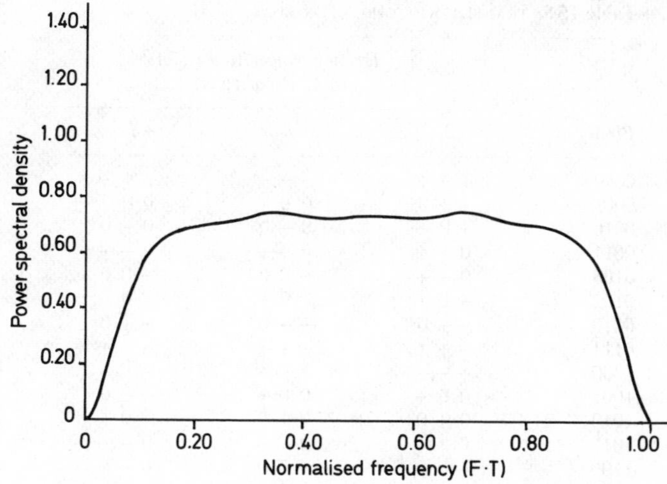

Fig. 15.6 Power spectral density for 4B3T-2 code

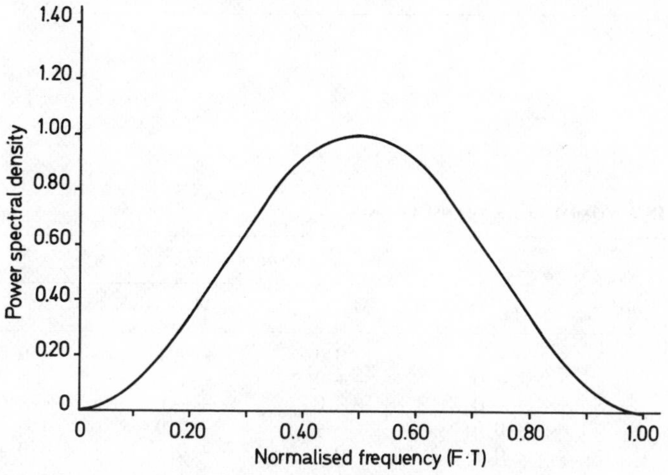

Fig. 15.7 Power spectral density for AMI code

15.3.2 General ternary *x*B*y*T codes

The first category was restricted to two alphabet codes; this is generalised here to include three or more alphabet codes. In this category the codewords are usually well defined in all alphabets and possibly in all states. The codeword selection is governed by the input word as well as the status of the encoder state.

Table 15.8. TRANSLATION TABLE FOR MS43

Binary	Ternary transmitted when total disparity is: −2	−1 or 0	+1
0000	+ + +	− + −	− + −
0001	+ + 0	0 0 −	0 0 −
0010	+ 0 +	0 − 0	0 − 0
0011	0 − +	0 − +	0 − +
0100	0 + +	− 0 0	− 0 0
0101	− 0 +	− 0 +	− 0 +
0110	− + 0	− + 0	− + 0
0111	− + +	− + +	− − +
1000	+ − +	+ − +	− − −
1001	0 0 +	0 0 +	− − 0
1010	0 + 0	0 + 0	− 0 −
1011	0 + −	0 + −	0 + −
1100	+ 0 0	+ 0 0	0 − −
1101	+ 0 −	+ 0 −	+ 0 −
1110	+ − 0	+ − 0	+ − 0
1111	+ + −	+ − −	+ − −

Table 15.9. FOMOT CODE TRANSLATION

Binary word	Ternary word M_1	M_2	M_3	M_4
0000	− + +	− 0 0	− + +	− 0 0
0001	− + 0	− + 0	− + 0	− + 0
0010	+ − 0	+ − 0	+ − 0	+ − 0
0011	+ 0 0	+ − −	+ 0 0	+ − −
0100	− 0 +	− 0 +	− 0 +	− 0 +
0101	+ + +	− + −	− + −	− + −
0110	+ 0 +	+ 0 +	− 0 −	− 0 −
0111	+ 0 −	+ 0 −	+ 0 −	+ 0 −
1000	0 + +	0 + +	− − 0	− − 0
1001	0 + 0	0 − 0	0 + 0	0 − 0
1010	+ − +	+ − +	+ − +	− − −
1011	+ + 0	+ + 0	0 − −	0 − −
1100	0 0 +	− − +	0 0 +	− − +
1101	0 + −	0 + −	0 + −	0 + −
1110	0 − +	0 − +	0 − +	0 − +
1111	+ + −	0 0 −	+ + −	0 0 −

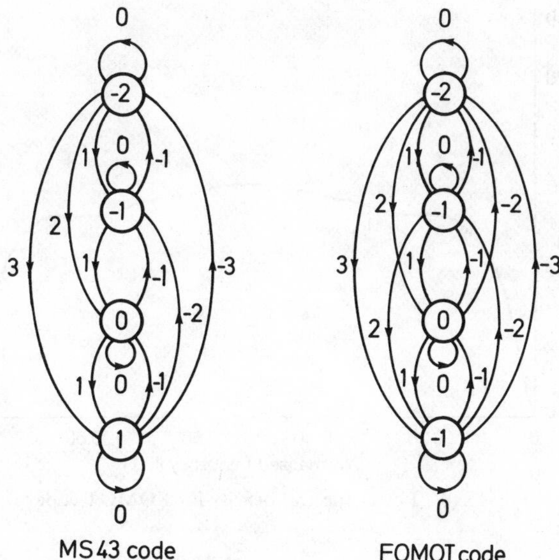

Fig. 15.8 Encoder state diagram for MS43 and FOMOT codes

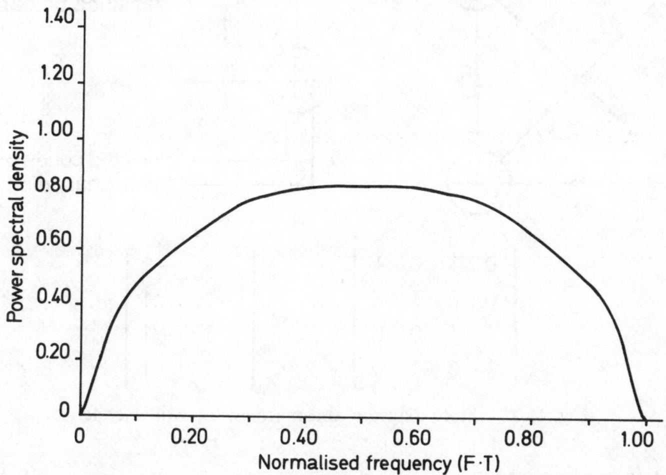

Fig. 15.9 Power spectral density for MS43 code

Two common examples, MS43 and FOMOT codes, are examined. The codeword dictionaries are defined in Tables 15.8 and 15.9 [10]. Both codes are variants of 4B3T code. State diagrams for these codes differ only slightly, as shown in Fig. 15.8; the corresponding code spectra are presented in Figs. 15.9 and 15.10.

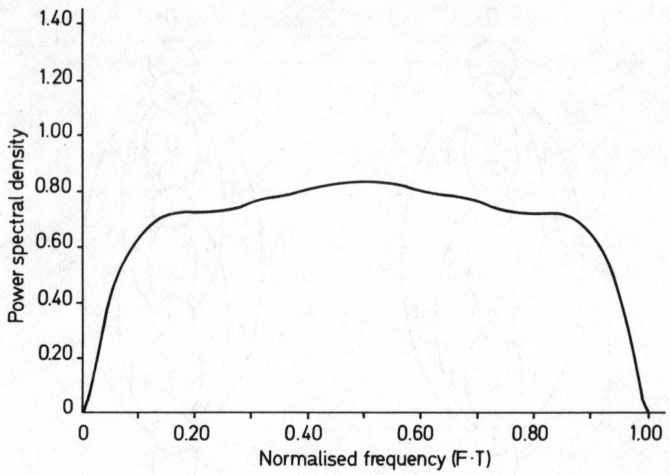

Fig. 15.10 Power spectral density for FOMOT code

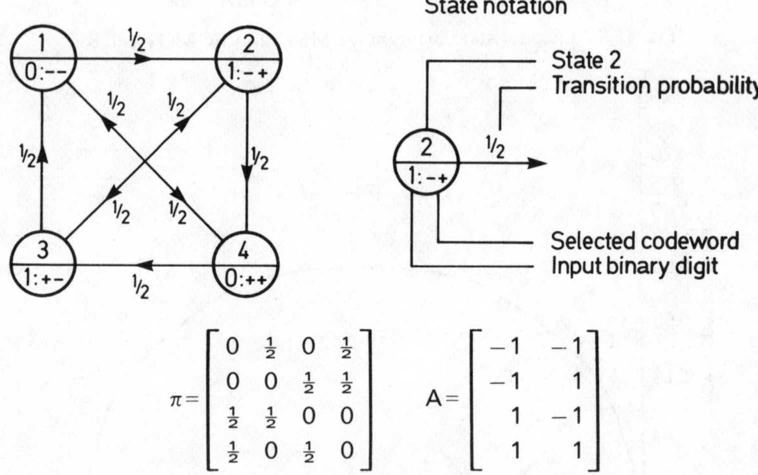

$$\pi = \begin{bmatrix} 0 & \frac{1}{2} & 0 & \frac{1}{2} \\ 0 & 0 & \frac{1}{2} & \frac{1}{2} \\ \frac{1}{2} & \frac{1}{2} & 0 & 0 \\ \frac{1}{2} & 0 & \frac{1}{2} & 0 \end{bmatrix} \qquad A = \begin{bmatrix} -1 & -1 \\ -1 & 1 \\ 1 & -1 \\ 1 & 1 \end{bmatrix}$$

Fig. 15.11 State diagram and matrices for Miller code

15.3.3 Adaptive codes

In this category, the codes are to be analysed by codeword state diagrams which can be constructed from coding rules. The transition matrix π (= E1) and the codeword matrix **A** can be derived directly from the state diagram.

The following examples are considered: Miller

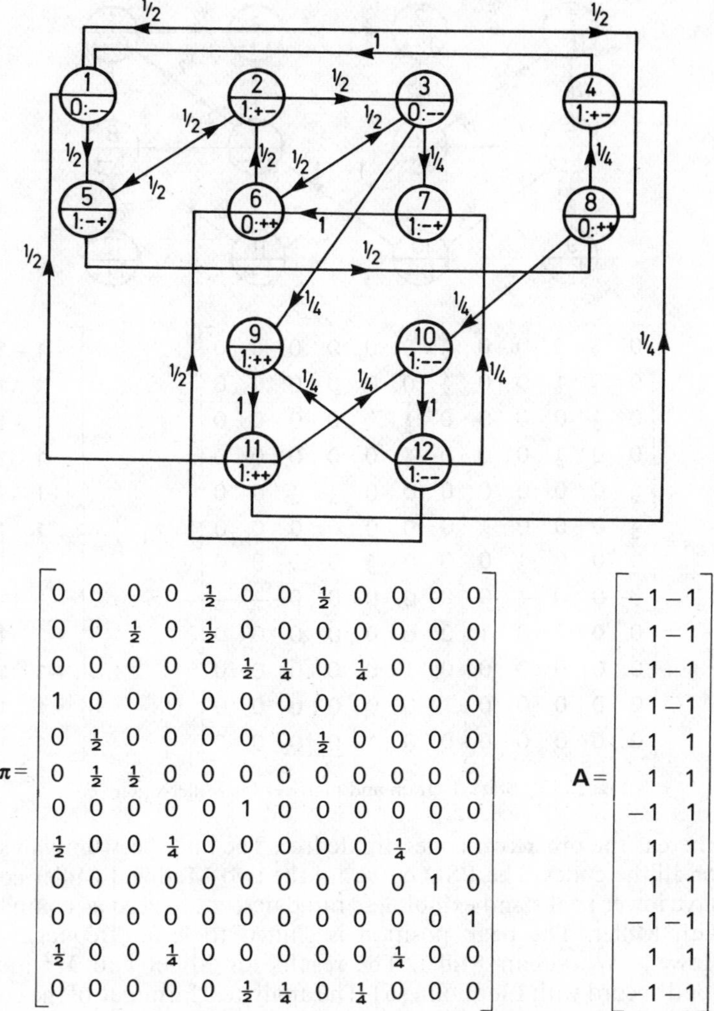

$$\pi = \begin{bmatrix} 0 & 0 & 0 & 0 & \frac{1}{2} & 0 & 0 & \frac{1}{2} & 0 & 0 & 0 & 0 \\ 0 & 0 & \frac{1}{2} & 0 & \frac{1}{2} & 0 & 0 & 0 & 0 & 0 & 0 & 0 \\ 0 & 0 & 0 & 0 & 0 & \frac{1}{2} & \frac{1}{4} & 0 & \frac{1}{4} & 0 & 0 & 0 \\ 1 & 0 & 0 & 0 & 0 & 0 & 0 & 0 & 0 & 0 & 0 & 0 \\ 0 & \frac{1}{2} & 0 & 0 & 0 & 0 & 0 & \frac{1}{2} & 0 & 0 & 0 & 0 \\ 0 & \frac{1}{2} & \frac{1}{2} & 0 & 0 & 0 & 0 & 0 & 0 & 0 & 0 & 0 \\ 0 & 0 & 0 & 0 & 0 & 1 & 0 & 0 & 0 & 0 & 0 & 0 \\ \frac{1}{2} & 0 & 0 & \frac{1}{4} & 0 & 0 & 0 & 0 & 0 & \frac{1}{4} & 0 & 0 \\ 0 & 0 & 0 & 0 & 0 & 0 & 0 & 0 & 0 & 0 & 1 & 0 \\ 0 & 0 & 0 & 0 & 0 & 0 & 0 & 0 & 0 & 0 & 0 & 1 \\ \frac{1}{2} & 0 & 0 & \frac{1}{4} & 0 & 0 & 0 & 0 & 0 & \frac{1}{4} & 0 & 0 \\ 0 & 0 & 0 & 0 & 0 & \frac{1}{2} & \frac{1}{4} & 0 & \frac{1}{4} & 0 & 0 & 0 \end{bmatrix} \qquad \mathbf{A} = \begin{bmatrix} -1 & -1 \\ 1 & -1 \\ -1 & -1 \\ 1 & -1 \\ -1 & 1 \\ 1 & 1 \\ -1 & 1 \\ 1 & 1 \\ 1 & 1 \\ -1 & -1 \\ 1 & 1 \\ -1 & -1 \end{bmatrix}$$

Fig. 15.12 State diagram and matrices for Howells-Woodman Miller code

Howells–Woodman–Miller and Miller-squared codes [11, 12]. Coding rules and properties of codes will not be detailed here but the state diagrams and the corresponding π and $\mathbf{A}$ matrices are given in Figs. 15.11–15.13.

The PSD of the three Miller codes are displayed in Figs. 15.14–15.16. As expected, a small dc content is detected in Miller

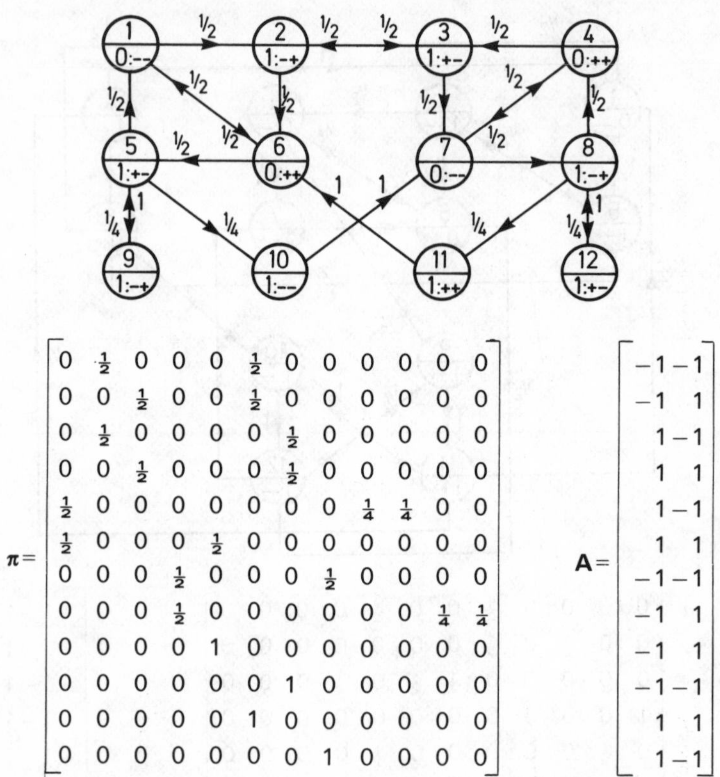

$$
\pi = \begin{bmatrix}
0 & \frac{1}{2} & 0 & 0 & 0 & \frac{1}{2} & 0 & 0 & 0 & 0 & 0 & 0 \\
0 & 0 & \frac{1}{2} & 0 & 0 & \frac{1}{2} & 0 & 0 & 0 & 0 & 0 & 0 \\
0 & \frac{1}{2} & 0 & 0 & 0 & 0 & \frac{1}{2} & 0 & 0 & 0 & 0 & 0 \\
0 & 0 & \frac{1}{2} & 0 & 0 & 0 & \frac{1}{2} & 0 & 0 & 0 & 0 & 0 \\
\frac{1}{2} & 0 & 0 & 0 & 0 & 0 & 0 & 0 & \frac{1}{4} & \frac{1}{4} & 0 & 0 \\
\frac{1}{2} & 0 & 0 & 0 & \frac{1}{2} & 0 & 0 & 0 & 0 & 0 & 0 & 0 \\
0 & 0 & 0 & \frac{1}{2} & 0 & 0 & 0 & \frac{1}{2} & 0 & 0 & 0 & 0 \\
0 & 0 & 0 & \frac{1}{2} & 0 & 0 & 0 & 0 & 0 & 0 & \frac{1}{4} & \frac{1}{4} \\
0 & 0 & 0 & 0 & 1 & 0 & 0 & 0 & 0 & 0 & 0 & 0 \\
0 & 0 & 0 & 0 & 0 & 0 & 1 & 0 & 0 & 0 & 0 & 0 \\
0 & 0 & 0 & 0 & 0 & 1 & 0 & 0 & 0 & 0 & 0 & 0 \\
0 & 0 & 0 & 0 & 0 & 0 & 0 & 1 & 0 & 0 & 0 & 0
\end{bmatrix}
\qquad
\mathbf{A} = \begin{bmatrix}
-1 & -1 \\
-1 & 1 \\
1 & -1 \\
1 & 1 \\
1 & -1 \\
1 & 1 \\
-1 & -1 \\
-1 & 1 \\
-1 & 1 \\
-1 & -1 \\
1 & 1 \\
1 & -1
\end{bmatrix}
$$

Fig. 15.13 State diagram and matrices for Miller-square code

whereas the other two codes are dc free. The line spectrum vanishes for all the codes. The PSD curves of the two Modified Miller codes have lower peaks and exhibit less rapid descent to zero as compared with Miller. The peak position is shifted to lower frequency for Howells–Woodman–Miller. The results for Miller and M^2 are in good accord with Lindholm [6]. The analytical formulae of the codes are shown in the Appendix. The expression of Miller code is identical to that found by Rousseau [13] but appears somewhat different from that derived by Hecht and Guida [14].

15.3.4 Unbounded block codes

Block codes 6B4T and 3B2T are typical examples of codes with unbounded total disparity. Codeword dictionaries for these codes are available in the literature [2, 7]; kernel segments for the state diagrams are displayed in Fig. 15.17.

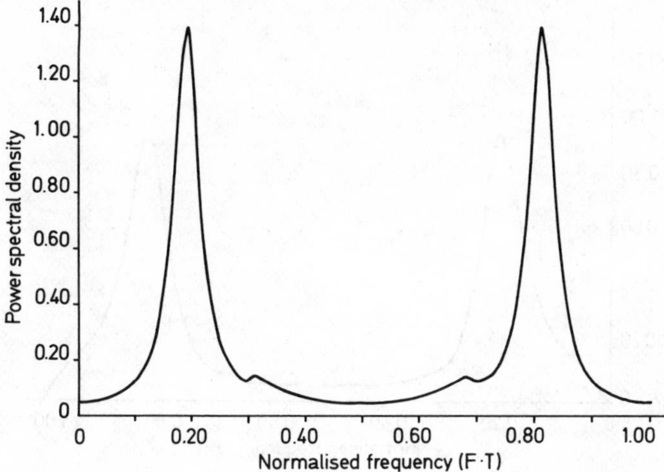

Fig. 15.14 Power spectral density for Miller code

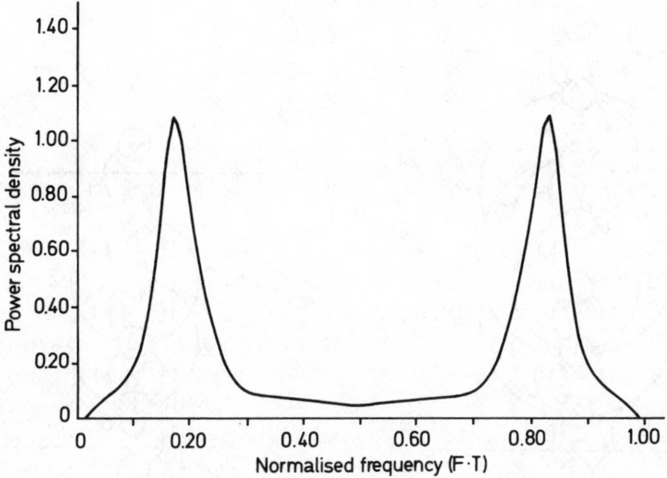

Fig. 15.15 Power spectral density for M Miller code

Consider first the 6B4T code. It possesses seven word disparities: ± 3, ± 2, ± 1, 0. The ± 1 disparity words are not paired and as a consequence the total disparity is not bounded. The number of possible states is thus unlimited. A kernel segment of the state diagram is displayed in Fig. 15.17a. To apply the successive approximation method one begins in this case with six states. The

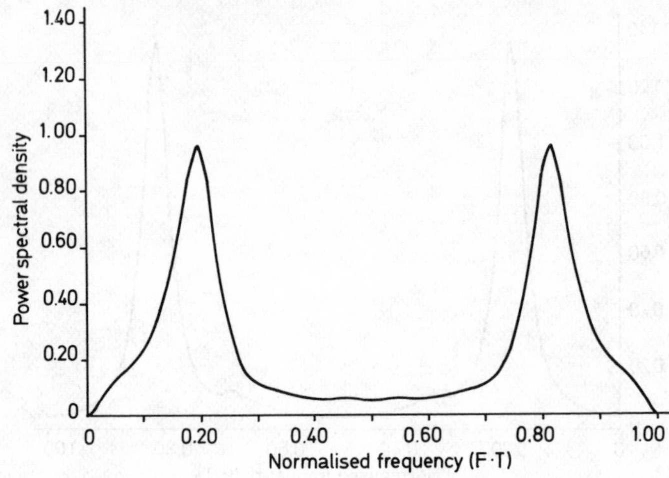

Fig. 15.16 Power spectral density for Miller-square code

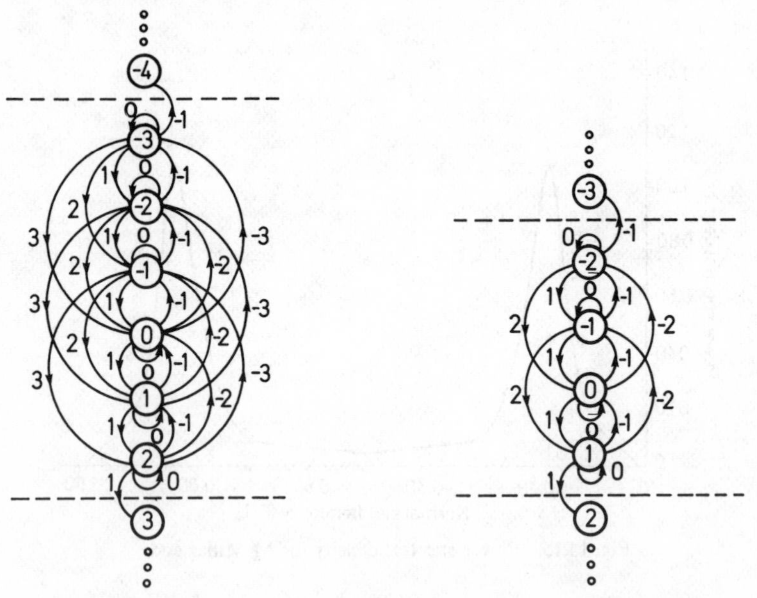

Fig. 15.17 Encoder state diagram for 6B4T and 3B2T codes

transitions of ± 1 disparity at the extreme ends are truncated giving the reduced state diagram of Fig. 15.18a. The spectrum is computed for this reduced-state encoder and the number of states is then increased evenly by two (or a multiple of two) to yield the new system

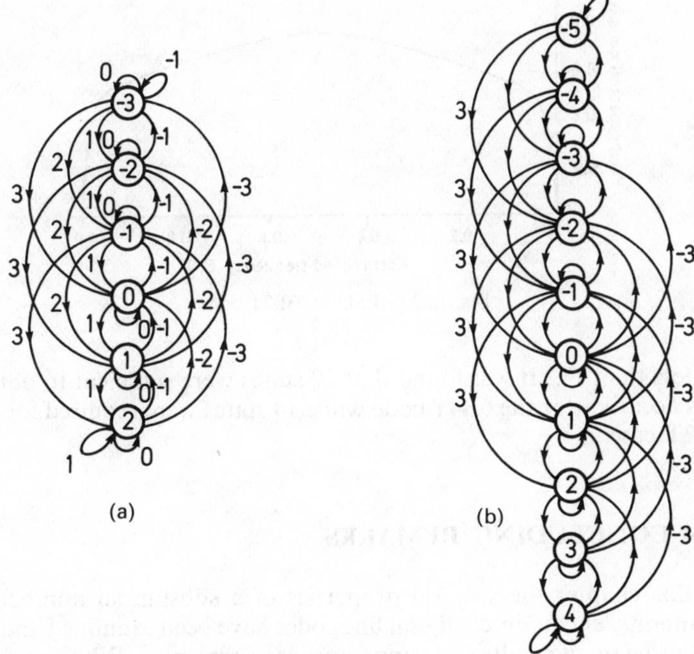

Fig. 15.18 (a) 6-State 6B4T encoder state diagram, (b) 10-state 6B4T encoder state diagram

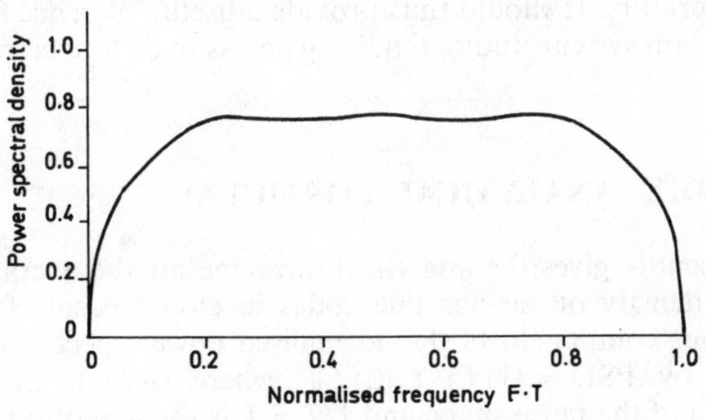

Fig. 15.19 PSD for 6B4T code

of Fig. 15.18b. The computation is then repeated. This process of state augmentation and spectrum calculation is repeated until the desired accuracy is obtained. Power spectral density functions for 6B4T and 3B2T codes are presented in Figs. 15.19 and 15.20. In each case iteration was terminated once successive values were found to differ

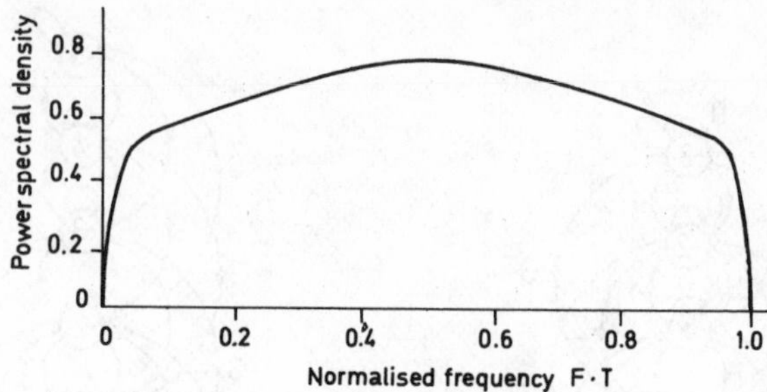

Fig. 15.20 PSD for 3B2T code

by less than 1%. It was found that 10 states were sufficient to obtain this accuracy for the 6B4T code while 14 states were required for the 3B2T case.

15.4 CONCLUDING REMARKS

In this chapter the spectral properties of a substantial number of commonly encountered digital line codes have been examined and an extensive set of results and supportive data presented. While no such collection can hope to be complete, that presented here is considerably more comprehensive than any known to be available elsewhere in the literature. It should thus provide a useful reference for digital transmission system studies requiring access to code spectral density data.

APPENDIX: ANALYTICAL FORMULAE

This appendix gives the analytical formulae for the average power spectral density of various line codes in closed cosine form. The continuous component of the normalised power spectral density is denoted by $\mathrm{PSD} = W(f)\cdot T/|G(f)|^2$ where $G(f)$ is the Fourier transform of the pulse shape and $1/T = f_r$ is the repetition rate. The parameter θ is given by $\theta = 2\pi f T$.

$$\begin{aligned}\mathrm{PSD}_{5B6B} = 1 &- 0.33333 \cos\theta - 0.20833 \cos 2\theta - 0.125 \cos 3\theta \\ &- 0.04166 \cos 4\theta - 0.04166 \cos 5\theta + (0.00911 \\ &- 0.00065 \cos\theta - 0.00911 \cos 2\theta - 0.00651 \cos 3\theta \\ &- 0.00520 \cos 4\theta - 0.01627 \cos 5\theta - 0.03645 \cos 6\theta\end{aligned}$$

$$-0.02213\cos 7\theta-0.01302\cos 8\theta-0.01302\cos 9\theta$$
$$-0.01562\cos 10\theta-0.01171\cos 11\theta)/(1.0625-0.5\cos 6\theta)$$

$$\begin{aligned}\mathrm{PSD}_{7B8B}=\;&0.99993-0.23817\cos\theta-0.19540\cos 2\theta\\&-0.09757\cos 3\theta-0.03912\cos 4\theta-0.01167\cos 5\theta\\&-0.02346\cos 6\theta-0.00389\cos 7\theta+(0.04551\\&+0.06380\cos\theta+0.04164\cos 2\theta+0.02475\cos 3\theta\\&+0.00764\cos 4\theta-0.00943\cos 5\theta-0.02620\cos 6\theta\\&-0.04820\cos 7\theta-0.07535\cos 8\theta-0.05820\cos 9\theta\\&-0.04394\cos 10\theta-0.03277\cos 11\theta-0.02154\cos 12\theta\\&-0.01040\cos 13\theta+0.00050\cos 14\theta+0.01442\cos 15\theta\\&+0.03140\cos 16\theta+0.02586\cos 17\theta+0.02109\cos 18\theta\\&+0.01715\cos 19\theta+0.01321\cos 20\theta+0.00930\cos 21\theta\\&+0.00545\cos 22\theta+0.00080\cos 23\theta-0.00467\cos 24\theta\\&-0.00401\cos 25\theta-0.00341\cos 26\theta-0.00287\cos 27\theta\\&-0.00233\cos 28\theta-0.00180\cos 29\theta-0.00126\cos 30\theta\\&-0.00066\cos 31\theta)/(2.34952-3.07286\cos 8\theta\\&+0.87516\cos 16\theta-0.08898\cos 24\theta)\end{aligned}$$

$$\begin{aligned}\mathrm{PSD}_{4B3T\text{-}1}=\;&0.6875-0.08333\cos\theta-0.04166\cos 2\theta+(0.25472\\&+0.19657\cos\theta-0.11630\cos 2\theta-0.42917\cos 3\theta\\&-0.21425\cos 4\theta+0.00066\cos 5\theta+0.21558\cos 6\theta\\&+0.12354\cos 7\theta+0.03150\cos 8\theta-0.06054\cos 9\theta\\&-0.03741\cos 10\theta-0.01428\cos 11\theta+0.00884\cos 12\theta\\&+0.00571\cos 13\theta+0.00259\cos 14\theta-0.00053\cos 15\theta\\&-0.000356\cos 16\theta-0.000178\cos 17\theta)/(2.99493\\&-4.30626\cos 3\theta+1.70784\cos 6\theta-0.37922\cos 9\theta\\&+0.04393\cos 12\theta-0.00201\cos 15\theta)\end{aligned}$$

$$\begin{aligned}\mathrm{PSD}_{4B3T\text{-}2}=\;&0.625-0.16666\cos\theta-0.08333\cos 2\theta+(0.07919\\&+0.05615\cos\theta-0.04608\cos 2\theta-0.14831\cos 3\theta\end{aligned}$$

$$-0.08129\cos 4\theta-0.01428\cos 5\theta+0.05273\cos 6\theta$$
$$+0.03320\cos 7\theta+0.01367\cos 8\theta-0.00585\cos 9\theta$$
$$-0.00390\cos 10\theta-0.00195\cos 11\theta)/(1.59789$$
$$-1.78710\cos 3\theta+0.39843\cos 6\theta-0.03124\cos 9\theta)$$

$$PSD_{AMI}=0.5-0.5\cos\theta$$

$$PSD_{CMI}=0.75-0.5\cos 2\theta-0.25\cos 3\theta$$

$$PSD_{MS43}=0.64732-0.24404\cos\theta-0.06845\cos 2\theta+(0.02788$$
$$+0.00567\cos\theta-0.02038\cos 2\theta-0.05493\cos 3\theta$$
$$-0.02891\cos 4\theta-0.01623\cos 5\theta-0.00348\cos 6\theta$$
$$-0.00003\cos 7\theta+0.00054\cos 8\theta+0.00119\cos 9\theta$$
$$+0.00045\cos 10\theta+0.00022\cos 11\theta)/(1.25008$$
$$-1.00773\cos 3\theta+0.01074\cos 6\theta+0.00976\cos 9\theta)$$

$$PSD_{FOMOT}=0.68749-0.16666\cos\theta-0.04166\cos 2\theta+(0.05055$$
$$+0.02690\cos\theta-0.3358\cos 2\theta)-0.10820\cos 3\theta$$
$$-0.06419\cos 4\theta-0.02525\cos 5\theta+0.01397\cos 6\theta$$
$$+0.01102\cos 7\theta+0.00557\cos 8\theta+0.00366\cos 9\theta$$
$$+0.00244\cos 10\theta+0.00101\cos 11\theta)/(1.25030$$
$$-0.98461\cos 3\theta-0.03906\cos 6\theta+0.01562\cos 9\theta)$$

$$PSD_{Miller}=1-(1.5-0.25\cos\theta+2.5\cos 2\theta+0.25\cos 3\theta$$
$$+\cos 4\theta)/(2.25+3\cos 2\theta+\cos 4\theta)$$

$$PSD_{H.W.\,Miller}=1+0.33333\cos\theta+(-1.1904-0.61783\cos\theta$$
$$-1.98501\cos 2\theta-0.92252\cos 3\theta-0.56119\cos 4\theta$$
$$-0.14713\cos 5\theta+0.62238\cos 6\theta+0.29166\cos 7\theta$$
$$+0.48957\cos 8\theta+0.11979\cos 9\theta+0.09374\cos 10\theta$$
$$+0.01041\cos 11\theta)/(2.19140+2.53124\cos 2\theta$$
$$-0.09374\cos 4\theta-1.03124\cos 6\theta-0.62499\cos 8\theta$$
$$-0.12499\cos 10\theta)$$

$$\begin{aligned}
\mathrm{PSD}_{M^2} = {} & 1 + 0.09090 \cos\theta + (-1.01526 + 1.15473 \cos\theta \\
& - 1.43575 \cos 2\theta - 0.85635 \cos 3\theta + 0.21270 \cos 4\theta \\
& - 0.89109 \cos 5\theta + 1.19177 \cos 6\theta + 0.14142 \cos 7\theta \\
& + 0.55157 \cos 8\theta + 0.28888 \cos 9\theta - 0.15189 \cos 10\theta \\
& + 0.02370 \cos 11\theta - 0.15926 \cos 12\theta - 0.03746 \cos 13\theta \\
& - 0.00639 \cos 14\theta - 0.00177 \cos 15\theta + 0.01491 \cos 16\theta \\
& + 0.00071 \cos 17\theta)/(2.70927 + 2.20702 \cos 2\theta \\
& - 2.10277 \cos 4\theta - 2.18162 \cos 6\theta - 0.28515 \cos 8\theta \\
& + 0.41796 \cos 10\theta + 0.16015 \cos 12\theta - 0.01562 \cos 14\theta \\
& - 0.01562 \cos 16\theta)
\end{aligned}$$

$$\begin{aligned}
\mathrm{PSD}_{6B4T} = {} & 0.65624 - 0.11718 \cos\theta - 0.07812 \cos 2\theta \\
& - 0.03906 \cos 3\theta + (0.08943 + 0.10027 \cos\theta \\
& + 0.02167 \cos 2\theta - 0.05692 \cos 3\theta - 0.13553 \cos 4\theta \\
& - 0.08823 \cos 5\theta - 0.04094 \cos 6\theta + 0.00635 \cos 7\theta \\
& + 0.05365 \cos 8\theta + 0.03863 \cos 9\theta + 0.02361 \cos 10\theta \\
& + 0.00859 \cos 11\theta - 0.00642 \cos 12\theta - 0.00522 \cos 13\theta \\
& - 0.00403 \cos 14\theta - 0.00283 \cos 15\theta - 0.00163 \cos 16\theta \\
& - 0.00115 \cos 17\theta - 0.00067 \cos 18\theta - 0.00019 \cos 19\theta \\
& + 0.00029 \cos 20\theta + 0.00022 \cos 21\theta + 0.00015 \cos 22\theta \\
& + 0.00008 \cos 23\theta + 0.00001 \cos 24\theta + 0.00001 \cos 25\theta) \\
& (5.29666 - 7.45738 \cos 4\theta + 2.31849 \cos 8\theta \\
& - 0.07404 \cos 12\theta - 0.08830 \cos 16\theta \\
& + 0.00534 \cos 20\theta + 0.00108 \cos 24\theta)
\end{aligned}$$

$$\begin{aligned}
\mathrm{PSD}_{3B2T} = {} & 0.625 - 0.125 \cos\theta + (0.18141 + 0.03971 \cos\theta \\
& - 0.28340 \cos 2\theta - 0.07597 \cos 3\theta + 0.13144 \cos 4\theta \\
& + 0.04880 \cos 5\theta - 0.03384 \cos 6\theta - 0.01471 \cos 7\theta \\
& + 0.00442 \cos 8\theta + 0.00208 \cos 9\theta - 0.00024 \cos 10\theta \\
& - 0.00012 \cos 11\theta)/(6.91892 - 10.29169 \cos 2\theta
\end{aligned}$$

$$+4.08642 \cos 4\theta - 0.75254 \cos 6\theta + 0.03796 \cos 8\theta$$

$$+0.00189 \cos 10\theta - 0.00009 \cos 12\theta)$$

REFERENCES

1. Valin, J., 'Codes used for the transmission of binary data', *Thomson-CSF Eng. Rev.*, **11**, 1–42 (1979)
2. Waters, D. B., 'Codes for digital line transmission', *Commun. Int.*, **5**, 19–27 (1978)
3. Cariolaro, G. L. and Tronca, G. P., 'Spectra of block coded digital signals', *IEEE Trans.*, **COM-22**, 1555–1563 (1974)
4. Poo, G. S., 'Computer aids for code spectra calculations', *Proc. IEEE*, **128**, Pt. F, No. 5, 323–330 (1981)
5. Poo, G. S., 'Power spectra of 6B4T and 3B2T codes', *Electron. Lett.*, **18**, 884–885 (1982)
6. Lindholm, D. A., 'Power spectra of channel codes for digital magnetic recording', *IEEE Trans.*, **MAG-14**, 321–323 (1978)
7. Catchpole, R. J., 'Efficient ternary transmission codes', *Electron. Lett.*, **11**, 482–484 (1975)
8. Alexandra, J. R. and Hagra, A. S. T., 'Transformation of binary coded signals into a form having lower disparity', U.K. Patent Specification 1540617
9. CCITT Cons. Comm., '4B3T codes for a 140 Mbit/s digital transmission system', **COM XVIII**, No. 27-E, 1–14 (1977)
10. Buchner, T. B., 'Ternary line signal codes', Zurich Seminar, F1, 1–9 (1974)
11. Mallinson, J. C. and Miller, J. W., 'Optimal codes for digital magnetic recording', *Radio Electron. Eng.*, **47**, No. 4, 172–176 (1977)
12. Howells, G. A. and Woodman, D. E., 'Modifications to Miller code to achieve a balanced mode of operation', U.K. Patent Application No. 52215/77.
13. Rousseau, M., 'Applying Markov chains to the calculation of the power spectral density of binary signals', *Annls. Telecommun.* (France, Anteau), Jan.–Feb., **31** (1–2), 8–16 (1976)
14. Hecht, M. and Guida, A., 'Delay modulation', *Proc. IEEE*, **57**, 1314–1316 (1969)

16

Design of a 7B8B Line Code with Improved Error Monitoring Capabilities

R. M. Brooks

16.1 INTRODUCTION

Line coding of digital information is usually desirable from the viewpoint of matching the transmitted signal to the transmission media and allowing certain benefits to arise in terms of transmission equipment realisation and system operation. Economic aspects and coder complexity also have to be taken into account and play a significant part in the choice of code for any particular system.

For optical fibre digital transmission systems binary line codes are usually the most attractive both in terms of system penalty and hardware realisation. The 7B8B low disparity alphabetic code [1] has many attractive features which make it a strong contender for 140 Mbit/s and higher bit rate optical fibre transmission systems. This is particularly the case for systems employing optical receivers with high impedance (integrating) front end stages [2, 3] where a balanced code [4] is more or less essential to achieve an adequate dynamic range.

The detection of errors gives an indication of the performance of a digital transmission system and whether or not the system is serviceable. It is therefore essential to monitor error ratio at the system terminals on an in-service basis. The monitoring of errors at dependent repeaters permits the identification of a faulty repeater section and so provides useful information for maintenance purposes. Whether or not the monitoring is undertaken on an in-service or out-of-service basis depends on the maintenance policy of the particular Telecommunication Administration or Operating Company.

A simple in service technique used by British Telecom for coaxial digital line systems is known as variable period mark parity [5,6] and has the following advantages:

(1) simple circuitry with only a small amount operating at high speed (mainly one D-type bistable)
(2) low power consumption
(3) no increase in bit rate
(4) no requirement for framing or word recognition
(5) good accuracy at low error ratios.

There is a certain loss of precision for high error ratios ($>10^{-6}$) and burst errors but this is not usually too important for fault location purposes and is more than offset by the advantages.

This chapter considers the use of a 7B8B line code in conjunction with the established mark parity technique to give an improved error monitoring capability [7–11].

16.2 VARIABLE PERIOD EVEN MARK PARITY ERROR DETECTION

The basic concept of this error detection technique is to control the number of marks transmitted to line in such a way that for specific time intervals an even number of marks are transmitted. The intervals are variable but delimited by the appearance at the encoder of a particular input word, known as the control word. The occurrence of a control word provides an opportunity to make the number of marks transmitted to line even since the previous opportunity. It is therefore required that the control word be mapped to a pair of output words of which one word should contain an even and the other an odd number of marks.

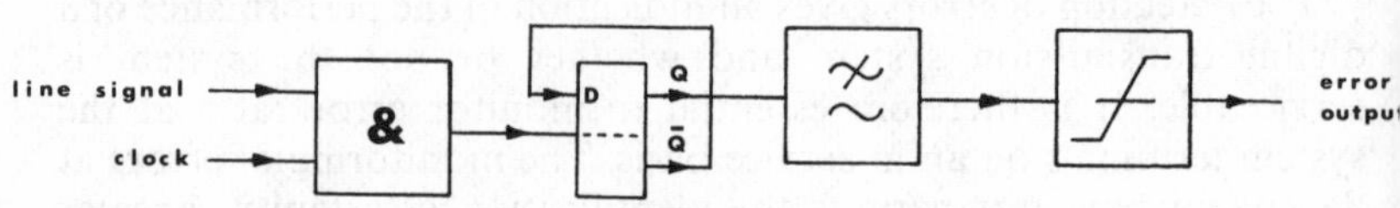

Fig. 16.1 Schematic of dependent repeater error detection

The repeater error detection circuitry shown in Fig. 16.1 consists of:

(1) an 'and' gate (if the data does not have a return to zero pulse format)
(2) a D-type bistable, arranged to change state on each received mark—referred to in this chapter as a 'toggle'
(3) a low-pass filter
(4) a method of determining changes in mean level.

Consider the toggle to be in a particular state after receipt of a mark parity word. Provided there are no transmission errors, the toggle will be in the same state after receipt of subsequent mark parity words and consequently can give rise to a small dc component. If, however, a single or odd number of transmission errors occur between any two mark parity words, the toggle dc voltage will be reversed and remain reversed until further transmission errors occur. Errors are detected by monitoring low frequency changes of the mean toggle voltage, by way of a threshold crossing detector.

16.3 7B8B STATE TRANSITION DIAGRAM

It is assumed for the purpose of this chapter that CCITT recommended vocabulary [4] will apply and that:

binary 0 will be assigned a disparity of $-\frac{1}{2}$

and

binary 1 will be assigned a disparity of $+\frac{1}{2}$

Under the normal coding rules, from an accumulated zero disparity condition there would seem no advantage in the transmission of positive disparity words compared with negative disparity words. However, by choosing a preferred polarity the digital sum variation can be reduced, negative will be arbitrarily chosen to serve as an illustration. In practice this is achieved by initialising the disparity counter with the disparity of binary 1.

Summary of pertinent 7B8B balanced code properties of the code in reference 1.

(1) word disparities: 0, ± 1, ± 2
(2) each word has no more than three leading or tailing like elements
(3) maximum number of consecutive like elements = 6
(4) digital sum variation = 6
(5) minimum number of transitions per word = 2

(6)

Word disparity	Number of words used	Spare words as regards above properties
0	67	10101010—normally reserved for realignment but spare if scramblers are employed
±1	48	None
±2	13	01000001, 10111110 10000001, 01111110 10000010, 01111101

Additionally, 0 and ±2 disparity words contain an even number of marks and ±1 disparity words contain an odd number of marks.

Therefore, for an integral number of adjacent transmitted words, if the number of marks is odd then the disparity is ±1 and if the number of marks is even then the disparity is 0 or −2 (assuming no transmission errors), that is, there is a fixed relationship between disparity and mark parity. Although arbitrary, if an even state of the toggle is considered to correspond to zero disparity then the state transition diagram is as shown in Fig. 16.2.

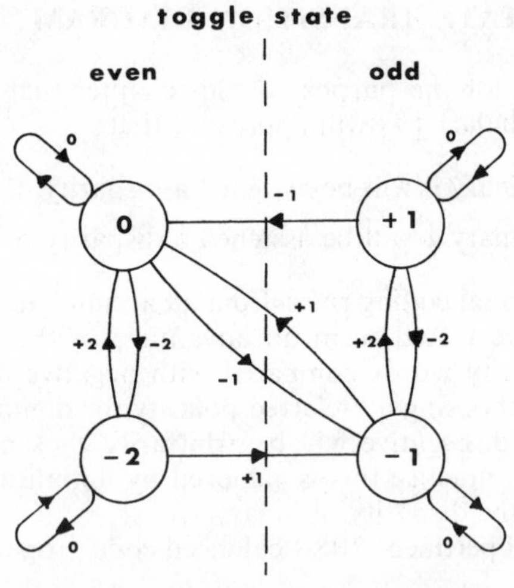

Fig. 16.2 7B8B transition diagram. Note: (1) Numbers in the circles denote the disparity states at the transmit terminal. (2) Other numbers indicate word disparity

Let

q_0 be the probability of a 0 disparity word

q_1 be the probability of a ±1 disparity word

q_2 be the probability of a ±2 disparity word

then

$$q_0 + q_1 + q_2 = 1 \tag{16.1}$$

If P_i is the probability of the ith disparity state where $i = +1, 0, -1, -2$, then assuming equilibrium, we obtain

$$\begin{bmatrix} q_0 & q_1 & q_1 & q_2 \\ 0 & q_0 & q_2 & 0 \\ q_1 & q_2 & q_0 & q_1 \\ q_2 & 0 & 0 & q_0 \end{bmatrix} \begin{bmatrix} P_0 \\ P_1 \\ P_{-1} \\ P_{-2} \end{bmatrix} = \begin{bmatrix} P_0 \\ P_1 \\ P_{-1} \\ P_{-2} \end{bmatrix} \tag{16.2}$$

We can obtain a solution by using any three rows of the matrix (note that the four equations are not independent) and the equation

$$P_0 + P_1 + P_{-1} + P_{-2} = 1 \tag{16.3}$$

Therefore we obtain

$$P_{-1} = P_0 = \frac{q_1 + q_2}{2(q_1 + 2q_2)} \tag{16.4}$$

$$P_{+1} = P_{-2} = \frac{q_2}{2(q_1 + 2q_2)} \tag{16.5}$$

$$P_{\text{even}} = P_0 + P_{-2} = P_{\text{odd}} = P_{-1} + P_{+1} = 0.5 \tag{16.6}$$

Now, assuming equiprobable input words

$$q_1 = \tfrac{48}{128} = 0.3750$$

$$q_2 = \tfrac{13}{128} = 0.1016$$

$$P_{-1} = P_0 = 0.4122$$

$$P_{+1} = P_{-2} = 0.0878$$

16.4 COMBINATION OF THE 7B8B CODE AND MARK PARITY ERROR DETECTION

16.4.1 Unmodified code

Having considered the state transition diagram it is now possible to calculate the mean voltage produced under equilibrium conditions in the toggle circuit and in the absence of transmission errors. As an example Fig. 16.3 shows the toggle response to the word 11011000 assuming the output swings between $\pm V_0$ and starts from an even state.

A complete list of the 7B8B code words and associated mean toggle

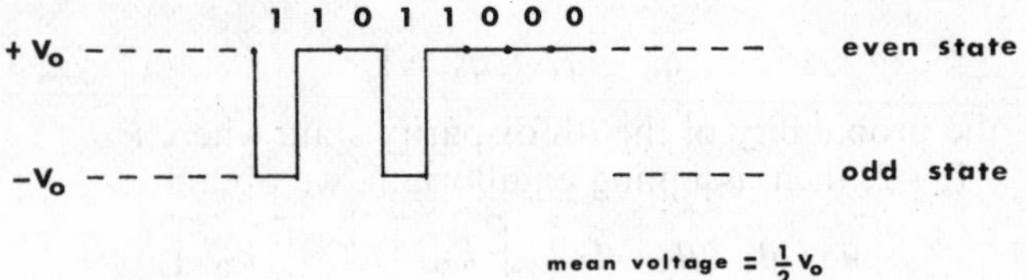

Fig. 16.3 Toggle response to 11011000

voltages is given in the appendix. An overall summary is given in Table 16.1.

Table 16.1

Word disparity (i)	Mean toggle voltage: Word (x_i)	Mean toggle voltage: Overall
0	$0.0896V_0$	$q_0x_0(P_{\text{even}} - P_{\text{odd}}) = 0$
+1	$-0.1250V_0$	$q_1x_1(P_{-2} - P_{-1}) = 0.015V_0$
−1	$-0.1250V_0$	$q_1x_{-1}(P_0 - P_1) = -0.015V_0$
+2	$-0.0192V_0$	$q_2x_2(P_{-2} - P_{-1}) = 0.0006V_0$
−2	$0.1346V_0$	$q_2x_{-2}(P_0 - P_1) = 0.0044V_0$

Notice that although +1 and −1 words each produce a non-zero mean toggle voltage these exactly balance, but +2 and −2 words reinforce.

The total overall mean toggle voltage $= 0.005V_0$. If an error occurs the toggle states would be interchanged as regards the transmit terminal disparity states and so the overall mean toggle voltage will change polarity and can be detected in the same way as for even mark parity. However, the voltage swing is too small for practical detection and does not give an adequate signal to noise ratio. The noise includes the continuous components of the line signal after processing by the error detection circuitry as well as additive noise.

16.4.2 Modified 7B8B code with enhanced toggle dc voltage

The overall mean toggle voltage can be significantly increased by permitting ±1 and ±2 words to have four like elements at the word beginning or end. Notice that this does not affect the digital sum variation since this is governed only by the permitted word disparities and the number of permitted like elements at the beginning or end of a zero disparity word.

If the ± 2 words are considered, then replacement of code words with any or all of the spare words within the present coding rules will not increase the magnitude of the toggle voltage. On inspecting the spare ± 2 words outside the present code rules it is apparent that to increase the toggle voltage swing the mean voltage of

-2 words should be made more positive

$+2$ words should be made more negative.

This effect can be enhanced by the mean voltage of

-1 words being made more positive

$+1$ words being made more negative.

On examining the ± 1 words it is also apparent that this is the best way to proceed. There are three complementary pairs of ± 1 words which are suitable for inclusion, denoted by $ in the appendix. The best words for exclusion are denoted by $\neq$ and there are two such pairs. There are many choices of second best words to exclude and this decision can be made on the basis of keeping words with better timing content and/or making the toggle voltage swing more uniform.

A similar exercise can be performed on the ± 2 words and so, in summary, the best changes to make are as shown in Table 16.2. If the changes are made the overall mean toggle voltage is increased to $0.0223\,V_0 \simeq V_0/45$ which gives a good signal to noise ratio with negligible spurious error ratio (better than 1 error per year).

Apart from word disparity and digital sum variation being unaffected so also is the minimum number of transitions per word. The maximum number of consecutive like elements has slightly increased from six to seven.

The calculated power spectral density characteristic for the 7B8B and modified 7B8B codes are shown in Fig. 16.4. The differences between the two codes is insignificant particularly at low frequencies (and close to harmonics of the symbol frequency).

Table 16.2

Word disparity	*Include*	*Exclude*
± 1	00001011, 11110100	10000110, 01111001
	00001101, 11110010	10011000, 01100111
	00001110, 11110001	10000011, 01111100
± 2	00001010, 11110101	00100001, 11011110
	00001100, 11110011	10000100, 01111011
	01010000, 10101111	00010001, 11101110
	00110000, 11001111	10001000, 01110111

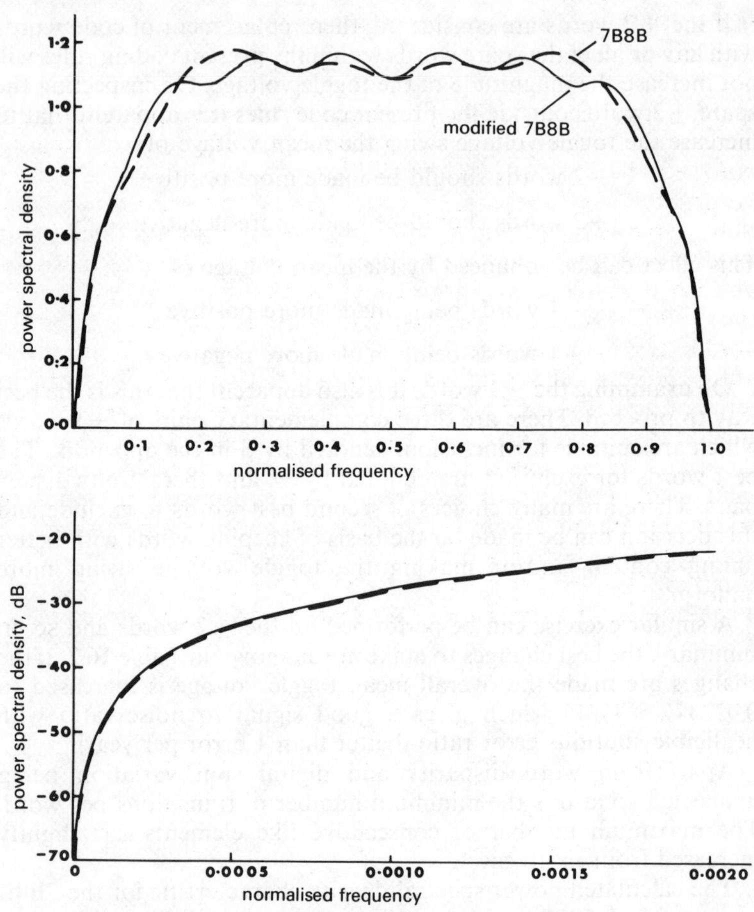

Fig. 16.4 Normalised power spectral density for 7B8B and modified 7B8B line codes (impulse stream with equiprobable 1's and 0's)

In summary, by a small modification to the 7B8B coding table it is possible to obtain an enhanced mean toggle voltage without significantly changing the code properties. The mean toggle voltage is sufficient to permit error detection in a similar way to even mark parity without the need for mark parity control logic at the terminals. It is important to ensure random data into the encoder to obtain the desired effect and a scrambler would therefore be required. Coders and error detection circuits have been built at British Telecom Research Laboratories for demonstration in 140 M bit/s land optical fibre system and 280 M bit/s submarine system experiments [10] and have performed satisfactorily.

16.4.3 7B8B code with variable period even mark parity error detection

In the normal coding scheme an input word is mapped to either a single zero disparity word or a balanced pair of non-zero disparity words e.g. 10011000, 01100111. Notice that in the case of a pair of words, both words contain either an even or odd number of marks. Now, variable period mark parity requires that at least one input word, known as the control word, be mapped to a pair of words of which one word contains an even and the other an odd number of marks. One way of satisfying both requirements is to map the control word to two balanced pairs of words, one pair containing even mark words and the other odd mark words. However, the relationship between disparity and mark parity outlined in Section 16.3 enables an unbalanced pair of words to be used without affecting the digital sum variation. An example will be given of this scheme which only uses valid spare code words and seems straightforward in terms of practical implementation. Let us consider that the even mark parity word will be of zero disparity and the odd mark parity word of −1 disparity. The modified state transition diagram is shown in Fig. 16.5. It can be seen that the single −1 disparity word used for even mark

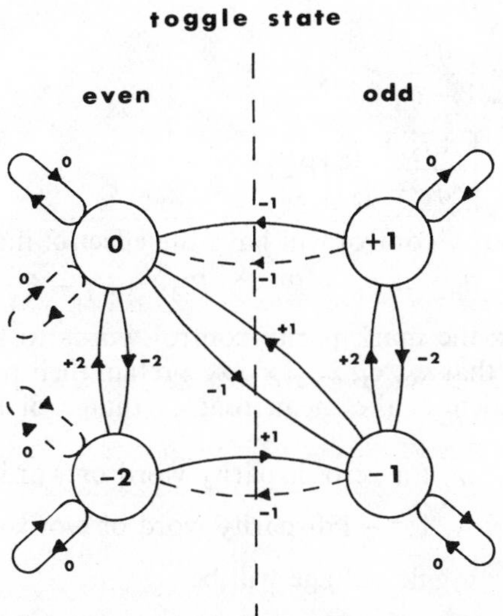

Fig. 16.5 Modified 7B8B state transition diagram with even mark parity. Note: (1) Numbers in the circles denote the disparity states at the transmit terminal. (2) Other numbers indicate word disparity. (3) --- Denotes a transition due to a mark parity word

parity control does not change the disparity states, just the associated probabilities, although a small modification to the method of disparity control would be required.

Using the previous notation, if q_p is the probability of a parity control word then

$$q_0 + q_1 + q_2 + q_p = 1 \tag{16.7}$$

and

$$\begin{bmatrix} q_0 + q_p & q_1 + q_p & q_1 & q_2 \\ 0 & q_0 & q_2 & 0 \\ q_1 & q_2 & q_0 & q_1 \\ q_2 & 0 & q_p & q_0 + q_p \end{bmatrix} \begin{bmatrix} P_0 \\ P_1 \\ P_{-1} \\ P_{-2} \end{bmatrix} = \begin{bmatrix} P_0 \\ P_1 \\ P_{-1} \\ P_{-2} \end{bmatrix} \tag{16.8}$$

If we let

$$A = \frac{1}{q_1}\left[\frac{q_1 + q_2}{q_1 + 2q_2}\right]\left[1 - q_0 - \frac{q_1 q_p}{q_1 + q_2} - \frac{q_2^2}{1 - q_0}\right] \tag{16.9}$$

then

$$P_{-1} = \left[1 + A + \frac{q_2}{1 - q_0} + \frac{Aq_2 + q_p}{q_1 + q_2}\right]^{-1} \tag{16.10}$$

$$P_0 = AP_{-1} \tag{16.11}$$

$$P_1 = \frac{q_2}{1 - q_0} P_{-1} \tag{16.12}$$

$$P_{-2} = \left[\frac{Aq_2 + q_p}{q_1 + q_2}\right] P_{-1} \tag{16.13}$$

Even mark parity control will have the effect of making

$$P_{\text{even}} = P_0 + P_{-2} > P_{\text{odd}} = P_1 + P_{-1}$$

If we consider the mark parity control words to be in a separate category such that $x_0, x_1, x_{-1}, x_2, x_{-2}$ retain their previous meaning then we can denote the mean toggle voltage of the mark parity words as

x_{0p} for zero disparity word or words

x_{-1p} for -1 disparity word or words

the total mean toggle voltage will be

$$q_0 x_0 (P_{\text{even}} - P_{\text{odd}}) + q_1[x_1(P_{-2} - P_{-1}) + x_{-1}(P_0 - P_1)] + q_2[x_2(P_{-2} - P_{-1}) + x_{-2}(P_0 - P_1)] + q_p[x_{0p}P_{\text{even}} - x_{-1p}P_{\text{odd}}] \tag{16.14}$$

Table 16.3

Even mark parity		Odd mark parity	
Word	*Toggle voltage*	*Word*	*Toggle voltage*
11011000	$\frac{1}{2}V_0$	10011000	$-\frac{3}{4}V_0$
11000110	$\frac{1}{2}V_0$	10000110	$-\frac{3}{4}V_0$
11000011	$\frac{1}{2}V_0$	10000011	$-\frac{3}{4}V_0$

Clearly x_{0p} should be of opposite polarity to x_{-1p} to maximise the mean toggle voltage.

It is possible to make indirect use of the three existing spare ± 2 disparity code words that are within the valid coding rules by substituting them for the ± 1 disparity words:

10011000, 01100111
10000110, 01111001
10000011, 01111100

If the mark parity pairs shown in (Table 16.3) are formed, then the calculated mean toggle voltage $\simeq \pm V_0/50$, should be sufficient for error detection purposes. Approximately 73% of the mean toggle voltage is directly attributable to the mark parity words the remainder being due to the residual unbalance of the toggle due to normal code words. It will be noticed that the even mark parity words only differ from the odd mark parity words in the second digit and this can give rise to the possible terminal implementation shown in Fig. 16.6.

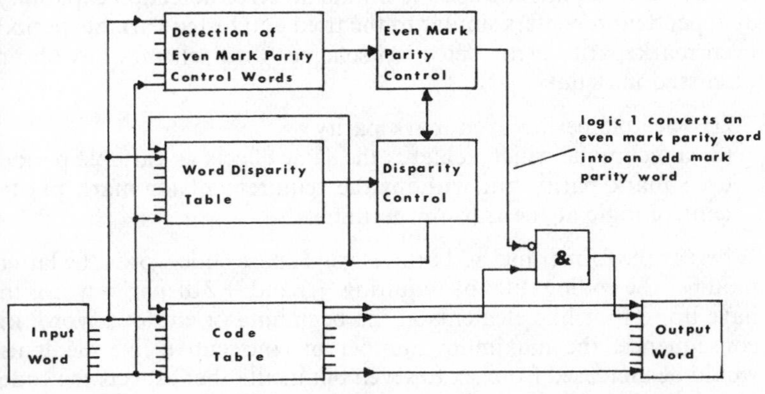

Fig. 16.6 A possible terminal configuration for even mark parity

The code conversion takes place by way of a 'look up' table (programmable read only memories) which contains both modes of the alphabet, the zero disparity words being identical for both modes. On the detection of an even mark parity control word, a decision is made by the even mark parity control, according to the disparity state, whether to transmit an even or odd mark parity word. By default the zero disparity word and disparity are 'looked up' and both need to be modified should an odd mark parity word be required. The output word is modified by changing the second digit to zero and the disparity counter is decremented by a disparity of 1.

An alternative approach would be to pair the even and odd mark parity words within the code table, modify the disparity table accordingly and detect the even mark parity control words one cycle in advance. The disparity control would then make a decision, on which mode of the code table to use, based on the digital sum and the detection of a mark parity control word. The correct code word and disparity would be 'looked up' and so require no further modification.

In summary, it can be seen that the mean toggle voltage can be enhanced by employing the technique of variable period even mark parity. This can be achieved by utilising valid spare code words thus preserving the main code properties. Usually the code 'noise effects are only slightly changed by the operation. The mean toggle voltages should be sufficient to permit an error detection capability with a very low spurious error rate.

16.5 CONCLUSION

Consideration has been given to the problem of modifying the 7B8B low disparity alphabetic code to obtain an error detection capability at dependent repeaters similar to the tried and tested variable period even mark parity error detection scheme. Two schemes have been discussed in detail:

(1) variable period even mark parity

(2) a technique which achieves the same effects as variable period even mark parity but without the requirement for mark parity control logic at the transmit terminals.

Whereas the former method utilises valid spare code words the latter modifies the coding rules by requiring ± 1 and ± 2 disparity words to have up to four like elements at the beginning or end of a word. In consequence, the maximum number of consecutive like elements would be increased from six to seven but in all other aspects the code properties would remain substantially unchanged. The latter technique has been realised within British Telecom Research Laboratories both for 140 Mbit/s land system and 280Mbit/s submarine system application and has worked well.

Subsequent to the original preparation of this work, a new 7B8B code has been developed which gives a slightly improved toggle voltage of $\pm V_0/17.1$ but at the expense of permitting up to 8 consecutive line elements [11].

Acknowledgements

I would like to thank Professor Ken Cattermole and my colleagues Ken Fitchew and Bob Carpenter for useful discussions. Acknowledgement is made to the Director of British Telecom Research Laboratories for permission to publish this work.

APPENDIX: 7B8B OUTPUT CODE WORDS [1] AND ASSOCIATED TOGGLE VOLTAGES

The toggle voltages are assumed to start from an even state, $+V_0$, with the odd state denoted as $-V_0$.

As far as Section 16.4.2 is concerned:

≠ indicates the best words to remove
* indicates the second choice best words to remove
$ indicates the best words to include.

Table A16.1. ZERO DISPARITY WORDS

Currently used output word	*Mean toggle word voltage* $\times V_0$
11101000	0.25
11100100	0
11100010	−0.25
11100001	−0.5
11011000	0.5
11010100	0.25
11010010	0
11010001	−0.25
11001100	0.5
11001010	0.25
11001001	0
11000110	0.5
11000101	0.25

Table A16.1 (*continued*)

Currently used output word	*Mean toggle word voltage* $\times V_0$
11000011	0.5
10111000	0.25
10110100	0
10110010	−0.25
10110001	−0.5
10101100	0.25
10101001	−0.25
10100110	0.25
10100101	0
10100011	0.25
10011100	0
10011010	−0.25
10011001	−0.5
10010110	0
10010101	−0.25
10010011	0
10001110	−0.25
10001101	−0.5
10001011	−0.25
10000111	−0.5
01111000	0.5
01110100	0.25
01110010	0
01110001	−0.25
01101100	0.5
01101010	0.25
01101001	0
01100110	0.5
01100101	0.25
01100011	0.5
01011100	0.25
01011010	0
01011001	−0.25
01010110	0.25
01010101	0
01010011	0.25
01001110	0
01001101	−0.25
01001011	0
01000111	−0.25
00111100	0.5
00111010	0.25
00111001	0
00110110	0.5
00110101	0.25
00110011	0.5
00101110	0.25

Table A16.1 *(continued)*

Currently used output word	*Mean toggle word voltage* $\times V_0$
00101101	0
00101011	0.25
00100111	0
00011110	0.5
00011101	0.25
00011011	0.5
00010111	0.25
Mean	$0.0896V_0$

Available spare words (outside present code rules):

11110000	0.5
00001111	0.5

(Note, 10101010 is reserved for realignment purposes except when scrambling is used.)

Table A16.2. ±1 DISPARITY WORDS

Mode 1 (−1)		*Mode 2 (+1)*		
Output word	*Mean toggle word voltage* $\times V_0$	*Output word*	*Mean toggle word voltage* $\times V_0$	
11001000	−0.25	00110111	0.25	*
11000100	0	00111011	0	
11000010	0.25	00111101	0.25	
11000001	0.5	00111110	0	
10101000	−0.5	01010111	0	*
10100100	−0.25	01011011	−0.25	
10100010	0	01011101	0	
10100001	0.25	01011110	−0.25	
10011000	−0.75	01100111	0.25	≠
10010100	−0.5	01101011	0	*
10010010	−0.25	01101101	0.25	*
10010001	0	01101110	0	
10001100	−0.75	01110011	−0.25	*
10001010	−0.5	01110101	0	*
10001001	−0.25	01110110	−0.25	
10000110	−0.75	01111001	0.25	≠
10000101	−0.5	01111010	0	*
10000011	−0.75	01111100	−0.25	*

Table A16.2 (*continued*)

Mode 1 (−1)		*Mode 2 (+1)*		
Output word	*Mean toggle word voltage* $\times V_0$	*Output word*	*Mean toggle word voltage* $\times V_0$	
01101000	−0.25	10010111	−0.25	
01100100	0	10011011	−0.5	
01100010	0.25	10011101	−0.25	
01100001	0.5	10011110	−0.5	
01011000	−0.5	10100111	0	*
01010100	−0.25	10101011	−0.25	
01010010	0	10101101	0	
01010001	0.25	10101110	−0.25	
01001100	−0.5	10110011	−0.5	
01001010	−0.25	10110101	−0.25	
01001001	0	10110110	−0.5	
01000110	−0.5	10111001	0	*
01000101	−0.25	10111010	−0.25	
01000011	−0.5	10111100	−0.5	
00111000	−0.25	11000111	0.25	*
00110100	0	11001011	0	
00110010	0.25	11001101	0.25	
00110001	0.5	11001110	0	
00101100	−0.25	11010011	−0.25	
00101010	0	11010101	0	
00101001	0.25	11010110	−0.25	
00100110	−0.25	11011001	0.25	*
00100101	0	11011010	0	
00100011	−0.25	11011100	−0.25	
00011100	0	11100011	−0.5	
00011010	0.25	11100101	−0.25	
00011001	0.5	11100110	−0.5	
00010110	0	11101001	0	
00010101	0.25	11101010	−0.25	
00010011	0	11101100	−0.5	
Mean	$-0.125V_0$	Mean	$-0.125V_0$	
Available spare words (outside present code rules):				
11100000	−0.75	00011111	0.25	
11010000	−0.5	00101111	0	
10110000	−0.75	01001111	−0.25	
01110000	−0.5	10001111	−0.5	
00000111	0.5	11111000	−0.5	
00001011	0.25	11110100	−0.25	\$
00001101	0.5	11110010	0	\$
00001110	0.25	11110001	0.25	\$

Table A16.3. ±2 DISPARITY WORDS

Mode 1 (−2)		*Mode 2 (+2)*		
Output word	*Mean toggle word voltage* $\times V_0$	*Output word*	*Mean toggle word voltage* $\times V_0$	
10001000	0	01110111	0	*
10000100	−0.25	01111011	0.25	≠
01001000	0.25	10110111	−0.25	
01000100	0	10111011	0	*
01000010	−0.25	10111101	−0.25	*
00101000	0.5	11010111	0	
00100100	0.25	11011011	0.25	*
00100010	0	11011101	0	*
00100001	−0.25	11011110	0.25	≠
00011000	0.75	11100111	−0.25	
00010100	0.5	11101011	0	
00010010	0.25	11101101	−0.25	
00010001	0	11101110	0	*
Mean	$0.1346V_0$	Mean	$-0.0192V_0$	
Available spare words: within present code rules:				
01000001	−0.5	10111110	0	
10000001	−0.75	01111110	0.25	
10000010	−0.5	01111101	0	
outside present code rules:				
00000011	0.75	11111100	0.25	
00000101	0.5	11111010	0	
00000110	0.75	11111001	−0.25	
00001001	0.25	11110110	0.25	
00001010	0.5	11110101	0	$
00001100	0.75	11110011	0.25	$
11000000	0.75	00111111	0.25	
10100000	0.5	01011111	0	
01100000	0.75	10011111	−0.25	
10010000	0.25	01101111	0.25	
01010000	0.5	10101111	0	$
00110000	0.75	11001111	0.25	$

Table A16.4. ±3 DISPARITY WORDS (OUTSIDE PRESENT CODE RULES)

Mode 1 (−3)		*Mode 2 (+3)*	
Output word	*Mean toggle word voltage* $\times V_0$	*Output word*	*Mean toggle word voltage* $\times V_0$
10000000	−1.0	01111111	0
01000000	−0.75	10111111	−0.25
00100000	−0.5	11011111	0
00010000	−0.25	11101111	−0.25
00001000	0	11110111	0
00000100	0.25	11111011	−0.25
00000010	0.5	11111101	0
00000001	0.75	11111110	−0.25

Table A16.5. ±4 DISPARITY WORDS (OUTSIDE PRESENT CODE RULES)

00000000	1.0	11111111	0

REFERENCES

1. British Patent 1 540 617: 'Transformation of Binary Coded Signals into a Form Having Lower Disparity'
2. Hooper, R. C., Smith, D. R. and White, B. R., 'PIN-FET for digital optical receivers', *30th IEEE Electronic Components Conference*, San Francisco (1980)
3. Personick, S. D., 'Receiver design for digital fiber optic communication systems, Pts. I and II', *Bell Syst. Tech. J.*, **52**, No. 6, 843–886 (1973)
4. CCITT Recommended Vocabulary on Codes: CCITT Orange Book, Line Transmission, Vol. III-2, Recommendation G702, 8001–8013 (1976)
5. Jessop, A., 'Line system error detection without frame alignment', *IEE Colloquium on Data Transmission Codes*, Digest No. 1978/51, 11/1–11/5 (1978)
6. British Patent 1 536 337: 'Error Detection in Digital Systems'.
7. Brooks, R. M., 'Line coding for optical fibre submarine cable digital transmission systems', Project Report for the Degree of MSc in Telecommunication Systems, University of Essex, 26–28, 43–59 (1979/80)
8. Brooks, R. M., '7B8B balanced code with simple error detection capability', *Electron Lett.*, **16**, No. 12, 458–459 (1980)
9. Sharland, A. J. and Stevenson, A., 'A simple in-service error detection scheme based on the statistical properties of line codes for optical fibre systems', *Int. J. Electronics*, **55**, No. 1, 151–158 (1983)
10. Faulkner, D. W. and O'Shea, C. D., 'A mark parity error detector integrated circuit for binary transmission systems', *Second Int. Conf. on 'The Impact of High Speed VLSI Technology on Communication Systems'*, IEE, London (1983)
11 Brooks, R. M. and Jessop, A., 'Line Coding for optical fibre systems', *Int. J. Electronics*, **55**, No. 1, 112 (1983)

17

A Hermite Series for Error Probability Calculation in Digital Line Systems

P. Cochrane

17.1 INTRODUCTION

Digital transmission systems are generally subject to a number of performance degrading effects that are introduced by the transmission channel and associated regenerator circuitry. In the specific case of cable systems the principal degrading phenomena are: thermal noise, cross-talk, echoes and intersymbol interference (ISI). Each of these components introduces a reduction in system performance and it is therefore necessary to predict the resultant effect of such phenomena from the outset of a particular system design.

Whilst thermal noise may be accounted for in a straightforward manner [1], cross-talk [2], echoes [3] and ISI [4, 5] are rather more problematic. The reported techniques generally adopted to account for such effects range from worst-case and other bounds [6, 7] through to the direct invocation of the central limit theorem [8]. Distinctions are also drawn, in the literature, between the various phenomena and the theoretical treatments used for each tend to be quite different.

An alternative method of predicting system error probability performance based on a Hermite series expansion is presented in this chapter. The approach adopted is a unified one in that cross-talk, echoes and ISI are accounted for by the one theory.

Throughout the chapter it is assumed that the transmitted data is statistically independent and additive Gaussian noise is present. For practical application of the technique it is also assumed that a time domain description of the various phenomena would be available, either by direct characterisation of a channel or via computation from other data.

The treatment and supporting practical evidence presented deal only with binary systems as the extension to M-ary signalling is straightforward.

17.2 THE PHENOMENA

The theory described in this chapter was born from the realisation that there is a marked similarity (in the time domain) between cross-talk, echoes and ISI (i.e. each is transitory in nature and tends to decay in an exponential manner). The major difference between them, as far as the described theory is concerned, lies in their relative amplitudes and duration.

Before we describe the theory it is therefore useful to highlight the similarity of the various phenomena to be considered by a number of practical examples. Figure 17.1 illustrates the temporal response of cross-talk, echoes and ISI introduced by pair and coaxial cables under actual system conditions.

17.3 THEORY

17.3.1 The general approach

For mathematical convenience we assume a binary bit stream exhibiting statistical independence between bits which may occupy the states ± 1 with equal probability. Let us also assume that some transient, spanning a number of bit periods, is associated with each bit transmitted. Such a condition is depicted in Fig. 17.2 where the launched signal is assumed to consist of $\delta(t)$ functions.

It will be appreciated that, whilst the bit stream and associated interferers shown in Fig. 17.2 have been depicted with the same polarity, for a bipolar signal with equiprobable states, they are governed by a bit probability density function (PDF) of the form shown in Fig. 17.3. Assuming bit independence, it therefore follows that, at some sample instant in the received waveform, the PDF of a single interferer is given by

$$p_i(v) = \tfrac{1}{2}\delta(v - a_i) + \tfrac{1}{2}\delta(v + a_i) \tag{17.1}$$

where

$$a_i = |e(t - \tau i)| \tag{17.2}$$

and $\sum_{i=1}^{m} a_i$ is a bounded sum.

Hence the addition of all the contributions implies the convolution of $p_i(v)$ for all i. This may be more conveniently expressed in terms of the characteristic function of eqn. (17.1).

$$P_i(\psi) = \mathscr{F} p_i(v) = \cos \psi a_i \tag{17.3}$$

$$P(\psi) = \prod_{i=1}^{m} P_i(\psi) = \mathscr{F} p_i(v)^*|_{i=1}^{m} \tag{17.4}$$

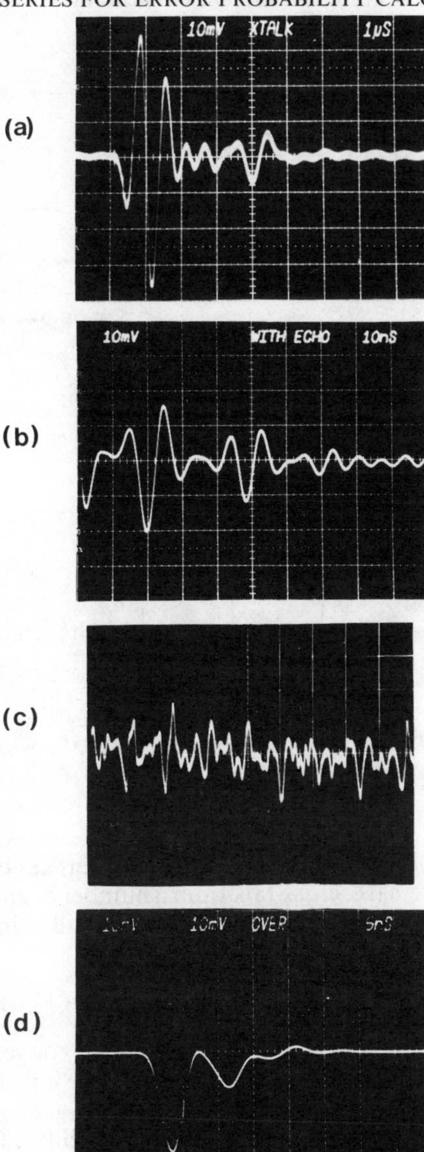

Fig. 17.1 Examples of transmission path impulse response imperfections. (a) Near-end cross-talk (pair-to-pair) on a 2 Mbit/s PCM systems operating over a PCQT cable, (b) echoes produced by closely spaced joints on an installed 1.2/4.4 mm coaxial cable, (c) periodic echoes produced by cyclic variations in the impedance with length characteristic of a 1.2/4.4 mm coaxial cable, (d) the intersymbol interference component of an over-equalised digital transmission channel

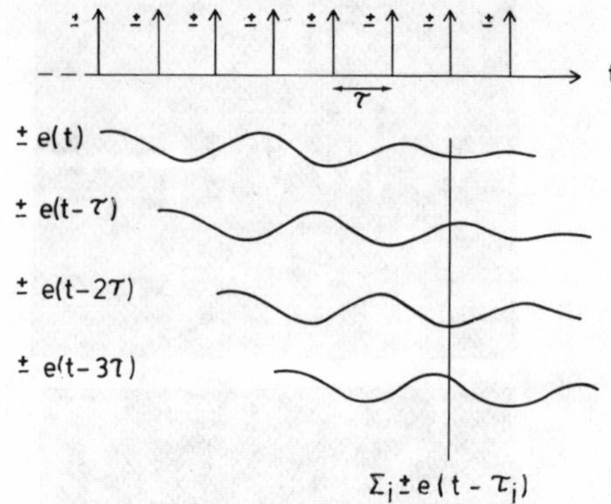

Fig. 17.2 The addition of bit-dependent interferers

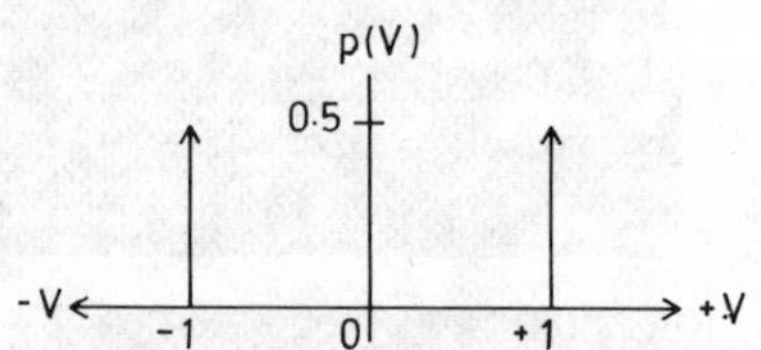

Fig. 17.3 The bit PDF of the transmitted message

where $*|_{i=1}^{m}$ implies the convolution of $p_i(v)$ over all i. Thus

$$P(\psi) = \prod_{i=1}^{m} \cos \psi a_i \tag{17.5}$$

It therefore follows that, if a number of different contributors are present on a system (i.e. cross-talk from a number of different pairs for example), eqn. (17.5) may be extended in the following manner

$$P(\psi) = \prod_{i=1}^{m} \cos \psi a_i \prod_{j=1}^{n} \cos \psi b_j \ldots \tag{17.6}$$

In some practical cases this may be a more convenient form for calculation or expansion, but in the general case it is still reducible to eqn. (17.5) which is thus assumed as the general description. Moreover, for practical system calculations, additive Gaussian noise may also be accounted for by the multiplication of the characteristic functions [9]. Let the thermal noise PDF be defined as

$$\eta(v) = \frac{\exp\left\{-\frac{1}{2}\left(\frac{v}{\sigma}\right)^2\right\}}{\sigma\sqrt{2\pi}} \tag{17.7}$$

where σ is the RMS value and v is the instantaneous amplitude, then

$$N(\psi) = \mathscr{F}\eta(v) = \exp\{-\tfrac{1}{2}(\psi\sigma)^2\} \tag{17.8}$$

and

$$N(\psi)P(\psi) = \exp\{-\tfrac{1}{2}(\psi\sigma)^2\} \prod_{i=1}^{m} \cos \psi a_i \tag{17.9}$$

The PDF of all the additive contributors is thus given by the following general solution

$$p(v) = \mathscr{F}^{-1}N(\psi)P(\psi) \tag{17.10}$$

$$= \int_{-\infty}^{+\infty} \exp\{j\psi v\} \exp\{-\tfrac{1}{2}(\psi\sigma)^2\} \prod_{i=1}^{m} \cos \psi a_i \, d\psi \tag{17.11}$$

Unfortunately there is no closed form analytic solution to this rather formidable general expression [10]. To find a solution we thus neglect, for a moment, the contribution of the Gaussian noise and return to eqn. (17.5). This is now expressed as a cumulant generating function

$$\log_e P(\psi) = \sum_{i=1}^{m} \log_e \{\cos \psi a_i\} \tag{17.12}$$

This in turn may be expanded as a series [11] in ψ containing Bernoulli numbers B_r so that

$$\log_e P(\psi) = -\sum_{i=1}^{m} \sum_{r=1}^{\infty} \left. \frac{2^{2r-1}(2^{2r}-1)B_r(\psi a_i)^{2r}}{r(2r)!} \right|_{|\psi a_i| < \pi/2} \tag{17.13}$$

also

$$\log_e P(\psi) = \sum_{r=1}^{\infty} \frac{\chi_{2r}(j\psi)^{2r}}{(2r)!} \tag{17.14}$$

where the even cumulants χ_{2r} are (by equating eqns. (17.13) and (17.14)) given by the relationship

$$\chi_{2r} = \frac{2^{2r-1}(2^{2r}-1)B_r}{r(j)^{2(r-1)}} \sum_{i=1}^{m} a_i^{2r} \tag{17.15}$$

The even cumulants may thus be shown to be generated by relationships of the form given below. Their numerical values are therefore easily computed directly from the transient responses being considered.

$$\chi_2 = \sum_{i=1}^{m} a_i^2 \tag{17.16}$$

$$\chi_4 = -2 \sum_{i=1}^{m} a_i^4 \tag{17.17}$$

$$\chi_6 = 16 \sum_{i=1}^{m} a_i^6 \tag{17.18}$$

$$\chi_8 = -272 \sum_{i=1}^{m} a_i^8 \tag{17.19}$$

In turn, these cumulants may be related to the even moments of the characteristic function in the following manner [9]

$$M_2 = \chi_2 \tag{17.20}$$

$$M_4 = \chi_4 + 3\chi_2^2 \tag{17.21}$$

$$M_6 = \chi_6 + 15\chi_2\chi_4 + 15\chi_2^3 \tag{17.22}$$

$$M_8 = \chi_8 + 28\chi_2\chi_6 + 210\chi_2^2\chi_4 + 105\chi_2^4 + 35\chi_4^2 \tag{17.23}$$

The characteristic function is thus

$$P(\psi) = \sum_{r=0}^{\infty} \frac{(j\psi)^{2r} M_{2r}}{(2r)!} \tag{17.24}$$

Combining this function with the thermal noise and taking the Fourier transform yields

$$\mathscr{F}^{-1} N(\psi)P(\psi) = \eta(v) * p(v) \tag{17.25}$$

$$= \eta(v) + \sum_{r=1}^{\infty} \frac{M_{2r}}{(2r)!} \eta(v)^{(2r)} \tag{17.26}$$

The resulting system error probability may thus be defined in the following manner (where the power in parenthesis denotes differentiation to that power)

$$P_e(x) = \int_x^{\infty} \eta(v)\, dv + \sum_{r=1}^{\infty} \frac{M_{2r}}{(2r)!} \int_x^{\infty} \eta(v)^{(2r)}\, dv \tag{17.27}$$

$$= P_{e0}(x) + \left[\sum_{r=1}^{\infty} \frac{M_{2r}}{(2r)!} \eta(v)^{(2r-1)} \right]_x^{\infty} \tag{17.28}$$

$$= P_{e0}(x) - \sum_{r=1}^{\infty} \frac{M_{2r}}{(2r)!} \eta(x)^{(2r-1)} \tag{17.29}$$

where $P_{e0}(x)$ is the error probability due entirely to thermal noise and the summation term represents the additional contribution of the interferer. This second term may be expanded as a Gram–Charlier series ([12] and Chapter 4 of this volume) in terms of the Hermite polynomials H_{2r-1} ([11] and Chapter 9 of this volume) as follows

$$P_e(x) = P_{eo}(x) + \frac{\exp(-x^2/2\sigma^2)}{\sqrt{\pi}} \sum_{r=1}^{\infty} \frac{M_{2r}}{(2r)!} \left(\frac{1}{2\sigma^2}\right)^r H_{2r-1}\left(\frac{x}{\sqrt{2}\sigma}\right) \tag{17.30}$$

As an aid to numerical calculation a further simplification is also acceptable for many applications. Let us approximate $P_{eo}(x)$ by a well-established and accurate upper bound [13]

$$P_{eo}(k) \approx \frac{1}{\sqrt{2\pi k}} \exp\left(-\frac{k^2}{2}\right) \tag{17.31}$$

where k^2 is the normalised peak signal-to-RMS-noise power (i.e. $\sigma = 1$) at the decision point. Substituting eqn. (17.31) in (17.30) gives

$$P_e(k) \approx P_{eo}(k)\left\{1 + k\sqrt{2} \sum_{r=1}^{\infty} \frac{M_{2r}}{(2r)!} \frac{1}{2^r} H_{2r-1}\left(\frac{k}{\sqrt{2}}\right)\right\} \tag{17.32}$$

The error probability may thus be computed to a predefined accuracy by the inclusion of sufficient terms in the series.

It is worth mentioning at this point that similar derivations have previously been published by Ho and Yeh [14] and Celebiler and Shimbo [15]. However, their solutions appear to have been formulated for application to ISI on data transmission links where the severity of the phenomena can be expected to be far greater than that experienced on PCM and higher bit-rate transmission systems. In particular the paper by Ho and Yeh describes a recursive relationship, between the even moments, that allows for a more protracted computational sequence than is likely to be necessary for the systems considered in this chapter.

7.3.2 Extension to multiple independent interferers

Unlike the class of interferer considered in Section 17.3.1, this case is concerned with interference that arises as a sum of a number of independent interferers that may, in general, be plesiochronous. An example of such a case would be a number of cross-talk interferers produced by several PCM systems operating over a pair-type cable.

We account for this additional degree of freedom by modifying eqn. (17.5) in the following manner

$$P(\psi) = \prod_{l=1}^{n} \prod_{i=1}^{m} \cos\{\psi a_i(t_l)\} \tag{17.33}$$

It should be noted that the possibility of a plesiochronous condition now dictates that our a_i values become time-dependent, i.e. we have to allow for the relative phase-slip between the numerous interferers. The cumulant-generating function is thus given by

$$\log_e P(\psi) = \sum_{l=1}^{n} \sum_{i=1}^{m} \log_e \cos(\psi a_{l,i}) \tag{17.34}$$

Note that the nomenclature of the a_i term has been modified for ease of manipulation.

$$\log_e P(\psi) = -\sum_{l=1}^{n} \sum_{i=1}^{m} \sum_{r=1}^{\infty} \frac{2^{2r-1}(2^{2r}-1)B_r(\psi a_{l,i})^{2r}}{r(2r)!} \tag{11.35}$$

$$= \sum_{r=1}^{\infty} \frac{\chi_{2r}(j\psi)^{2r}}{(2r)!} \tag{17.36}$$

Hence, by equating eqns. (17.35) and (17.36) it may be shown that the cumulants of the random function are given by

$$\chi_{2r} = \frac{2^{2r-1}(2^{2r}-1)B_r}{r(j)^{2(r-1)}} \sum_{l=1}^{n} \sum_{i=1}^{m} a_{l,i}^{2r} \tag{17.37}$$

The moments of the characteristic function may therefore be derived as indicated in eqns. (17.20)–(17.23) and the resulting error probability may be computed using eqn. (17.30).

Because we are generally interested only in the long-term error probability of a system, and because $a_{l,i}$ is a function of the uniform random variable t_l, it is generally more convenient to form, numerically, an average value for each of the M_{2r} moments. This may be achieved by modifying eqns. (17.16) through (17.23) in the following manner

$$\bar{M}_2 = \frac{1}{n} \sum_{l=1}^{n} \sum_{i=1}^{m} a_{l,i}^2 \tag{17.38}$$

$$\bar{M}_4 = \frac{1}{n} \sum_{l=1}^{n} \left[-2 \sum_{i=1}^{m} a_{l,i}^4 + 3\left\{ \sum_{i=1}^{m} a_{l,i}^2 \right\}^2 \right] \tag{17.39}$$

$$\bar{M}_6 = \frac{1}{n} \sum_{l=1}^{n} \left[16 \sum_{i=1}^{m} a_{l,i}^6 - 30 \sum_{i=1}^{m} a_{l,i}^4 \sum_{i=1}^{m} a_{l,i}^2 + 15\left\{ \sum_{i=1}^{m} a_{l,i}^2 \right\}^3 \right] \tag{17.40}$$

Alternatively, upper and lower bound values may be computed by assuming the contributors to be stationary and in either pessimistic or optimistic states. The problem may thus be reduced in complexity to become comparable to those described in the previous sections.

17.4 PRACTICAL EXAMPLES

17.4.1 Introduction

As far as possible the examples cited in this section have been selected to be representative of real systems conditions [10, 17]. However, it has proved necessary, in most cases, to scale the magnitudes of the phenomena (in a pessimistic manner) in order to produce a measurable effect on the transmission systems employed.

17.4.2 A single discrete echo

This is probably the simplest form of system performance degradation encountered in practice and is by far the easiest to deal with theoretically. In order to generate a discrete echo, the transmission path of a 140 Mbit/s binary system was modified in the manner depicted in Fig. 17.4.

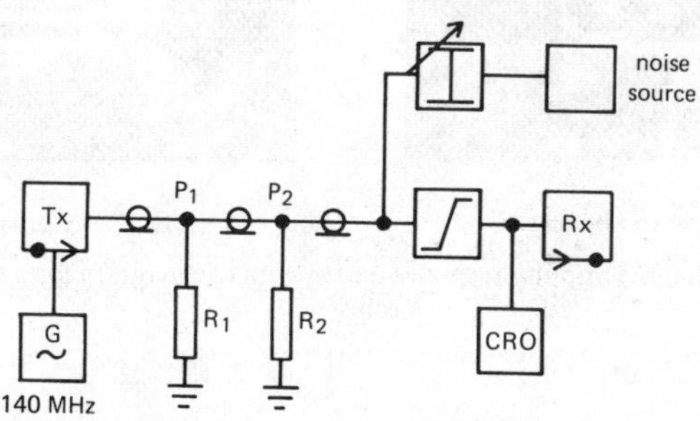

Fig. 17.4 The modified transmission path used to generate noise

A pair of non-reactive discontinuities were purposely selected for this example in order to simplify the theoretical treatment. In addition, the transmission time delay between the discontinuities was arranged to be an integer number of bit periods. The echo therefore introduced the maximum eye closure whilst only influencing a single decision epoch. This condition is depicted in Fig. 17.5 which shows the impulse response of the channel as well as the system eye diagrams both with and without the echo introduced.

Because we have only a single echo contributor our basic characteristic function (eqn. (17.5)) reduces to

$$P(\psi) = \cos \psi a \tag{17.41}$$

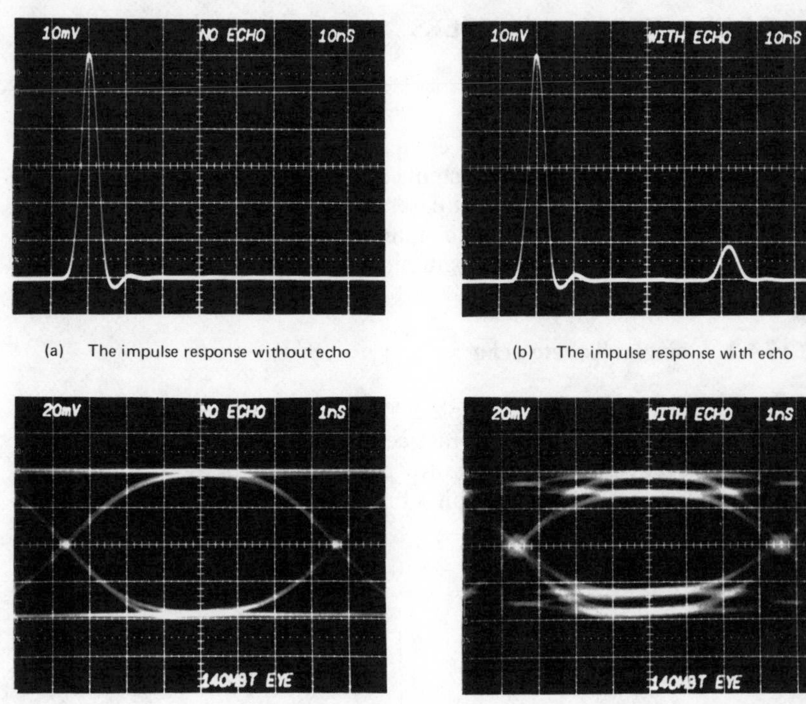

(a) The impulse response without echo

(b) The impulse response with echo

(c) The eye diagram without echo

(d) The eye diagram with echo

Fig. 17.5 The channel impulse response and system eye diagram for a single discrete echo

Consequently eqn. (17.32) becomes

$$P_e(k) \approx P_{e0}(k)\left\{1 + \frac{(ak)^2}{2!} + \frac{(ak)^4}{4!}\left(1 - \frac{3}{k^2}\right) + \frac{(ak)^6}{6!}\left(1 - \frac{10}{k^2} + \frac{15}{k^6}\right) + \cdots\right\} \tag{17.42}$$

where a = the echo amplitude and k = the signal-to-noise ratio.

From Fig. 17.5 it will be seen that the normalised amplitude of the discrete echo is -17.2 dB, the absolute value of a is thus

$$a = 0.138k \tag{17.43}$$

For practical transmission systems it is likely that discrete echoes would be constrained to much smaller values whilst the signal-to-noise ratio would almost certainly be in excess of 16–20 dB. Thus, a

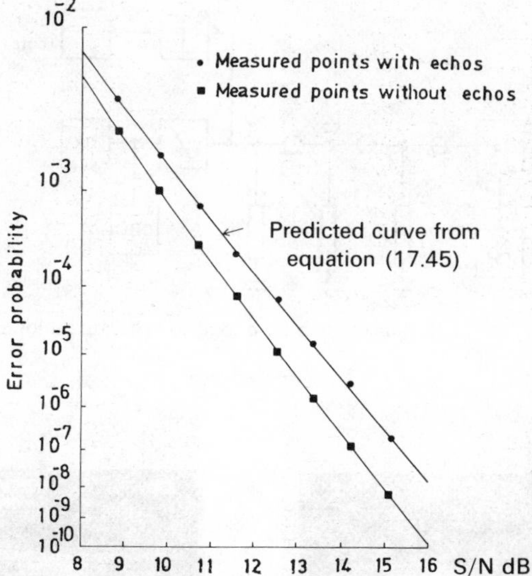

Fig. 17.6 Predicted and measured error probability results for a single discrete echo

useful approximation to eqn. (17.42) will be seen to be

$$P_e(k) = P_{e0}(k) \cosh (ak) \tag{17.44}$$

which, for our specific example, becomes

$$P_e(k) = P_{e0}(k) \cosh (0.138k^2) \tag{17.45}$$

Predicted and measured results are shown in Fig. 17.6.

17.4.3 Multiple discrete echoes

For this case the transmission path was modified by the inclusion of three non-reactive discontinuities as indicated in Fig. 17.7. Again the position of the discontinuities was arranged to give echoes at integer intervals of the bit period in order to achieve the maximum eye closure.

The transient response of the transmission channel and the associated eye diagram of the system with the echoes present are shown in Fig. 17.8. The transient response and eye diagram without the echoes present are as previously depicted in Figs. 17.5a and 17.5c.

From Fig. 17.8 it will be seen that the relative amplitudes of the

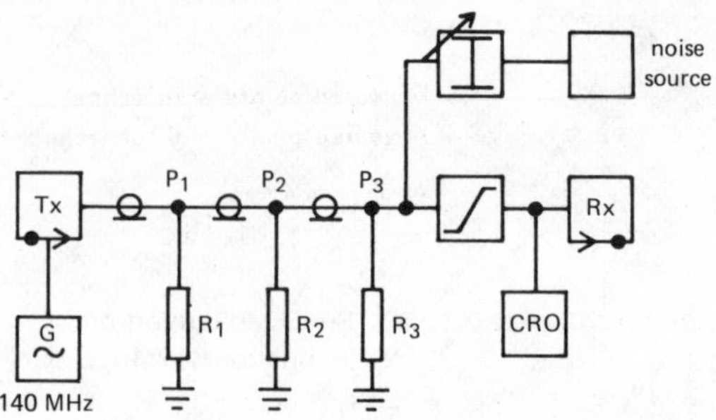

Fig. 17.7 The modified transmission path used to generate multiple discrete echoes

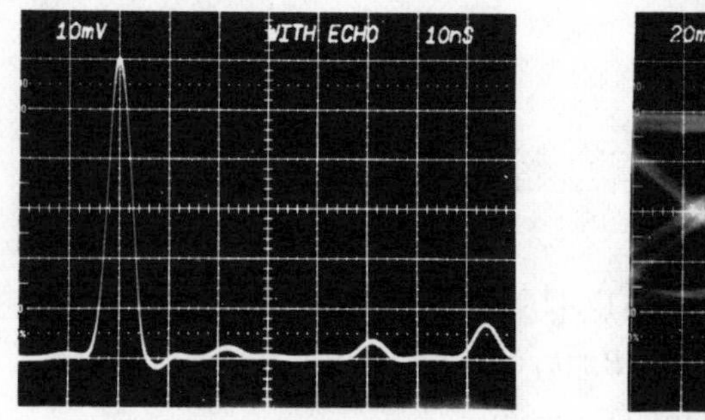

(a) The impulse response with echoes

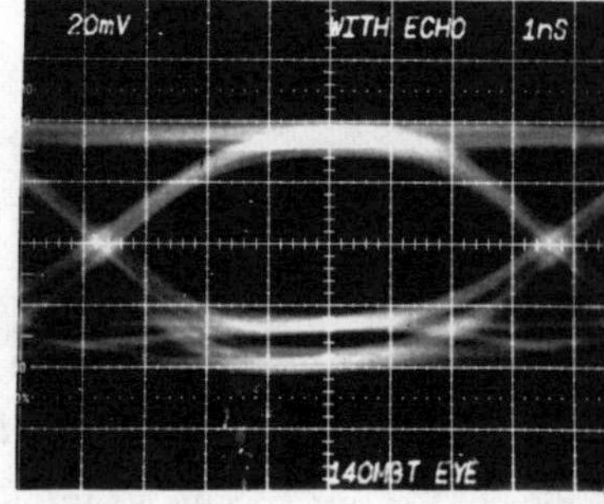

(b) The eye diagram with echoes

Fig. 17.8 The channel impulse response and eye diagram for multiple discrete echoes

respective echoes are

$$\hat{e}_1 = -28.5 \text{ dB}, \quad \hat{e}_2 = -24.1 \text{ dB}, \quad \hat{e}_3 = -19.0 \text{ dB}$$

The absolute values are thus

$$a_1 = 3.76 \times 10^{-2}k, \quad a_2 = 6.24 \times 10^{-2}k,$$
$$a_3 = 1.12 \times 10^{-1}k$$

Using eqns. (17.16) through (17.23) the even moments may be shown to be

$$M_2 = 1.79 \times 10^{-2}k^2 \quad M_4 = 6.12 \times 10^{-4}k^4$$
$$M_6 = 2.49 \times 10^{-5}k^6 \quad M_8 = 7.43 \times 10^{-6}k^8$$

Expanding eqn. (17.32) for this case gives

$$P_e(k) = P_{e0}(k)\left\{1+\frac{M_2k^2}{2!}+\frac{M_4k^4}{4!}\left(1-\frac{3}{k^2}\right)+\frac{M_6k^6}{6!}\left(1-\frac{10}{k^2}+\frac{15}{k^4}\right)\right.$$
$$\left.+\frac{M_8k^8}{8!}\left(1-\frac{21}{k^2}+\frac{105}{k^4}-\frac{105}{k^6}\right)+\cdots\right\} \tag{17.46}$$

Substituting the calculated values for M_{2r} gives

$$P_e(k) = P_{e0}(k)\left\{1+8.95\times10^{-3}k^4+2.55\times10^{-5}k^8\left(1-\frac{3}{k^2}\right)\right.$$
$$+3.46\times10^{-8}k^{12}\left(1-\frac{10}{k^2}+\frac{15}{k^4}\right)$$
$$\left.+1.84\times10^{-10}k^{16}\left(1-\frac{21}{k^2}+\frac{105}{k^4}-\frac{105}{k^6}\right)+\cdots\right\} \tag{17.47}$$

The error probability predicted by this equation is shown together with measured values in Fig. 17.9.

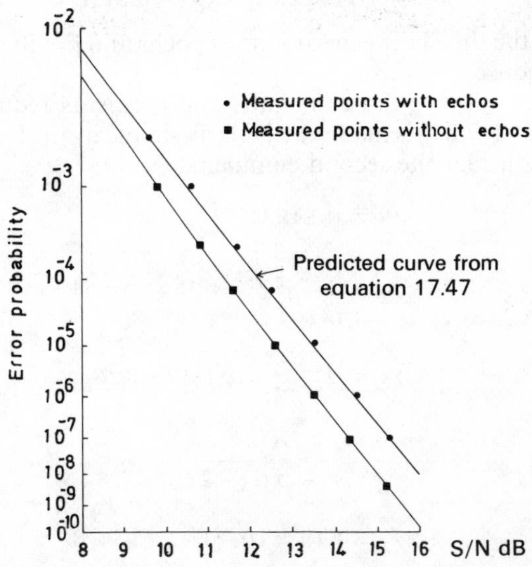

Fig. 17.9 Predicted and measured error probability results for three discrete echoes

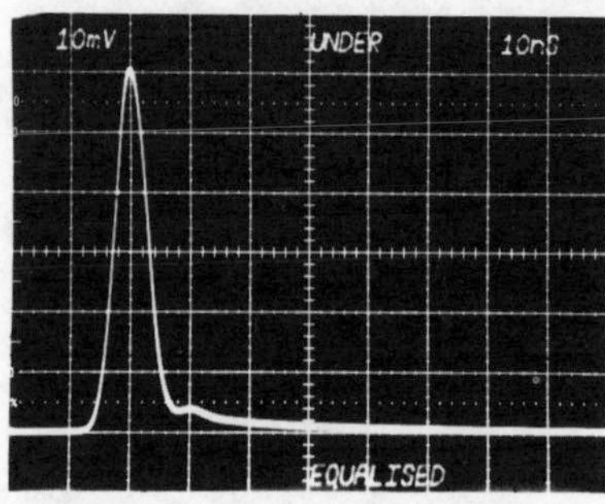

(a) The impulse response with ISI

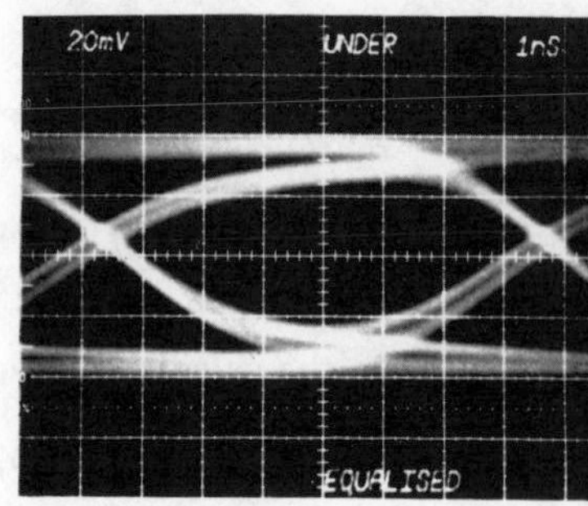

(b) The eye diagram with ISI

Fig. 17.10 The channel impulse response and eye diagram with ISI due to under-equalisation

17.4.4 Intersymbol interference

For this particular example the transmission path was modified by an additional length of coaxial cable to simulate an under-equalised channel. The resulting transient response of the channel and eye diagram of the system are shown in Fig. 17.10.

The exponential tail of the under-equalised channel is well approximated by the relative amplitude function

$$a_i = e(\tau i) \approx 0.074 \exp(-0.313i) \tag{17.48}$$

where τ = the time between sampling epochs and i = the ith position of the response.

The calculation of the cumulants and moments required (as per eqns. (17.16)–(17.23)) is made particularly simple by the form of a_i. For example, consider the second cumulant

$$a_i = A \exp(-\alpha i) \tag{17.49}$$

and

$$\chi_2 = \sum_{i=1}^{\infty} a_i^2 k^2 \tag{17.50}$$

$$\chi_2 = A^2 \sum_{i=1}^{\infty} \exp(-2\alpha i) k^2 \tag{17.51}$$

$$= \frac{A^2 k^2}{1 - \exp(-2\alpha)} \tag{17.52}$$

where

$$\chi_2 = 1.18 \times 10^{-2} k^2 \tag{17.53}$$

The higher-order cumulants and hence the moments may be

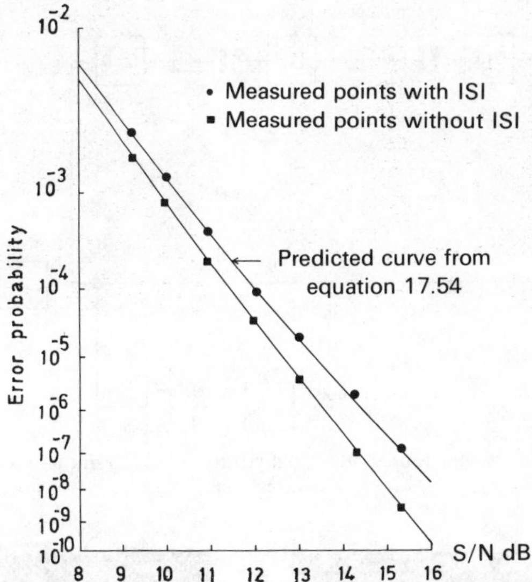

Fig. 17.11 Predicted and measured error probability results for the ISI case

expressed in a similar manner and eqn. (17.32) is shown to be

$$P_e(k) = P_{e0}(k)\left\{1 + 5.89\times10^{-3}k^4 + 1.38\times10^{-5}k^8\left(1-\frac{3}{k^2}\right)\right.$$
$$+1.77\times10^{-8}k^{12}\left(1-\frac{10}{k^2}+\frac{15}{k^4}\right)$$
$$\left.+1.43\times10^{-11}k^{16}\left(1-\frac{21}{k^2}+\frac{105}{k^4}-\frac{105}{k^6}\right)+\cdots\right\} \tag{17.54}$$

This function is depicted in Fig. 17.11 together with the measured error probability characteristic.

17.4.5 Cross-talk from a single interferer

The transmission path of the 140 Mbit/s system was modified as shown in Fig. 17.12 to model the cross-talk conditions commonly found [16, 18, 19] to exist between the metallic pairs used for PCM systems.

Figure 17.13 shows the impulse response of the cross-talk path frozen in time relative to the measurement channel and also the

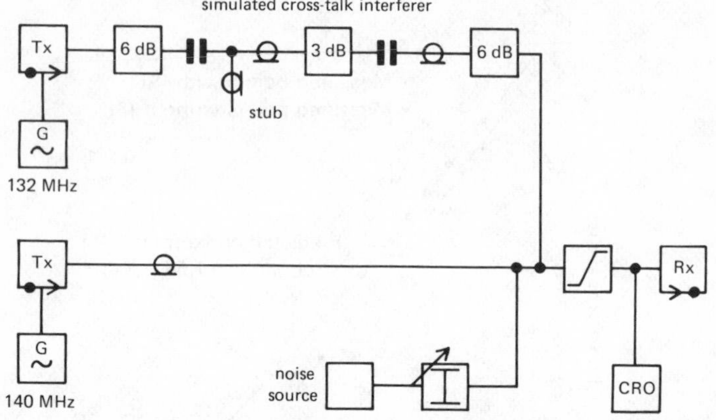

Fig. 17.12 The transmission system configuration used to simulate a single cross-talk interferer

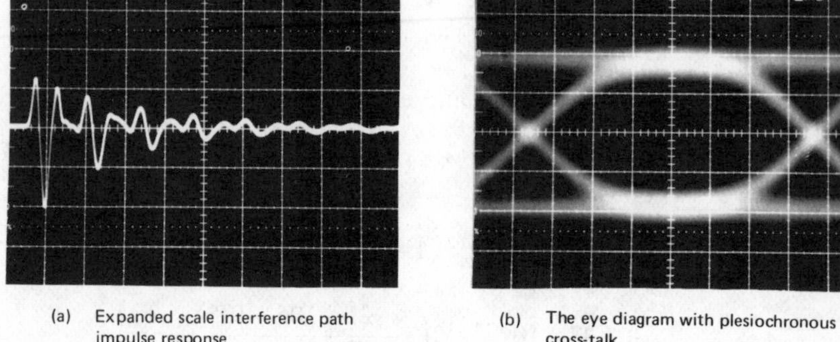

(a) Expanded scale interference path impulse response

(b) The eye diagram with plesiochronous cross-talk

Fig. 17.13 The channel impulse response and eye diagram with a single cross-talk interferer

resulting eye diagram, with the cross-talk slipping in time relative to the operational bit stream. From the impulse response given we calculate the average even moments (via eqns. (17.38) through (17.40)) to be

$$\bar{M}_2 = 1.51 \times 10^{-2}k^2 \quad \bar{M}_4 = 5.03 \times 10^{-4}k^4$$
$$\bar{M}_6 = 2.23 \times 10^{-5}k^6 \quad \bar{M}_8 = 1.15 \times 10^{-6}k^8$$

(A more detailed account of this process is given in Reference 20.) The mean error probability is thus defined by inserting these values in eqn.

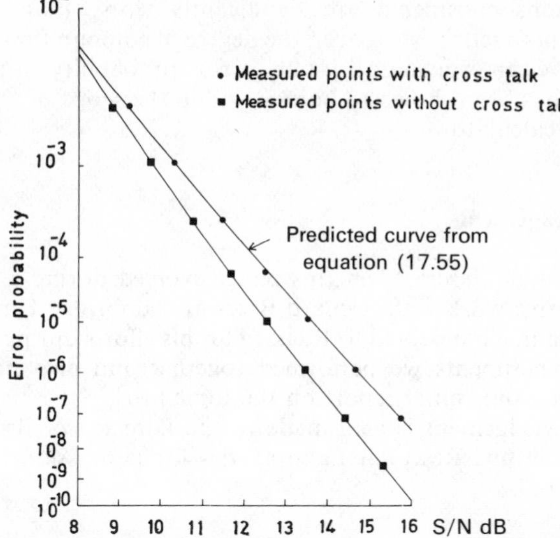

Fig. 17.14 Predicted and measured error probability results for a single cross-talk interferer

(17.32), which gives

$$\overline{P_e(k)} = P_{e0}(k)\left\{1+7.53\times10^{-3}k^4+2.10\times10^{-5}k^8\left(1-\frac{3}{k^2}\right)\right.$$
$$+3.10\times10^{-8}k^{12}\left(1-\frac{10}{k^2}+\frac{15}{k^4}\right)$$
$$\left.+2.86\times10^{-11}k^{16}\left(1-\frac{21}{k^2}+\frac{105}{k^4}-\frac{105}{k^6}\right)+\cdots\right\} \tag{17.55}$$

The experimental results and theoretical predictions are shown in Fig. 17.14.

17.5 CONCLUSIONS

The theory presented in this chapter has been demonstrated, by direct experimental evidence, to provide accurate estimates of the actual system error probability performance when operating over a non-ideal channel exhibiting ISI, echoes and cross-talk. All of the experimental cases cited are extreme in the sense that the channel degradations considered are significantly worse than would be expected in practice. Moreover, the degree of computation necessary to achieve an adequate system error probability performance estimate has been shown to be well within the scope of a pocket or desk-top calculator.

Acknowledgements

The technique described in this paper evolved during a period of collaborative work with Richard Bates at Cambridge University in 1979. I am much indebted to Richard for his efforts during the many system experiments we performed together and also his help in formulating our initial report on the topic [10].

Acknowledgement is also made to the Director of British Telecommunications Research Laboratories for permission to publish this material.

REFERENCES

1. Schwartz, M., *Information Transmission, Modulation, and Noise*, McGraw-Hill, pp. 330–339 (1970)
2. Cravis, H. and Crater, T. B., 'Engineering of T1 carrier system repeatered lines', *Bell Syst. Tech. J.*, **42**, 431–486 (1963)
3. Staff of BTL, *Transmission Systems for Communications*, Bell Telephone Laboratories, p. 727 (1971)
4. Jenq, Y. C., 'Probability of error in PAM systems with intersymbol interference and additive noise', *IEEE Trans. Inf. Theory*, **IT-23**, No. 5, pp. 575–582 (1977)
5. Sawai, A., 'Evaluation of intersymbol interference in multilevel dc balanced code transmission systems', *Electronics and Communications in Japan*, **57-A**, No. 3, pp. 31–39 (1974)
6. Glave, E. F., 'An upper bound on the probability of error due to intersymbol interference for correlated digital signals', *IEEE Trans. Inf. Theory*, **IT-18**, No. 3, 356–363 (1972)
7. Yao, K., *Moment Space Error Bounds in Digital Communication Systems*, Communications Systems and Random Process Theory, NATO ASI Series, Applied Science No. 25, pp. 605–618 (1978)
8. Jacobson, B. B., 'Cable cross-talk limits on low capacity pcm systems', *Electr. Commun.*, **48**, Nos. 1 and 2, 98–107 (1973)
9. Cramer, H., *Mathematical Methods of Statistics*, Princeton University Press, pp. 89, 185, 186, 222, 258 (1974)
10. Cochrane, P. and Bates, R. J. S., *A Unified Approch to the Problems of Cross-Talk Echoes and Intersymbol Interference on Digital Transmission Systems*, Post Office Research Department Report 797 (1979)
11. Spiegel, M. R., *Mathematical Handbook of Formulas and Tables*, McGraw-Hill, pp. 112 and 300 (1962)
12. Fry, T. C., *Probability and its Engineering Uses*, Van Nostrand, pp. 251–260 (1928)

13. Abramowitz, M. and Stegun, A., *Handbook of Mathematical Functions*, Dover Publications, p. 298 (1972)
14. Ho, E. Y. and Yeh, Y. S., 'A new approach for evaluating the error probability in the presence of intersymbol interference and additive Gaussian noise', *Bell Syst. Tech. J.*, **49**, 2249–2266 (1970)
15. Celebiler, M. I. and Shimbo, O., 'The probability of error due to interference and Gaussian noise in digital communication systems', *IEEE Trans. Commun.*, **COM-19**, 113–120 (1971)
16. Bylanski, P. and Ingram, D. G. W., *Digital Transmission Systems*, IEE Telecommun. Series 4, Peter Peregrinus, pp. 147–164 (1976)
17. Crank, G. J., 'High-frequency characteristics and measurement techniques for wideband coaxial cables', *Post Office Electr. Eng. J.*, **71**, Pt. 3, 167–174 (1978)
18. Bates, R. J. S., 'Error probability due to Gaussian noise and near end cross-talk', *Electron. Lett.*, **15**, No. 4, 116–117 (1979)
19. Bates, R. J. S., Probability of Error Rate due to Cross-Talk Interference, Cambridge University Engineering Department Report (CUED/ElectTR55) (1979)
20. Bates, R. J. S. and Cochrane, P., 'Method of estimating the capacity of multipair cables to carry low-speed pcm transmission systems', *Proc. IEE*, **127**, No. 1, 16–21 (1980)

Index